Transformation Wave Physics

Transformation Wave Physics

Electromagnetics, Elastodynamics, and Thermodynamics

edited by
Mohamed Farhat
Pai-Yen Chen
Sebastien Guenneau
Stefan Enoch

CRC Press
Taylor & Francis Group
Boca Raton London New York

CRC Press is an imprint of the
Taylor & Francis Group, an **informa** business

Transformation Wave Physics: Electromagnetics, Elastodynamics, and Thermodynamics

First published 2016 by Pan Stanford Publishing Pte. Ltd.

Published 2022 by CRC Press
Taylor & Francis Group
6000 Broken Sound Parkway NW, Suite 300
Boca Raton, FL 33487-2742

ISBN-13: 978-981-4669-95-5 (hbk)

Visit the Taylor & Francis Web site at
http://www.taylorandfrancis.com

and the CRC Press Web site at
http://www.crcpress.com

British Library Cataloguing-in-Publication Data
A catalogue record for this book is available from the British Library.

Contents

Preface

The theory of composites is a vast and highly interdisciplinary topic that can be traced back to Lord Rayleigh's work on effective medium formulae. John William Strutt, 3rd Baron Rayleigh, who earned the Nobel Prize in Physics in 1904 for his discovery of argon, is most famous among the wave community for Rayleigh scattering (for the layperson, this explains why the sky is blue) and Rayleigh waves that are, with Love waves, responsible for much of earthquake disasters in human infrastructures. Rayleigh's textbook, *The Theory of Sound*, has been, together with Augustus Edward Hough Love's monograph *Some Problems of Geodynamics,* an invaluable source of inspiration for generations of physicists, engineers, and mathematicians. Rayleigh waves are both longitudinal and transverse motions that decrease exponentially in amplitude as distance from the surface increases, so they are also known as surface Rayleigh waves, while Love waves are horizontally polarized surface waves.

Metamaterials are composites with extraordinary properties and were introduced in the context of electromagnetic waves by Sir John Pendry and his colleagues toward the end of the twentieth century, following the advent of photonic crystals. To achieve this tour de force Pendry proposed to combine split ring resonators and thin, straight wires, which exhibit, respectively, a negative permeability and a negative permittivity, the square root of the product of which turns negative upon resonance (hence the negative refractive index!). The first experimental proof of negative refraction came in 2000 with the team of David Smith. Inspired by Victor Veselago's 1968 proposal of a flat convergent lens via negative refraction, Pendry further proposed at the turn of the millennium a lens whose resolution is not limited by the wave wavelength, that is, with a resolution not limited by Rayleigh's criterion on resolution of optical instruments. Finally, 10 years ago, Pendry proposed, together with David Smith and David Schurig, a route toward invisibility, using coordinate changes in Maxwell's equations, which leads to anisotropic heterogeneous tensors of permittivity and permeability. Ulf Leonhardt independently proposed to use conformal mappings to design invisibility cloaks without resorting to anisotropy. The paradigms of negative refraction, transformation optics, and conformal optics have revolutionized the field of photonics, as we have known it since the time of Snell and Descartes.

The present book explores theories and applications of metamaterials not limited to the control of electromagnetic waves. It is a collection of works by leading international experts in the fields of electromagnetics, plasmonics, hydrodynamics, elastodynamics, and diffusion waves.

It starts by a survey (by the world-famous theoretical physicist Leonhardt) of space–time transformations as a design tool for metamaterials underpinning intimate connections between Maxwell's equations and Einstein's theory of relativity, and then the book moves on to practical applications in the control of radio frequency and microwaves, water waves, mechanical waves, and even heat and mass diffusion. We stress that all these contributions promise to revolutionize ways of controlling the propagation of sound, light, and any particular form of waves at macroscopic and microscopic scales. Indeed, potential applications range from subwavelength lensing and time reversal, to underwater camouflaging and electromagnetic invisibility, to enhanced biosensors and protection from harmful physical waves (e.g., tsunamis and earthquakes). This volume covers theoretical as well as experimental aspects in these different areas that include nanoscale (plasmonics) and meter-scale (geophysics) media.

The outline of the chapters is as follows:

As mentioned earlier, the book starts with a chapter by Leonhardt, which describes how theoretical ideas arising from Einstein's general theory of relativity in optical and electrical engineering for designing devices can do the (almost) impossible: invisibility cloaks, perfect imaging, levitation, and the creation of analogues of the event horizon. This chapter gives an introduction to this field requiring minimal prerequisites.

The second chapter, by the group of Liu, reviews the fundamentals and applications of conformal mapping in transformation optics. It first introduces the basics of conformal mapping and how this subset of transformations in the complex plane can eliminate the undesired anisotropy in 2D systems. It then presents metamaterials with minimized anisotropy derived by quasi-conformal mapping; it addresses the promising applications for plasmonics, where conformal mapping manifests its versatility when dealing with the surface plasmon polaritons (SPPs) and localized surface plasmons (LSPs). Interestingly, an attempt is made at applying conformal mapping to the construction of 3D devices.

The third chapter, by the group of Li, further investigates quasi-conformal map as a useful tool for minimizing anisotropy, while the index range can be further minimized by avoiding sharp corners at boundaries. Analytic and numerical approaches show that the corresponding map generation can be simplified by taking an electrostatic analogy. Extension to acoustic and elastic waves is proposed for carpet cloaks. These considerations are useful in a wide class of applications such as invisibility cloaks and optimized integrated devices at the optical wavelengths.

The fourth chapter, by the group of Chan, introduces a mechanism to control the spatial distribution of either an electric or a magnetic field instead of both fields, which can be controlled in an almost arbitrary manner in wavelength and subwavelength scales. Interestingly, the principle of flux control does not rely on surface waves as in the field of plasmonics, but relies on the evanescent waves induced by the strong anisotropy and designed inhomogeneity of media. Such inhomogeneous anisotropic media exhibit surprisingly robust high transmittance. Combined with transformation optics, such high transmittance property can be utilized to build waveguide devices with almost arbitrary shapes and bending angles.

The fifth chapter, by the group of Itoh, reviews progress on radio frequency and microwave beam-forming techniques using planar metamaterials and metamaterial surfaces (metasurfaces). Principles and physics of guided- and leaky-wave characteristics of periodic planar structures are discussed, such as transmission-line-based metamaterials, followed by their active versions that enable novel smart antennas with beam-steering functions. Practical realizations are discussed, as well as the integration with active elements and circuits, which enables adaptively tailoring electromagnetic waves. Planar metamaterial-/metasurface-based beam-forming techniques are clarified through a critical assessment and comparative analysis in the radio-frequency and microwave bands. These techniques offer promising applications in modern wireless communication, radar, remote sensing, and medical and security imaging.

The sixth chapter, by the group of Alù, aims at providing a comprehensive insight into recent developments and applications of gradient metasurfaces to control and engineer the propagation of electromagnetic waves. Various aspects of this technology are explored, starting from miniaturized metasurface building blocks at the lowest level, moving to primary optical elements for radiation patterning, and ultimately incorporating graded metasurfaces into more complex optical devices.

The seventh chapter, by Douglas and Werner, presents scattering manipulation of objects using anisotropic metasurfaces in the microwave range and plasmonic loadings at optical wavelengths. Different from previously reported transformation optics–enabled coatings, they are achieved on the basis of modifying the complex Mie scattering coefficients of an object. The nonvanishing radial response of an anisotropic metasurface is exploited to accomplish near-perfect cloaking and angle-tolerant illusion for objects beyond the quasi-static limit. It is also demonstrated how plasmonic core–shell particles can provide a compact and robust solution toward the realization of nanocircuit loads that offer unprecedented flexibility in tuning the response of a nanodipole-type antenna. Indicative examples are provided that demonstrate the tuning range that core–shell particles are capable of offering.

The eighth chapter, by some of this book's editors, emphasizes that metamaterials can mimic a transformed space in many wave physics areas. The light rays follow trajectories according to Fermat's principle in transformed electromagnetic, acoustic, hydrodynamic, or elastic space, instead of the laboratory space. Homogenization techniques are used to approach such media. This allows one to ultimately manipulate electromagnetic, water, and mechanical wave behaviors, with various exotic characteristics, such as (but not limited to) invisibility cloaks and flat convergent lenses.

The ninth chapter, by the group of Puvirajesinghe, investigates transformational techniques applied to diffusion phenomena. Coordinates transform in the Fourier and Fick's equations bridges transformational thermodynamics to control of mass diffusion. Potential applications range from invisibility cloaks and concentrators for control of heat flux in electronics to biocloaks enabling delayed drug delivery for medical applications.

The tenth chapter, by the group of Wegener, points out that cloaking can be seen as a look-alike contest: the goal is to make some object A appear like another object B with respect to some physical observable. Early mathematical literature has indeed spoken of the nonuniqueness of the tomography (inverse) problem with the works of Calderon (1980), Kohn and Vogelius (1984), and Greenleaf et al. (2003). The experiments discussed in this chapter concern the observables light, sound, elastic waves, static elasticity, electric conduction, heat conduction, and particle diffusion.

The eleventh chapter, by the group of Sato, refreshes our mind with geometrical interpretation to required material parameters through linear-algebraic operations, before moving on to a systematic approach for designing a single functionality in a given physical domain. In this chapter, an extension of the transformation optics formalism is proposed to go beyond a single functionality to independently manipulate multiple physical phenomena simultaneously. A multifunctional shell behaving as an electrical invisibility cloak and a thermal concentrator opens a route to transformation multiphysics.

The twelfth, and last, chapter of the book, by the group of Maurel and Fink, takes us on a journey to the wonderland of time reversal of acoustic, elastic, and electromagnetic waves. In a standard time reversal experiment, waves generated by a source are first measured by an array of antennas positioned around the source and then time-reversed and simultaneously rebroadcasted by the same antenna array. Due to the time invariance of the wave process, the reemitted energy will focus back on the original source, whatever the complexity of the propagation medium. This chapter concentrates on the application of time reversal to the focusing and manipulation of water waves both in linear and nonlinear regimes. Applications are sought in water waves that are scalar waves referring to the evolution of small perturbation of the height of fluid under the action of gravity and surface tension. They are dispersive by nature, nonlinear when generated with standard wave makers, and they experience strong damping at the scale of laboratory experiments. The evolution dynamics in time and space of nonlinear wave trains in deep water can be modeled using the focusing nonlinear Schrödinger equation. The implication of the time reversal invariance on the nonlinear Schrödinger equation is discussed and a way to experimentally focus, both in time and space, rogue waves using the principles of time reversal mirrors is demonstrated.

As you can see, this book therefore touches upon many hot subjects in the physics of metamaterials, which were discovered less than 20 years ago. The first direct experimental evidence of gravitational waves that are ripples in the curvature of space–time that propagate like waves traveling outward from a source (say, two black holes falling onto one another) came in January 2016, that is, one century after Einstein foresaw their existence as a solution to his equations of general relativity. The experimental evidence of Higg's boson came just over two years ago. Graphene was discovered just over a decade ago. We have therefore lived an exciting new millennium of advances in physics thus far, and needless to say that this would already be even to fill with joy the life of any human being, but we believe that metamaterials offer a playground for many new discoveries. We hope that the present book will help foster theoretical and experimental efforts toward a brave new physics world!

We would like to convey our warmest thanks to all chapter authors for their excellent scientific contributions and their willingness to share their knowledge of metamaterials with a general readership. The assistance and professionalism of the Pan Stanford Publishing team is also greatly acknowledged.

We hope that you'll enjoy reading these chapters and find them as informative as we did!

Mohamed Farhat
Pai-Yen Chen
Stefan Enoch
Sebastien Guenneau
Thuwal, Detroit, and Marseille

Chapter 1

Transformation Optics

Ulf Leonhardt
Weizmann Institute of Science, Herzl St. 234, Rehovot, 7610001, Israel

1.1 Introduction

According to Einstein's general theory of relativity, the geometry of space–time is curved by the momentum and energy of macroscopic objects. This curvature is what we perceive as gravity, because it influences the motion of particles such as Newton's apple falling from a tree in the space–time geometry curved by Earth or the planets circling around in the space–time geometry curved by the sun. Gravity also influences the propagation of waves, the most striking demonstration of which is gravitational lensing where light from distant stars or galaxies is deflected and focused in the space–time geometry created by other stars or galaxies. Gravity is universal, because the geometry of space and time sets the scene for everything, particle and wave alike.

Transformation Wave Physics: Electromagnetics, Elastodynamics, and Thermodynamics
Edited by Mohamed Farhat, Pai-Yen Chen, Sebastien Guenneau, and Stefan Enoch

ISBN 978-981-4669-95-5 (Hardcover), 978-981-4669-96-2 (eBook)
www.panstanford.com

Analogues of gravity occur when the geometry of space–time appears to be altered by other means than momentum and energy. The most natural example of analogue gravity is the propagation of light in media. The medium—a piece of glass, the water in a vase, or any other transparent substance—distorts images much the same way stars and galaxies distort light. We may say, with some justification to be given later in this chapter, that media apparently alter the geometry of space–time for light. This geometry differs from the natural space–time geometry of gravity: the medium establishes a virtual geometry different from the natural geometry of physical space. The virtual geometry is created by a completely different physical process than the geometry of physical space and it is also not universal, but restricted to certain physical phenomena. In the case of light in media, the virtual geometry differs from the real geometry only for light (and, typically, only for light within a certain frequency range).

In this chapter, we show how and when virtual geometries arise for light, or electromagnetic waves in general. For this we combine ideas from two of the most beautiful theories of physics, Maxwell's electromagnetism and Einstein's general relativity, such that they become transformable into each other (Fig. 1.1). This connection between general relativity and electromagnetism in media is not new; it dates back to ideas by Gordon [1] published in 1923 and Tamm [2, 3] that appeared around that time and further back to Newton, who allegedly toyed with the idea that gravity is mediated by a medium before settling for Newtonian gravity [4], and also to the enigmatic genius of Fermat [5]. New are applications of these ideas in electrical and optical engineering, as design concepts for novel devices that do the (almost) impossible, for example, invisibility cloaking and perfect imaging. This new research area with old and deep roots, called transformation optics [6–11], has been regarded as one of the most fascinating research insights of the last decade [12]. In this chapter, we derive the foundations of this area and explain a few of the key applications. We also show how transformation optics is related to one of the recurring themes of this book, the physics of the event horizon.

Figure 1.1 Einstein, Einwell, and Maxwell. Transformation optics combines ideas from Einstein's general relativity and Maxwell's electromagnetism. In particular, it uses transformations of space—like the transformation between Einstein and Maxwell shown in the picture.

1.2 Maxwell's Electromagnetism

1.2.1 *Maxwell's Equations*

Let us begin at the beginning, with Maxwell's equations in empty, flat space in Cartesian coordinates:

$$\begin{aligned}
\nabla \cdot \boldsymbol{E} &= 0 && \text{Gauss's law,} \\
\nabla \times \boldsymbol{B} &= \frac{1}{c^2}\frac{\partial \boldsymbol{E}}{\partial t} && \text{Ampére's law with Maxwell's displacement current,} \\
\nabla \times \boldsymbol{E} &= -\frac{\partial \boldsymbol{B}}{\partial t} && \text{Faraday's law of induction,} \\
\nabla \cdot \boldsymbol{B} &= 0 && \text{Absence of magnetic monopoles.}
\end{aligned} \tag{1.1}$$

As usual, c denotes the speed of light in vacuum. Throughout this chapter we use SI units for the electromagnetic fields. Now suppose we change the spatial coordinates, for example, we use spherical coordinates r, θ, ϕ instead of the Cartesian x, y, z. The differentials of the new coordinates appear in a different way in the line element ds than the Cartesian differentials, for example, as

$$\mathrm{d}s^2 = \mathrm{d}x^2 + \mathrm{d}y^2 + \mathrm{d}z^2 = \mathrm{d}r^2 + r^2\,\mathrm{d}\theta^2 + r^2\sin^2\theta\,\mathrm{d}\phi^2 \tag{1.2}$$

in spherical coordinates. In general, curved coordinates x^i contribute to the line element as

$$\mathrm{d}s^2 = g_{ij}\,\mathrm{d}x^i\mathrm{d}x^j \tag{1.3}$$

where we sum over repeated indices (running from 1 to 3). The g_{ij} usually depend on the x^i, as the line element (Eq. 1.2) of the spherical coordinates shows. They constitute the metric tensor with determinant g and matrix inverse g^{ij},

$$g \equiv \det\left(g_{ij}\right), \quad g^{ij} \equiv \left(g_{ij}\right)^{-1}. \tag{1.4}$$

Differential geometry [11] tells us how to express the divergences and curls in Maxwell's equations (Eq. 1.1) in terms of curved coordinates:

$$\begin{aligned}
\frac{1}{\sqrt{g}}\,\partial_i\sqrt{g}\,g^{ij}E_j &= 0 && \text{Gauss's law,} \\
\epsilon^{ijk}\partial_j B_k &= \frac{1}{c^2}\frac{\partial g^{ij}E_j}{\partial t} && \text{Ampére's law with Maxwell's displacement current,} \\
\epsilon^{ijk}\partial_j E_k &= -\frac{\partial g^{ij}B_j}{\partial t} && \text{Faraday's law of induction,} \\
\frac{1}{\sqrt{g}}\,\partial_i\sqrt{g}\,g^{ij}B_j &= 0 && \text{Absence of magnetic monopoles,}
\end{aligned} \tag{1.5}$$

where the Levi–Civita tensor ϵ^{ijk}, appearing in the curls, is given in terms of the completely antisymmetric symbols $[ijk]$ as [11]:

$$\epsilon^{ijk} = \pm\frac{1}{\sqrt{g}}\,[ijk]\,. \tag{1.6}$$

The $\pm$ sign depends on the handedness of the coordinate system: $+$ in right-handed systems and $-$ in left-handed systems.

Maxwell's equations (Eq. 1.5) are not only valid in a flat space expressed in curved coordinates, but also in genuine curved spaces.[a] The reason is the following: Maxwell's equations (Eq. 1.5) are first-order partial differential equations containing maximally first derivatives of the metric tensor g_{ij}. Now, a theorem from differential geometry [11, 13] says that, for any given point, we can always construct a local Cartesian coordinate system where $g_{ij} = \delta_{ij}$ and $\partial_k g_{ij} = 0$ at that point, regardless how curved the geometry is. If the space is curved, these local Cartesian systems do not form a single, global Cartesian frame, but rather represent a patchwork of frames that are not consistent. The inconsistency is caused by the spatial curvature. However, there is always a local coordinate transformation from each local frame to the global frame of the curved manifold. As Maxwell's equations depend maximally on first derivatives of g_{ij} we can assume them in the form (Eq. 1.1) in each local Cartesian frame and then transform to the general form (Eq. 1.5) in the global frame. Therefore the form (Eq. 1.5) describes electromagnetism in curved space as well.

1.2.2 *The Medium of a Geometry*

Consider now a case similar to gravitational lensing[b] where we assume a given spatial geometry. In the following section we show how this geometry appears as a medium. For this we express Maxwell's equations (Eq. 1.5) for the quantities $\boldsymbol{E}$ (electric field strength), $\boldsymbol{D}$ (dielectric displacement), $\boldsymbol{H}$ (magnetic field), and $\boldsymbol{B}$

---

[a]We consider purely spatial geometries first and then, in Section 1.5, we generalize our theory to space–time geometries.

[b]In gravitational lensing, gravity alters primarily the measure of time, but due to the conformal invariance of electromagnetism (see Section 1.5.2) this is equivalent to altering the measure of space.

(magnetic induction) familiar from the macroscopic electromagnetism in media:

$$\begin{aligned}
\nabla \cdot \boldsymbol{D} &= 0 && \text{Gauss's law,} \\
\nabla \times \boldsymbol{H} &= \frac{\partial \boldsymbol{D}}{\partial t} && \text{Ampère's law with Maxwell's displacement current,} \\
\nabla \times \boldsymbol{E} &= -\frac{\partial \boldsymbol{B}}{\partial t} && \text{Faraday's law of induction,} \\
\nabla \cdot \boldsymbol{B} &= 0 && \text{Absence of magnetic monopoles.}
\end{aligned} \tag{1.7}$$

Considering Gauss's law, we can write it as $\nabla \cdot \boldsymbol{D} = \partial_i D^i = 0$ with

$$D^i = \varepsilon_0 \, \varepsilon^{ij} E_j \tag{1.8}$$

and $\varepsilon^{ij} \propto \sqrt{g}\, g^{ij}$. Here ε_0 denotes the electric permittivity of the vacuum. Let us see whether this definition of the dielectric displacement $\boldsymbol{D}$ is consistent with the other place $\boldsymbol{D}$ occurs in Maxwell's equations (Eq. 1.7), Ampère's law with Maxwell's displacement current:

$$\epsilon^{ijk} \partial_j B_k = \pm \frac{1}{\sqrt{g}} [ijk] \, \partial_j B_k = \frac{1}{c^2} \frac{\partial g^{ij} E_j}{\partial t} . \tag{1.9}$$

If we write

$$H_k = \varepsilon_0 c^2 B_k \tag{1.10}$$

and

$$\varepsilon^{ij} = \pm \sqrt{g}\, g^{ij} \tag{1.11}$$

we obtain Ampère's law for $\boldsymbol{D}$ given by definition (Eq. 1.8). Using the same arguments for the remaining two Maxwell equation we get

$$B^i = \mu_0 \mu^{ij} H_j \tag{1.12}$$

with the magnetic permeability of the vacuum

$$\mu_0 = \frac{1}{\varepsilon_0 c^2} \tag{1.13}$$

and the relative magnetic permeability

$$\mu^{ij} = \pm \sqrt{g}\, g^{ij} . \tag{1.14}$$

In the formalism developed here [11], $\boldsymbol{E}$, and $\boldsymbol{H}$ are the fundamental fields, whereas $\boldsymbol{D}$ and $\boldsymbol{B}$ are derived from E_i and H_i by raising the index with g^{ij} and multiplication by $\sqrt{g}$. Mathematically, the $\boldsymbol{E}$ and $\boldsymbol{H}$ are one-forms, whereas the $\boldsymbol{D}$ and $\boldsymbol{B}$ are vector densities with respect to the spatial geometry [11]. The fact that the ε^{ij} and μ^{ij} are matrices that depend on two indices indicates that the medium representing the geometry g_{ij} is anisotropic in general. We also see that a spatial geometry appears as a medium with

$$\varepsilon^{ij} = \mu^{ij} . \tag{1.15}$$

In electrical engineering, such media are called impedance matched (to the vacuum).

1.2.3 *The Geometry of a Medium*

The converse is also true: Impedance-matched media, that is, media satisfying the condition (Eq. 1.15), can be understood as spatial geometries for the following argument. We calculate the determinant $\det\varepsilon$ of the matrix ε^{ij} from Eq. 1.11 and obtain

$$\det\varepsilon = \pm\sqrt{g} . \tag{1.16}$$

Consequently, we can write g^{ij} in terms of ε^{ij} and, by virtue of impedance matching (Eq. 1.15), of μ^{ij} as well, as

$$g^{ij} = \frac{\varepsilon^{ij}}{\det\varepsilon} = \frac{\mu^{ij}}{\det\mu} . \tag{1.17}$$

Spatial geometries appear as impedance-matched media and impedance-matched media make virtual geometries. Impedance matching establishes an exact virtual geometry where electromagnetic fields are identical in all aspects to such fields in a real geometry. Without impedance matching, the geometric picture is not exact. In particular, the propagation of light in non-impedance-matched anisotropic media depends on the polarization, a phenomenon called birefringence. However, non-impedance-matched media can still be used for establishing geometries in planar media for specific polarizations (see, for example, the experiments [14–18]), because in such cases not all components of the ε^{ij} and μ^{ij} tensors are needed. Furthermore, in optically isotropic media (where $\varepsilon^{ij} = \varepsilon\,\delta_{ij}$ and $\mu^{ij} = \mu\,\delta_{ij}$) the geometric picture gives an excellent approximation for electromagnetic waves within the validity range of geometrical optics [11].

1.3 Spatial Transformations

In general, the virtual geometry of light in media is curved. To give a simple example, a lens focuses parallel light rays in the focal point; parallels are thus no longer parallel, but meet, which violates Euclid's parallel axiom of flat space. Curved space is common place in optics. It is much harder to create a virtual geometry that is flat. What would it do? As any flat geometry can be reduced to Cartesian coordinates by a coordinate transformation, the material would just perform a transformation of space where each point of physical space appears to be at a position in virtual space that may deviate from real one. If the new coordinates agree with the old ones outside of the device, which is made of the medium, we would not see the difference between propagation in the medium and empty, flat space. In short, the device would be completely invisible.

1.3.1 *Invisibility Cloaking*

Such invisible devices could be used to make other things invisible, too: they can be turned into invisibility devices as follows. Suppose the device performs the following transformation (Fig. 1.2): an extended region in physical space (Fig. 1.2 right) is condensed into a single point in virtual space (Fig. 1.2 left). Anything inside this region has thus become as small as a single point, invisibly small. Everything inside is hidden and, as the device itself is invisible, the very act of hiding is hidden as well. The transformation (Fig. 1.2) makes a perfect cloaking device [8].

Such a cloaking device has been demonstrated for microwaves [14] or, to be more precise, an approximation without impedance matching was made [14]. However, this device did only operate for one polarization, because it used a medium that is not impedance-matched, and for one frequency only, for a fundamental reason [9] that applies to all purely transformation-based cloaking devices [8]. This reason is easy to understand: consider a light ray that just straddles the invisible point in virtual space. As the device performs a spatial transformation from virtual to physical space, the light propagation in both spaces is synchronized. Therefore the light must go around the invisible region in precisely the time it takes to pass

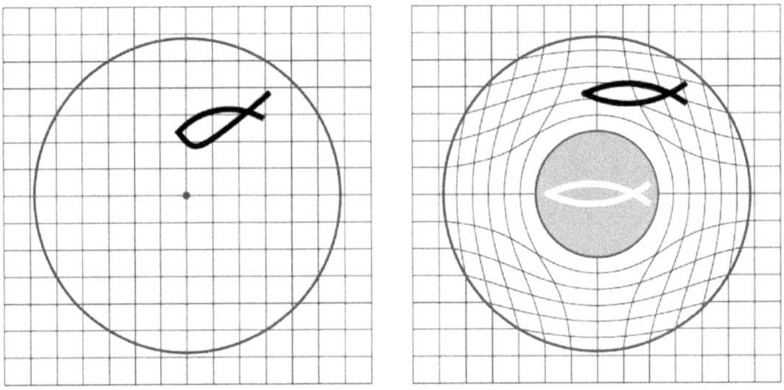

Figure 1.2 Transformation of space. Optical materials appear to change the perception of space; objects (fish) in physical space (right picture) appear at positions (left picture) different from where they actually are. Suppose the medium performs a coordinate transformation from physical space (right) to virtual space (left) and vice versa. Virtual space is empty and so light propagates along straight lines. In virtual space, we may draw a coordinate system as a rectangular grid of light rays (left grid). In physical space, the light rays are curved; the coordinate grid of virtual space is transformed into a curved coordinate system in physical space (right grid). As the coordinate transformation only changes space within a circle, this circle marks the boundary of the optical material used to transform space. We see that the images of the fish are distorted in virtual space because the coordinate transformation illustrated here is not uniform. Moreover, the white fish has completely disappeared because it was swimming within a region of physical space (gray) that, in virtual space, is contracted to a single, invisible point. Such an optical material makes an invisibility device.

a single point in virtual space: zero time. Consequently, the speed of light must tend to infinity at the inner surface of the cloaking device [9, 19]. By speed of light we mean the phase velocity here, whereas the group velocity turns out to approach zero [19]. Relativistic causality does not prohibit an infinity phase velocity in a medium, but it allows it only for a single frequency—a single color—and thus in a purely stationary regime where nothing changes and no new information is transferred. Any change would cause distortions, which defeats the point of a cloaking device;[c] one might as well use a hologram of the background.

[c]It might be reassuring to know that perfect deception is impossible; the truth will always appear in the end.

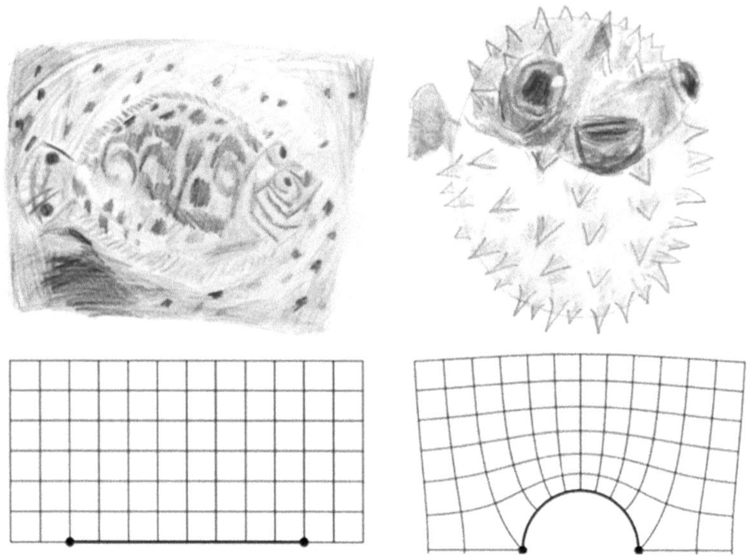

Figure 1.3 From fugu to flatfish. A coordinate transformation may turn a voluminous object in physical space (fugu) into a flat object (flatfish). Reproduced with friendly permission of Maria Leonhardt.

Perfect invisibility is impossible, but this does not prevent invisibility that is good enough. One might be content with deforming a surface by conformal [7] or quasiconformal transformations [20], which does not make an object disappear altogether, but makes it optically flat; the fugu in Fig. 1.3 is turned into a flatfish. One could then use conventional camouflage to disguise the flat object (as flatfish are masters of). Or one might use non-Euclidean cloaking devices where virtual space is not flat, but curved in an appropriate way [7, 21, 22]. In this case the speed of light is finite in the device (and can be made slower than c [22]) but the price to be paid is a time delay of the light making the detour in the cloaking device. Furthermore, relativistic causality is of little concern to the cloaking of sound waves, because the normal speed of sound is several orders of magnitude slower than c. Near-perfect acoustical cloaking over a broad band of frequencies is possible and has been demonstrated [23].

1.3.2 *Transformation Media*

What does it take to build a cloaking device and similar transformation devices? Let us work out the electromagnetic properties of devices required for transforming space. We use x^i to denote the coordinates of physical space and $x^{i'}$ for virtual space (the prime at the index does not mean that we simply use a different index variable but shall indicate the different coordinates of virtual space). As the device performs a transformation from physical to virtual space and vice versa, the $x^{i'}$ of virtual space are thus functions $x^{i'}(x^i)$ of the physical space coordinates x^i—the device performs a coordinate transformation. Suppose that virtual space is flat and empty, and that we describe it in Cartesian coordinates. In this case we obtain for the line element:

$$ds^2 = \delta_{i'j'}\, dx^{i'} dx^{j'} = \delta_{i'j'} \frac{\partial x^{i'}}{\partial x^i} \frac{\partial x^{j'}}{\partial x^j} dx^i dx^j \tag{1.18}$$

and thus, according to definition (Eq. 1.3),

$$g_{ij} = \Lambda^{i'}_{\ i}\, \delta_{i'j'}\, \Lambda^{j'}_{\ j} \quad \text{with} \quad \Lambda^{i'}_{\ i} \equiv \frac{\partial x^{i'}}{\partial x^i}. \tag{1.19}$$

From this expression we get for the matrix inverse of g_{ij}:

$$g^{ij} = \Lambda^{i}_{\ i'}\, \delta^{i'j'}\, \Lambda^{j}_{\ j'} \quad \text{with} \quad \Lambda^{i}_{\ i'} \equiv \frac{\partial x^{i}}{\partial x^{i'}}, \tag{1.20}$$

or, in matrix notation,

$$\mathrm{g}^{-1} = \boldsymbol{\Lambda}\, \boldsymbol{\Lambda}^T \tag{1.21}$$

where g denotes the matrix g_{ij} and

$$\boldsymbol{\Lambda} \equiv \left(\frac{\partial x^i}{\partial x^{i'}} \right). \tag{1.22}$$

Now we can calculate the matrices $\boldsymbol{\varepsilon}$ and $\boldsymbol{\mu}$ of the ε^{ij} and μ^{ij} according to the recipe (Eq. 1.11) and (Eq. 1.14). There we need the determinant g of g_{ij}, which is $g = (\det \boldsymbol{\Lambda})^{-2}$ according to formula (Eq. 1.21). We thus obtain

$$\boldsymbol{\varepsilon} = \boldsymbol{\mu} = \frac{\boldsymbol{\Lambda}\, \boldsymbol{\Lambda}^T}{\det \boldsymbol{\Lambda}}. \tag{1.23}$$

Equation 1.23 formulates a simple recipe for calculating the required electromagnetic properties of a spatial transformation device. This recipe is valid in Cartesian coordinates, where both virtual space and physical space are described by Cartesian grids, but it can be easily extended to curved coordinates [9, 11].

1.3.3 *Perfect Imaging with Negative Refraction*

Apart from invisibility cloaking, another prominent application of transformation optics is perfect imaging [9]. Imagine a device that performs the following transformation in Cartesian coordinates:

$$x = x(x'), \quad y = y', \quad z = z', \tag{1.24}$$

where $x(x')$ is folded, as shown in Fig. 1.4. We see in Fig. 1.4 that in the fold of the function $x(x')$ each point x' in virtual space has three faithful images in physical space, so the electromagnetic field at these three points is the same as that at the point x'. Electromagnetic fields at each of the three points in physical space are therefore perfectly imaged at the other two: the device is a perfect lens. The transformation (Eq. 1.24) has the transformation matrix $\mathbf{\Lambda} = \mathrm{diag}\,(\mathrm{d}x/\mathrm{d}x',\ 1,\ 1)$ and we find from the recipe (Eq. 1.23):

$$\boldsymbol{\varepsilon} = \boldsymbol{\mu} = \mathrm{diag}\left(\frac{\mathrm{d}x}{\mathrm{d}x'}, \frac{\mathrm{d}x'}{\mathrm{d}x}, \frac{\mathrm{d}x'}{\mathrm{d}x}\right). \tag{1.25}$$

Inside the device, that is, inside the fold in the transformation of Fig. 1.4, the derivative $\mathrm{d}x'/\mathrm{d}x$ becomes negative and the coordinate system changes handedness. The electromagnetic left-handedness of such a material appears through a transformation to a left-handed coordinate system; the material is called a left-handed material and also a material with negative refraction. When the negative slope in the transformation is $\mathrm{d}x'/\mathrm{d}x = -1$, Eq. 1.25 gives a perfect lens made of an isotropic material with $\varepsilon = \mu = -1$ (otherwise the material (Eq. 1.25 is anisotropic). As Fig. 1.4 shows, the imaging range is equal to the thickness of the lens in this case.

Perfect lensing was first analyzed through the imaging of evanescent waves in a slab of negatively refracting material [24]. These are waves that may carry images finer than the optical resolution limit. Various aspects of this idea have been subject to a considerable theoretical debate (see Ref. [25]) but experiments have confirmed negative refraction (see, for example, Ref. [26]). Subresolution imaging was observed for a poor man's perfect lens [27] where the lens is effectively implemented by a subwavelength sheet of silver. Our pictorial argument leads to a simple intuitive explanation of why such lenses are indeed perfect. It also reveals some of the practical limitations of perfect imaging by negative refraction.

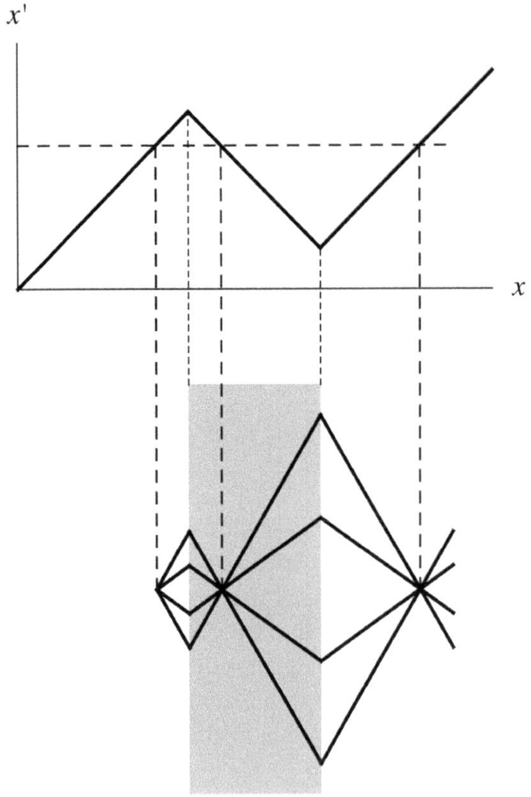

Figure 1.4 Perfect lens. Negatively refracting perfect lenses employ transformation media. The top figure shows a suitable coordinate transformation from the physical x axis to the electromagnetic x', the lower figure illustrates the corresponding device. The inverse transformation from x' to x is either triple or single valued. The triple-valued segment on the physical x axis corresponds to the focal region of the lens: any source point has two images, one inside the lens and one on the other side. Since the device facilitates an exact coordinate transformation, the images are perfect.

We have seen that if the imaging device performs the spatial transformation (Eq. 1.24) illustrated in Fig. 1.4 the electromagnetic field is identical in three separate regions of space. How is this possible? The electromagnetic field cannot instantly hop from one region to another—this is forbidden by relativistic causality, but it can settle to identical field structures over time in a stationary regime. But this implies that the stationary response of the electromagnetic material must be very different from the instantaneous response: the material must be dispersive. Dispersion is always accompanied by dissipation and the dissipation turns out [28] to severely reduce the resolution of imaging by negative refraction in left-handed materials. Therefore only poor man's perfect lenses have worked [27] where the imaging distance is a mere fraction of the wavelength. Nevertheless, the tantalizing ideas of negative refraction have inspired the entire research area of metamaterials and transformation optics [12].

1.4 Curved Space

Spatial transformations are simple and intuitive in theory, but often difficult to implement in practice. In particular, for the most interesting transformations such as invisibility cloaking and perfect imaging, fundamental problems prevent their practical realization in a meaningful way, as we have seen. The alternative to spatial transformations is the implementation of a curved virtual space, which is the standard case in isotropic media anyway (even non-impedance-matched isotopic media appear to electromagnetic waves as geometries, as long as the approximation of geometrical optics is valid [11]). The theory of light in curved space is harder, but the experiments are much easier than in implementations of flat space. In some cases, curved spaces have extraordinary optical properties that make their practical use in optical devices highly desirable. Let us discuss such a case that also represents the simplest curved space in 3D: the surface of the 4D hypersphere. There we can use a lot of the intuition we have about the surface of the ordinary 3D sphere to predict interesting physical phenomena without calculation, just by drawing pictures.

1.4.1 *Einstein's Universe and Maxwell's Fish Eye*

Our case is closely related to a famous cosmological model due to Einstein [30] that happens to be wrong for the universe but, when turned into an electromagnetic device, may be very useful in down-to-Earth applications. Einstein assumed that the universe is static and for this case derived an exact solution of the equations of general relativity that describe how matter curves space–time. Astronomical observations have shown, however, that the universe is not static but expanding, yet this should not deter us from turning Einstein's solution into a practical device. In Einstein's static universe light propagates as if it were confined to the 3D surface of a 4D hypersphere in $\{X, Y, Z, W\}$ space with

$$X^2 + Y^2 + Z^2 + W^2 = a^2 \tag{1.26}$$

where the constant a describes the radius of the hypersphere. Suppose that the hypersurface (Eq. 1.26) is the virtual space of an optical device. Furthermore, the Cartesian coordinates $\{x, y, z\}$ of physical space shall be connected to the virtual hypersphere by stereographic projection (Fig. 1.5):

$$x = \frac{X}{1 - W/a}, \quad y = \frac{Y}{1 - W/a}, \quad z = \frac{Z}{1 - W/a}. \tag{1.27}$$

One easily verifies that the virtual space coordinates are given by the following inverse stereographic projection:

$$X = \frac{2x}{1 + r^2/a^2}, \quad Y = \frac{2y}{1 + r^2/a^2},$$
$$Z = \frac{2z}{1 + r^2/a^2}, \quad W = a\,\frac{r^2 - a^2}{r^2 + a^2} \tag{1.28}$$

where r denotes the radius in physical space with

$$r^2 = x^2 + y^2 + z^2 . \tag{1.29}$$

To deduce the effective geometry in physical space and hence the medium required to implement it, we express the line element in virtual space in terms of the differentials in physical space with the help of the inverse stereographic projection (Eq. 1.28). We obtain

$$\mathrm{d}s^2 = \mathrm{d}X^2 + \mathrm{d}Y^2 + \mathrm{d}Z^2 + \mathrm{d}W^2 = n^2 \left(\mathrm{d}x^2 + \mathrm{d}y^2 + \mathrm{d}z^2\right) \tag{1.30}$$

with the radius-dependent prefactor

$$n = \frac{2}{1 + r^2/a^2} . \tag{1.31}$$

From the line element (Eq. 1.30) we read off the metric tensor g_{ij}, its determinant g and its matrix inverse g^{ij} as

$$g_{ij} = n^2\, \delta_{ij}\,, \quad g = n^6\,, \quad g^{ij} = n^{-2}\, \delta_{ij}\,. \tag{1.32}$$

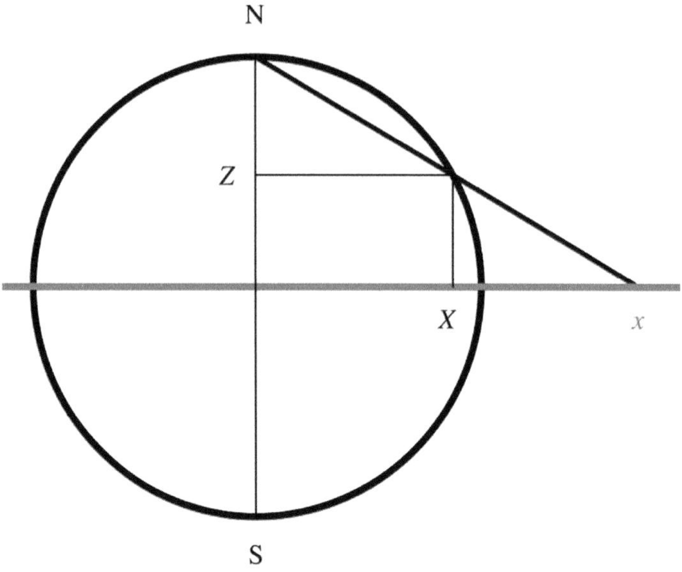

Figure 1.5 Stereographic projection. Points on the sphere (or hypersphere) are projected to the plane (or hyperplane) as follows. A line is drawn through the North Pole, N, and the point on the sphere. Where this line intersects the plane cut through the equator lies the projected point. The picture shows a cut through the sphere and plane where a point on the sphere is characterized by the coordinates $\{X, Z\}$ and the point on the plane by x.

According to Eqs. 1.11 and 1.14 this spatial geometry corresponds to a medium with

$$\boldsymbol{\varepsilon} = \boldsymbol{\mu} = n\mathbb{1} \,. \tag{1.33}$$

As $\boldsymbol{\varepsilon}$ and $\boldsymbol{\mu}$ are proportional to the unity matrix $\mathbb{1}$ the medium is optically isotropic, and it has the refractive index profile (Eq. 1.31). Isotropic media are usually the easiest to implement, which is the reason why we have chosen the stereographic projection (Eq. 1.27) and not any other mapping from virtual to physical space.

The index profile (Eq. 1.31) of Einstein's static Universe [30] was written down by Maxwell as a student at Trinity College Cambridge [31]. Maxwell was not aware of its relation to the stereographic projection of a sphere—that was discovered in optics by Luneburg [32] much later—Maxwell was simply fascinated by the extraordinary optical properties of a device with the profile (Eq. 1.31). It reminded him of the eye of fish and therefore such a device is called Maxwell's fish eye.

1.4.2 *Perfect Imaging with Positive Refraction*

Two properties of the fish eye Maxwell found particularly fascinating: (1) light goes in circles and (2) all light rays from any point meet at a corresponding image point. These properties turn out to be simple mathematical consequences of the light propagation on the virtual hypersphere and the stereographic projection to physical space. Let us, instead of the 4D hypersphere, imagine an ordinary 3D sphere—the hypersphere is not much different (Fig. 1.6). The light rays on the sphere are the geodesics, the great circles. Now, the stereographic projection always maps circles on the sphere to circles in physical space [11] (some degenerate into lines, that is, circles with infinite radius). From this follows property 1 of Maxwell's fish eye—light goes in circles. Property 2 is the easiest to understand. Consider the great circles of light rays emitted from a point P on the virtual sphere. All great circles from a given point P must intersect at the antipodal point. To see this, just rotate P to the North Pole of the sphere. In this case the great circles are the lines of longitude, and they all meet at the two poles. Therefore, the geodesics from the North Pole intersect at the South Pole. If we rotate the point P back to its original position the rotated South Pole turns into the antipodal point. Now, as the stereographic projections maps the surface of the hypersphere to physical space, the trajectories of light rays are simply the projections of the great circles. Therefore they all must meet in physical space as well, at the stereographic projection of the antipodal point. All light rays emitted from any point in Maxwell's fish eye meet at the corresponding image point. Devices where all light rays from any point within an object region intersect at the points of the image region are called absolute optical instruments [33].

Maxwell's fish eye is an absolute optical instrument and it has other curious optical properties, too—for example, light goes in circles—but it has never been built in its original form (Eq. 1.31) for two good reasons that are connected to each other. First, the profile

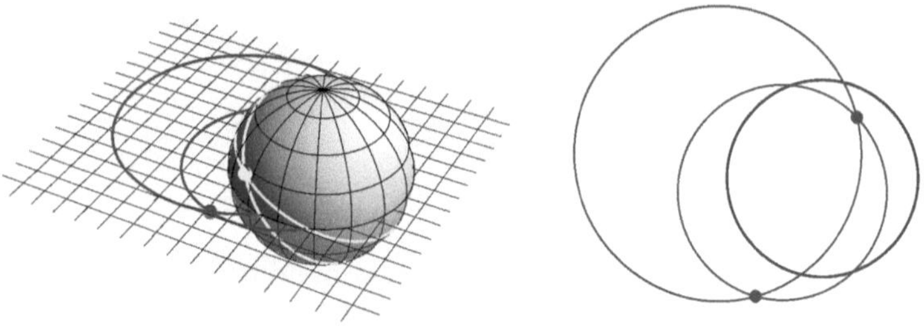

Figure 1.6 Light propagation in Einstein's universe and Maxwell's fish eye. The virtual space (left) is a sphere or hypersphere—Einstein's universe. Light follows the geodesics, the great circles here. All light rays emitted from one point must come together again at the antipodal point. The physical space (right) is Maxwell's fish eye with the refractive index profile (Eq. 1.31). As the stereographic projection (Fig. 1.5) maps circles into circles, light goes in circles in physical space. Moreover, all light rays emitted at an arbitrary given point must focus at a corresponding image point. Maxwell's fish eye makes a perfect imaging device.

of Maxwell's fish eye fills the entire physical space. Second, for $r \to \infty$ the refractive index (Eq. 1.31) tends to zero, that is, the speed of light becomes infinite at infinity. The two reasons are connected, because Maxwell's fish eye represents the geometry of a finite space, the virtual hypersphere, in an infinitely extended space. There the speed of light must go to infinity for keeping propagation times finite. However, there is a remedy [34, 35] that solves both problems in one stroke. Imagine we place a mirror around the equator of the virtual sphere. The mirror would create the illusion that the light propagates in the entire virtual sphere, whereas in reality it is confined to one of the Hemispheres, say the Southern Hemisphere. In physical space, the Southern Hemisphere corresponds to the region with $r \leq a$ where the refractive index ranges only from 1 at $r = a$ to 2 in the centre and the device is finite now. Therefore, Maxwell's fish eye with a mirror can be built, and recently it has been built [36, 37] in 2D. In 3D space the mirror should be a spherical shell at $r = a$ that encloses the index profile (Eq. 1.31) of the fish eye.

As long as ray optics is concerned, absolute optical instruments like Maxwell's fish eye create perfect images because all light rays from all object points faithfully arrive at the corresponding image points. However, the resolution of optical instruments is normally restricted by the wave nature of light [33] and cannot be made much finer than the wavelength. Is perfect imaging possible with Maxwell's fish eye? It is wise to consider this problem in virtual space, on the sphere (representing the virtual hypersphere for the 3D fish eye). Any source can be regarded as a collection of point sources, so it suffices to investigate the wave produced by a single-point source of arbitrary position on the sphere. A wave propagates from the point of emission round the sphere and focuses at the antipodal point; this corresponds with emission from a point in the plane of the actual device and focusing at the image point in physical space. The wave propagating round the virtual sphere would come to the antipodal point and focus there. Because of the symmetry of the sphere, the initially outgoing wave from the source points turns into an ingoing wave at the image point, that is, a time-reversed outgoing wave. However, the time reversal is only complete if one essential element is present at the image point: a reversed source, a drain. The drain at the image point is something natural in imaging where one wishes to detect an image, for example, by photochemical reactions or in a charge-coupled device (CCD) array. The drain represents a detector. Without the detector the image is not infinitely sharp, but limited by the wavelength. The perfect image may appear, but only if one looks [37].

The crucial point of perfect imaging is that, given a choice of detectors in the image area, the light localizes at the correct ones. The fact that the correct light localization naturally happens in Maxwell's fish eye is also understandable if we imagine the absorption in an array of detectors as the time reverse of the emission by a collection of point sources. Given perfect time symmetry, the light must settle down at the image points that correspond to the actual source points and avoid the ones corresponding to potential source points that did not emit. In this way, a sharp image is formed with a resolution given by the cross section of the detectors and not by the wave nature of light. Perfect imaging is possible with positive refraction [34, 35]. This idea has stirred up controversy [38] but theory [37] and experiment [39] indicate that it works.

1.5 Space–Time Media

So far we considered only spatial geometries for light—we showed that they appear as impedance-matched media and that impedance-matched media appear as spatial geometries. We discussed applications of spatial transformations and curved virtual space such as invisibility cloaking and perfect imaging. Let us now include time and expend our theory to space–time geometries.

1.5.1 *Space–Time Geometries*

We distinguish space–time coordinates x^α by Greek indices running for 0 to 3, where the 0-th coordinate refers to time, whereas purely spatial coordinates are indicated by latin indices. For example, $x^\alpha = \{ct, x, y, z\}$ are the Galileian coordinates of Minkowski space–time. The lines in space–time are world lines; they describe the trajectories of particles in space and time. The line element $\mathrm{d}s/c$ with

$$\mathrm{d}s^2 = -c^2\mathrm{d}t^2 + \mathrm{d}x^2 + \mathrm{d}y^2 + \mathrm{d}z^2 \tag{1.34}$$

in Minkowski space–time, measures the proper time experienced by a particle on its way. The line element $\mathrm{d}s$ characterizes the space–time geometry. The square of $\mathrm{d}s$ is a quadratic form of the increments of the space–time coordinates,

$$\mathrm{d}s^2 = g_{\alpha\beta}\,\mathrm{d}x^\alpha\mathrm{d}x^\beta\,. \tag{1.35}$$

As in the case of spatial geometries, we denote the determinant of $g_{\alpha\beta}$ by g and the matrix inverse by $g^{\alpha\beta}$; note that g is usually negative. For example, in Minkowski space–time, $g_{\alpha\beta} = \eta_{\alpha\beta}$ with

$$\eta_{\alpha\beta} = \mathrm{diag}\,(-1, 1, 1, 1) = \eta^{\alpha\beta} \tag{1.36}$$

and $g^{\alpha\beta} = \eta^{\alpha\beta}$ and $g = -1$. We know that purely spatial geometries act as impedance-matched media on electromagnetic fields; how do genuine space–time geometries appear?

1.5.2 *Magnetoelectric Media*

Plebanski [40] deduced a description of electromagnetism in space–time geometries that closely resembles the familiar form of constitutive equations. Here we will simply state Plebanski's result; readers interested in its derivation are referred to the appendices of Refs. [9] and [11]. Plebanski's constitutive equations are

$$\boldsymbol{D} = \varepsilon_0 \boldsymbol{\varepsilon} \boldsymbol{E} + \frac{\boldsymbol{w}}{c} \times \boldsymbol{H} \,, \quad \boldsymbol{B} = \mu_0 \boldsymbol{\mu} \boldsymbol{H} - \frac{\boldsymbol{w}}{c} \times \boldsymbol{E} \tag{1.37}$$

with the electromagnetic properties

$$\varepsilon^{ij} = \mu^{ij} = \mp \frac{\sqrt{-g}}{g_{00}} \, g^{ij} \,, \quad w_i = \frac{g_{0i}}{g_{00}} \,. \tag{1.38}$$

The constitutive equations (Eqs. 1.37 and 1.38) show that space–time geometries with $g_{0i} \neq 0$ mix electric and magnetic fields; they appear as magnetoelectric media (also known as bianisotropic media [41]). The magnetoelectric coupling vector $\boldsymbol{w}$ has the physical dimensions of a velocity. We will show in the next subsection that $\boldsymbol{w}$ is related to the local velocity of a moving medium. In a moving medium, the material responds to the electric and magnetic fields in locally co-moving frames where the medium appears to be at rest. There it is described by the constitutive equations (Eqs. 1.8 and 1.12). The Lorentz transformations to such locally co-moving frames mix electric and magnetic fields, which gives Plebanski's constitutive equations.

The dielectric tensors $\boldsymbol{\varepsilon}$ and $\boldsymbol{\mu}$ of Eq. 1.38 closely resemble the tensors (Eqs. 1.11 and 1.14) of purely spatial geometries, except that g and g_{00} are negative. Furthermore, only the ratio of $\sqrt{-g}\,g^{ij}$ and g_{00} matters to electromagnetic fields, and so does the ratio of g_{0i} and g_{00} in the magnetoelectric coupling, which reflects an important property of light (and electromagnetic radiation in general) known as conformal invariance. It originates from the fact that the proper time of a light ray is zero—light does not experience time. Therefore light does not recognize the magnitude of the line element (Eq. 1.35), but only the relative contributions of the increments $\mathrm{d}x^\alpha$ to $\mathrm{d}s^2$. This implies that we can multiply the line element (Eq. 1.35) by an arbitrary nonvanishing function of the space–time coordinates without any effect on the propagation of light. Thus suppose we make the following change:

$$g_{\alpha\beta} \to \Omega^2 g_{\alpha\beta} \,. \tag{1.39}$$

As $g^{\alpha\beta} \to \Omega^{-2} g^{\alpha\beta}$ and $g \to \Omega^8 g$, Plebanski's constitutive equations (Eq. 1.38) are invariant; hence electromagnetism is conformally invariant.

1.5.3 *Moving Media*

Let us discuss an instructive example of a medium that corresponds to a space–time geometry, the moving isotropic impedance-matched medium. We denote the local velocities of the medium by $\boldsymbol{u}$ where $\boldsymbol{u}$ may vary. For a given space–time point x^α we can always erect a locally co-moving frame where $\boldsymbol{u}$ vanishes at that point (where the medium is locally at rest). As the medium is impedance matched we can describe it in the locally co-moving frame by a space–time geometry with

$$g_{\alpha\beta} = \mathrm{diag}\left(-1, n^2, n^2, n^2\right) . \tag{1.40}$$

In view of the conformal invariance of electromagnetism we can replace this $g_{\alpha\beta}$ by

$$g_{\alpha\beta} = \mathrm{diag}\left(-n^{-2}, 1, 1, 1\right) . \tag{1.41}$$

Using the definition (Eq. 1.36) of the metric tensor of Minkowski space–time and introducing

$$u_\alpha = (-1, 0, 0, 0) \tag{1.42}$$

we can write $g_{\alpha\beta}$ as

$$g_{\alpha\beta} = \eta_{\alpha\beta} + \left(1 - n^{-2}\right) u_\alpha u_\beta . \tag{1.43}$$

Now, the Lorentz transformation from the locally co-moving frame back to the laboratory frame maintains the space–time geometry (Eq. 1.34) of Minkowski space and hence leaves $\eta_{\alpha\beta}$ invariant. Additionally, we write u_α in terms of quantities with a simple geometrical meaning in space–time:

$$u_\alpha = \eta_{\alpha\beta}\, u^\beta , \quad u^\alpha = \frac{\mathrm{d}x^\alpha}{\mathrm{d}s} \tag{1.44}$$

where $\mathrm{d}s$ refers to the Minkowski line element (Eq. 1.34). The u^α form the local four-velocity of the medium. In the co-moving frame $x^\alpha = (ct, 0, 0, 0)$ and $\mathrm{d}s = c\,\mathrm{d}t$ and so the four-velocity agrees with expression (Eq. 1.42). As $\mathrm{d}s$ is a Lorentz invariant, the four-velocity behaves like x^α under a Lorentz transformations to the laboratory frame. Applying Eq. 1.35 and $\boldsymbol{u} = \mathrm{d}\mathbf{x}/\mathrm{d}t$ we obtain the explicit expressions

$$u^\alpha = \frac{(1, \boldsymbol{u}/c)}{\sqrt{1 - u^2/c^2}} , \quad u_\alpha = \frac{(-1, \boldsymbol{u}/c)}{\sqrt{1 - u^2/c^2}} . \tag{1.45}$$

In this way we can easily express the tensor (Eq. 1.43) in the laboratory frame. This $g_{\alpha\beta}$ describes the space–time geometry established for light by the moving medium. It was discovered by Gordon [1] in 1923 and independently rediscovered several times [42–44]. The matrix inverse $g^{\alpha\beta}$ of $g_{\alpha\beta}$ is

$$g^{\alpha\beta} = \eta^{\alpha\beta} + \left(1 - n^2\right) u^\alpha u^\beta\,, \tag{1.46}$$

as one easily verifies by calculating the matrix product $g_{\alpha\gamma}\, g^{\gamma\beta}$ that gives δ_α^β, the unity matrix. With these expressions we can calculate the dielectric properties of moving media. They are particularly instructive in the limit of low velocities. In this case we obtain from Plebanski's constitutive equations (Eq. 1.38):

$$\boldsymbol{\varepsilon} = \boldsymbol{\mu} \approx n\,\mathbb{1}\,, \quad \boldsymbol{w} \approx (n^2 - 1)\boldsymbol{u}\,. \tag{1.47}$$

We see that the magnetoelectric coupling vector $\boldsymbol{w}$ is proportional to the velocity. The proportionally factor is the susceptibility $n^2 - 1$ that vanishes in empty space when $n = 1$. In the case of large velocities, $\boldsymbol{w}$ and also $\boldsymbol{\varepsilon} = \boldsymbol{\mu}$ depend in a more complicated way on the velocity of the moving medium [11].

1.5.4 *Space–Time Transformations*

Suppose the medium moves in one direction only, say, the z direction (but possibly with varying velocity), and that n may also only vary in z. We will show that the light propagation in the z direction is equivalent to a transformation in space and time; the one-dimensionally moving medium appears as a space–time transformation medium [9]. Our starting point is Gordon's space–time geometry (Eq. 1.46). If $\boldsymbol{u}$ has only a z component u and all other components vanish we have

$$g^{\alpha\beta} = \begin{pmatrix} \dfrac{u^2 - c^2 n^2}{c^2 - u^2} & 0 & 0 & \dfrac{(1-n^2)\,c\,u}{c^2 - u^2} \\ 0 & 1 & 0 & 0 \\ 0 & 0 & 1 & 0 \\ \dfrac{(1-n^2)\,c\,u}{c^2 - u^2} & 0 & 0 & \dfrac{c^2 - n^2 u^2}{c^2 - u^2} \end{pmatrix}. \tag{1.48}$$

We introduce the new coordinates t' and z' defined by

$$t' \mp \frac{z'}{c} = t - \int \frac{\mathrm{d}z}{v_\pm}\,. \tag{1.49}$$

Here $v_\pm$ denotes the relativistic addition of the velocity of light in the medium in either the positive or the negative direction, $\pm c/n$, and the velocity of the medium, u:

$$v_\pm = \frac{u \pm c/n}{1 \pm u/(c\,n)}\,. \tag{1.50}$$

Gordon's $g^{\alpha\beta}$ tensor appears in the new coordinates as

$$g^{\alpha'\beta'} = \Lambda^{\alpha'}_{\ \alpha}\, g^{\alpha\beta}\, \Lambda^{\beta'}_{\ \beta} \tag{1.51}$$

with the transformation matrix

$$\Lambda^{\alpha'}_{\ \alpha} = \begin{pmatrix} 1 & 0 & 0 & \dfrac{(n^2-1)\,c\,u}{c^2 - n^2u^2} \\ 0 & 1 & 0 & 0 \\ 0 & 0 & 1 & 0 \\ 0 & 0 & 0 & \dfrac{n(c^2-u^2)}{c^2-n^2u^2} \end{pmatrix}. \tag{1.52}$$

The result is the diagonal matrix

$$g^{\alpha'\beta'} = \mathrm{diag}\left(-\frac{n^2(c^2-u^2)}{c^2-n^2u^2},\ 1,\ 1,\ \frac{n^2(c^2-u^2)}{c^2-n^2u^2}\right) \tag{1.53}$$

with the inverse

$$g_{\alpha'\beta'} = \mathrm{diag}\left(-\frac{c^2-n^2u^2}{n^2(c^2-u^2)},\ 1,\ 1,\ \frac{c^2-n^2u^2}{n^2(c^2-u^2)}\right). \tag{1.54}$$

The determinant of the metric is

$$g' = -\frac{(c^2-n^2u^2)^2}{n^4(c^2-u^2)^2}\,. \tag{1.55}$$

The metric $g_{\alpha'\beta'}$ describes the geometry in virtual space–time. To find out how this geometry appears as a medium we use Plebanski's constitutive equations (Eq. 1.38) in virtual space–time, with primed instead of unprimed tensors. Since $g_{\alpha'\beta'}$ is diagonal, the magnetoelectric coupling vector $\boldsymbol{w}'$ vanishes: in virtual space–time the medium is at rest. For the dielectric tensors we obtain

$$\boldsymbol{\varepsilon}' = \boldsymbol{\mu}' = \mathrm{diag}\left(1,\ 1,\ \varepsilon'_{zz}\right) \tag{1.56}$$

with some ε'_{zz} we do not need to specify here. Since electromagnetic waves propagating in the z direction are polarized in the x, y plane their electromagnetic fields only experience the x and y components of the dielectric tensors. Consequently, for 1D wave propagation virtual space–time is empty, waves are free here. In virtual space-time, left- and right-moving wave packets are functions of either $t' + z'/c$ or $t' - z'/c$; in physical space–time they are modulated wave packets according to the transformation (Eq. 1.49). Instead of the velocity of light in vacuum they experience the relativistic velocity addition (Eq. 1.50) of the speed of light in the medium and the velocity of the medium.

Suppose that at some place, say $z = 0$, the velocity of the moving medium reaches the speed of light in the medium, as illustrated in the aquatic analogue of Fig. 1.7:

$$|u(0)| = \frac{c}{n(0)} \,. \tag{1.57}$$

Without loss of generality, we assume that $u < 0$ around $z = 0$ (the medium moves from the right to the left) and consider electromagnetic waves propagating against the flow as wave packets with velocity v_+ that, according to the addition (Eq. 1.50) of velocities, vanishes at $z = 0$. Linearizing v_+ at $z = 0$,

$$v_+ = \alpha z \,, \tag{1.58}$$

we see that the integral in the virtual coordinates (Eq. 1.49) develops a logarithmic singularity. Virtual space–time appears as

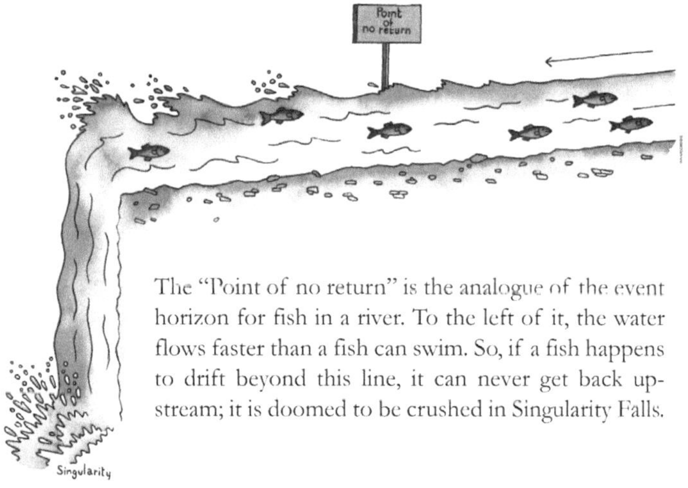

Figure 1.7 Aquatic analogue of the event horizon. Reproduced with friendly permission of Yan Nascimbene.

two branches, one corresponding to $z < 0$ and the other to $z > 0$, both being two independent Minkowski spaces. From this follows that right-propagating light confined to either the region where $z < 0$ or $z > 0$ will remain there; right-propagating light cannot cross $z = 0$, the horizon, where the velocity of the medium reaches the speed of light.

How should we create a horizon for light in practice? It seems rather hopeless for a realistic medium to reach the speed of light, because although normal media reduce the velocity of light, they do not reduce c by much; light is still rather fast. However [46], one can mimic a moving medium with an intense pulse of light in a suitable material with nonlinear optical response [45]. In such a case, the refractive index of the material n may get an additional contribution that is proportional to the intensity of the pulse (which is called the Kerr effect). The pulse enhances the refractive index as if an additional piece of material were added to the medium. As this effective piece of material is made by light, it naturally moves at the speed of light in the material. Consider two light fields now: the pulse and the probe light that experiences the pulse as a contribution to the refractive index. One can control the speed of both the pulse and the probe using the frequency and polarization dependence of the refractive index [46] and by the lateral profiles [47] of the light fields; different frequencies, polarizations, and lateral profiles propagate at different speeds. The place where the velocity of the probe is sufficiently reduced by the pulse to reach the speed of the pulse makes a horizon [46, 47].

Acknowledgments

I am most grateful for the discussions I had with many distinguished scientists about geometry, light, and a wee bit of magic. In particular, I thank my former group at St. Andrews and wish them well: Simon Horsley, Susanne Kehr, Thomas Philbin, Sahar Sahebdivan, and William Simpson. My work is supported by the European Research Council, the Israeli Science Foundation, and the Weizmann Institute of Science.

References

1. Gordon, W. (1923). *Ann. Phys. (Leipzig)*, **72**, 421 [in German].
2. Tamm, I.Y. (1924). *J. Russ. Phys.-Chem. Soc.*, **56**(2–3), 248 [in Russian].
3. Tamm, I.Y. (1925). *J. Russ. Phys.-Chem. Soc.*, **56**(3–4), 1 [in Russian].
4. R. Penrose (private communication).
5. Mahoney, M.S. (1994). *The Mathematical Career of Pierre de Fermat*, pp. 1601–1665 (Princeton University Press, Princeton).
6. Greenleaf, A., Lassas, M., and Uhlmann, G. (2003). *Math. Res. Lett.*, **10**, 685.
7. Leonhardt, U. (2006). *Science*, **312**, 1777–1780.
8. Pendry, J.B., Schurig, D., and Smith, D.R. (2006). *Science*, **312**, 1780.
9. Leonhardt, U. and Philbin, T.G. (2006). *New J. Phys.*, **8**, 247.
10. Chen, H., Chan, C.T., and Sheng, P. (2010). *Nat. Mater.*, **9**, 387.
11. Leonhardt, U., and Philbin, T.G. (2010). *Geometry and Light: The Science of Invisibility* (Dover, Mineola).
12. Service, R., and Cho, A. (2010). *Science*, **330**, 1622.
13. Schutz, B.F. (2009). *A First Course in General Relativity* (Cambridge University Press, Cambridge).
14. Schurig, D., Mock, J.J., Justice, B.J., Cummer, S.A., Pendry, J.B., Starr, A.F., and Smith, D.R. (2006). *Science*, **314**, 977.
15. Ma, Y.G., Ong, C.K., Tyc, T., and Leonhardt, U. (2009). *Nat. Mater.*, **8**, 639.
16. Zhang, B., Luo, Y., Liu, X., and Barbastathis, G. (2011). *Phys. Rev. Lett.*, **106**, 033901.
17. Chen, X., Luo, Y., Zhang, J., Jiang, K., Pendry, J.B., and Zhang, S. (2011). *Nat. Commun.*, **2**, 176.
18. Ma, Y.G., Sahebdivan, S., Ong, C.K., Tyc, T., and Leonhardt, U. (2011). *New. J. Phys.*, **13**, 033016.
19. Chen, H., and Chan, C.T. (2008). *J. Appl. Phys.*, **104**, 033113.
20. Li, J., and Pendry, J.B. (2008). *Phys. Rev. Lett.*, **101**, 203901.
21. Leonhardt, U. and Tyc, T. (2009). *Science*, **323**, 110.
22. Perczel, J., Tyc, T., and Leonhardt, U. (2011). *New J. Phys.*, **13**, 083007.
23. Zhang, S., Xia, C., and Fang, N. (2011). *Phys. Rev. Lett.*, **106**, 024301.
24. Pendry, J.B. (2000). *Phys. Rev. Lett.*, **85**, 3966.
25. Minkel, J.R. (2002). *Phys. Rev. Focus*, **9**, 23.
26. Valentine, J., Zhang, S., Zentgraf, T., Ulin-Avila, E., Genov, D.A., Bartal, G., and Zhang, X. (2008). *Nature*, **455**, 376.

27. Fang, N., Lee, H., Sun, C., and Zhang, X. (2005). *Science*, **308**, 534.
28. Stockman, M.I. (2007). *Phys. Rev. Lett.*, **98**, 177404.
29. Dzyaloshinskii, I.E., Lifshitz, E.M., and Pitaevskii, L.P. (1961). *Adv. Phys.*, **10**, 165.
30. Einstein, A. (1917). *Sitzungsberichte der Königlich Preußischen Akademie der Wissenschaften*, p. 142 (Berlin).
31. Maxwell, J.C. (1854). *Camb. Dublin Math. J.*, **8**, 188.
32. Luneburg, R.K. (1964). *Mathematical Theory of Optics* (University of California Press, Berkeley, Los Angeles).
33. Born, M., and Wolf, E. (1999). *Principles of Optics* (Cambridge University Press, Cambridge).
34. Leonhardt, U. (2009). *New J. Phys.*, **11**, 093040.
35. Leonhardt, U., and Philbin, T.G. (2010). *Phys. Rev. A*, **81**, 011804; Leonhardt, U., and Philbin, T.G. (2011). *Phys. Rev. A*, **84**, 049902.
36. Gabrielli, L.H., and Lipson, M. (2011). *J. Opt.*, **13**, 024010.
37. Leonhardt, U., and Sahebdivan, S. (2015). *Phys. Rev. A*, **92**, 053848.
38. Tyc, T., and Zhang, X. (2011). *Nature*, **480**, 42.
39. Miñano, J.C., Sánchez-Dehesa, J., González, J.C., Benítez, P., Grabovičkić, D., Carbonell, J., and Ahmadpanahi, H. (2014). *New J. Phys.*, **16**, 033015.
40. Plebanski, J. (1960). *Phys. Rev.*, **118**, 1396.
41. Serdyukov, A., Semchenko, I., Tretyakov, S., and Sihvola, A. (2001). *Electromagnetics of Bi-anisotropic Materials* (Gordon and Breach, Amsterdam).
42. Quan, P.M. (1957). *C. R. Acad. Sci. (Paris)*, **242**, 465.
43. Quan, P.M. (1957/1958). *Archive Rat. Mech. Anal.*, **1**, 54.
44. Leonhardt, U., and Piwnicki, P. (1999). *Phys. Rev. A*, **60**, 4301.
45. Boyd, R.W. (1992). *Nonlinear Optics* (Academic Press, San Diego).
46. Philbin, T.G., Kuklewicz, C., Robertson, S., Hill, S., Konig, F.,and Leonhardt, U. (2008). *Science*, **319**, 1367.
47. Belgiorno, F., Cacciatori, S.L., Clerici, M., Gorini, V., Ortenzi, G., Rizzi, L., Rubino, E., Sala, V.G., and Faccio, D. (2010). *Phys. Rev. Lett.*, **105**, 203901.

Chapter 2

Conformal Mapping in Transformation Optics

Kan Yao[a] and Yongmin Liu[a,b]
[a]*Department of Electrical and Computer Engineering, Northeastern University, Boston, MA 02115, USA*
[b]*Department of Mechanical and Industrial Engineering, Northeastern University, Boston, MA 02115, USA*
k.yao@neu.edu, y.liu@neu.edu

2.1 Introduction

The emerging technique of transformation optics (TO) has stimulated widespread interest in electromagnetics, optics and other fields related to wave physics [1–5]. Its mathematical foundations can be traced back to the 1920s when the theory of general relativity was intensively explored, or even to the earlier age as Fermat's principle was formulated [6, 7]. The investigation of Maxwell's equations in curved space–times lasted through the past century, and the form invariance of these governing equations of electromagnetic (EM) phenomena was then revealed by different means [8–10]. However, the early investigation mainly stayed at the

Transformation Wave Physics: Electromagnetics, Elastodynamics, and Thermodynamics
Edited by Mohamed Farhat, Pai-Yen Chen, Sebastien Guenneau, and Stefan Enoch

ISBN 978-981-4669-95-5 (Hardcover), 978-981-4669-96-2 (eBook)
www.panstanford.com

level of mathematics and theoretical physics, and the applications of this elegant property were rarely discussed. In 1996, aiming at simplifying the computation of light propagation in complex structures, Ward and Pendry studied Maxwell's equations in a general coordinate system [11]. It was found that the equations retain their expressions in the Cartesian coordinates, but the tensors of permittivity and permeability need to take a different form. The new tensors can be derived from the transformation that maps Cartesian frames into a new coordinate system, which defines the "push forward" of the original medium following mathematical terminology [3]. This equivalence between coordinate transformations and changes of material properties establishes a shortcut for computing fields in curved geometries. Soon afterward, the transformation approach was applied to the study of the perfect lens [12, 13], another theoretical milestone of metamaterials.

The real blossom of TO came out in 2006, manifested by two articles published in *Science* that discuss the protocol of invisibility cloaks [1, 2]. Despite a slightly earlier work by Greenleaf et al. employing the same transformation to formulate the nondetectable conductivities in electrostatics [14, 15], Pendry and coworkers developed a general design strategy in the context of EM fields at all frequencies [2]. The central idea of their cloaking device is to compress a solid volume into a shell by blowing up a point inside the volume. According to the aforementioned principle, the shell filled with the new medium, albeit anisotropic and inhomogeneous, will produce zero scattering regardless of the inclusion enclosed by the shell. In other words, a cloak creates the illusion of an empty space, although an object physically exists in that region. Subsequent simulations [16–18] and a proof-of-concept experiment at microwave frequencies [19] confirmed the cloaking effect, and this design was widely adopted as a paradigm in later research on cloaking of other waves [20–26]. The successful demonstration of the invisibility cloak opened up unprecedented opportunities to control EM fields via the visual and intuitive manipulation of coordinate frames. Vast devices were proposed following the TO framework within just a few years [27–30].

An alternative recipe suggested by Leonhardt restricts the scope of invisibility within the accuracy of geometric optics [1].

Consequently, the formula to be worked with is the Helmholtz equation and the constitutive tensors reduce to scalar refractive indices. In the 2D case, Helmholtz equation is form-invariant under a specific class of transformations, namely conformal mapping if the refractive index is reshaped accordingly. Whereas the choice of transformations is limited to a certain subset, the advantage of Leonhardt's proposal is the preservation of the isotropic property of media, which can be implemented by gradient index (GRIN) materials [31]. Furthermore, because conformal mapping can be expressed in the complex plane, it is possible to employ abstract operations of light rays between multiple Riemann sheets. This allows the usage of advanced tools in mathematical physics and differential geometry to design and study analytically many complicated optical devices [32, 33].

The main focus of this chapter is to review the fundamentals and applications of conformal mapping in TO. The first section introduces the basics of conformal mapping and how this subset of transformations can eliminate the undesired anisotropy in 2D systems. Representative examples from the literature are presented subsequently, covering both the most fascinating prediction—the invisibility cloak and other practical devices with minimized anisotropy derived by quasi-conformal mapping (QCM). In Section 2.3, we discuss the usage of an analogy strategy for conformal-mapping-based TO design. By drawing intuitive analogies between optics and other subjects of physics, most anisotropic devices difficult in realization can find their isotropic counterparts with reduced singularities. The fourth section addresses the promising applications for plasmonics, where conformal mapping manifests its versatility when dealing with the surface plasmon polaritons (SPPs) and localized surface plasmons (LSPs). These examples demonstrate that the TO framework is even valid for subwavelength control of light in the near-field regime. Section 2.5 makes an attempt to apply conformal mapping to the construction of 3D devices. In spite of being anisotropic, the resulting material parameters are optimized to possess practically feasible values and are thus less difficult to implement. Lastly, a brief summary and an outlook are presented in Section 2.6.

2.2 The Basics of Optical Conformal Mapping

To illustrate the design procedure based on conformal mapping, let us first limit our discussion to 2D cases. The coordinates of the two planes in the physical and the virtual space are denoted by (x, y) and (u, v), respectively. Since complex analysis is a powerful tool for solving 2D problems, normally we consider the two planes as complex planes and adopt the complex notation $z = x + \mathrm{i}y$ and $w = u + \mathrm{i}v$ for the representation of coordinates.

A conformal map $w = w(z)$ is an analytic function on the complex plane that preserves local angles between any line elements. The real and imaginary parts of $w(z)$ need to satisfy the well-known Cauchy–Riemann conditions [34]

$$\frac{\partial u}{\partial x} = \frac{\partial v}{\partial y}, \frac{\partial v}{\partial x} = -\frac{\partial u}{\partial y}. \tag{2.1}$$

These two equations guarantee that w depends only on z but not on its conjugate z^*, and the Laplacian operator in the z plane is transformed into

$$\nabla_z^2 = |\mathrm{d}w/\mathrm{d}z|^2\, \nabla_w^2. \tag{2.2}$$

Therefore, the Helmholtz equation becomes

$$(|\mathrm{d}w/\mathrm{d}z|^2\, \nabla_w^2 + n_z^2 k^2)\psi = 0 \tag{2.3}$$

in the w plane. Equation 2.3 can be rewritten into the standard form

$$(\nabla_w^2 + n_w^2 k^2)\psi = 0 \tag{2.4}$$

if the indices follow

$$n_z = n_w\, |dw/dz|\,. \tag{2.5}$$

Apparently, the medium n_w in the virtual space is connected to the transformed index n_z by a scale factor given by the modulus of the map. This describes the design strategy of TO in the accuracy of geometric optics when the transformation is conformal [1].

On the other hand, since the Helmholtz equation is rigorously valid for certain polarizations in 3D, the same conclusion can be drawn from the process applicable to general transformations [35, 36]. In cases where the original medium is isotropic and characterized by scalar permittivity ε and permeability μ, the transformed constitutive tensors are given by

$$\varepsilon'(\mu') = \frac{\Lambda\varepsilon(\mu)\Lambda^T}{|\Lambda|}$$

$$= \frac{\varepsilon(\mu)}{\frac{\partial u}{\partial x}\frac{\partial v}{\partial y} - \frac{\partial u}{\partial y}\frac{\partial v}{\partial x}} \begin{pmatrix} \left(\frac{\partial u}{\partial x}\right)^2 + \left(\frac{\partial u}{\partial y}\right)^2 & \frac{\partial u}{\partial x}\frac{\partial v}{\partial x} + \frac{\partial u}{\partial y}\frac{\partial v}{\partial y} & 0 \\ \frac{\partial u}{\partial x}\frac{\partial v}{\partial x} + \frac{\partial u}{\partial y}\frac{\partial v}{\partial y} & \left(\frac{\partial v}{\partial x}\right)^2 + \left(\frac{\partial v}{\partial y}\right)^2 & 0 \\ 0 & 0 & 1 \end{pmatrix} \quad (2.6)$$

with Λ the Jacobian matrix $\Lambda = \frac{\partial(u,v,s)}{\partial(x,y,s)}$. Note that the out-of-plane dimension s is untouched. When the Cauchy–Riemann conditions are satisfied, the off-diagonal elements vanish and the tensor reduces to

$$\varepsilon'(\mu') = \varepsilon(\mu)\begin{pmatrix} 1 & 0 & 0 \\ 0 & 1 & 0 \\ 0 & 0 & \Delta^{-1} \end{pmatrix}, \quad (2.7)$$

where $\Delta = (\partial u/\partial x)^2 + (\partial u/\partial y)^2$ equal to the square of $|dw/dz|$in the 2D case. The only nonunity element is responsible for the management of the electric (magnetic) fields polarized perpendicular to xy plane [19], resulting in the same distribution of refractive index as described by Eq. 2.5. Because naturally occurring materials are almost nonmagnetic, a pure dielectric implementation of the transformed medium works perfectly for only transverse electric (TE)-polarized waves at any length scales. For any other arbitrary polarization, however, the validity is limited to the accuracy of geometric optics.

We begin with the invisibility cloak to discuss the application of conformal mapping. Unlike other recently proposed cloaking devices, such as plasmonic cloaks [37, 38] and mantle cloaks [39] based on scattering cancellation [40], TO-based invisibility relies on the detour of light to avoid a certain region in space [1, 2]. To achieve this by conformal mapping, one may take advantage of the analytic functions on the complex plane, which allows the connection of several Riemann sheets according to the Riemann

Mapping Theorem. Although the physical space contains only one plane, the multiple sheets in the virtual space provide large degrees of freedom to manipulate light rays [1, 41]. For example, consider a simple but important map

$$w = z + \frac{a^2}{z}, \tag{2.8}$$

where a is a constant. Named after the Russian physicist Nikolai Joukowsky for his tremendous contributions to the early development of aerodynamics, Eq. 2.8 is called the Joukowsky transform and is historically used in the study of airfoil [42].

In the TO design of invisibility cloaks, the virtual space, here the two connecting Riemann sheets denoted by w coordinates, is empty in terms of nontrivial EM properties. Under this assumption, light on each Riemann sheet propagates along straight lines as in the vacuum. However, as the two layers are connected by a branch cut between $w = \pm 2a$, light rays propagating in different directions have different termini: Those who avoid the cut travel on the same layer all the way straight to infinity, while the rest crossing the cut enter the other layer and end at a different unbounded edge, as shown in Fig. 2.1a. Seen in the physical space with z coordinates, the branch cut is transformed into a circle of radius a, which separates the images of the upper and lower Riemann sheet as its exterior and interior region, respectively. In particular, the infinity of the upper layer corresponds to the true unlimited boundary in the physical space, and the other infinity on the lower sheet corresponds to a dielectric singularity when z approaches 0. Therefore, the constant a in Eq. 2.8 characterizes the extension of the branch cut and the resulting index $n = |1 - a^2/z^2|$.

So far the new medium is not invisible yet. The ray traces in Fig. 2.1b illustrate the reason. For the rays shown in blue and green, they avoid the branch cut and can reach the true infinity as if the space is occupied by a vacuum. However, rays like the one in red go across the branch cut and are eventually absorbed by the singularity at the origin, forming a shadow behind. To overcome this problem, one needs to not only remove the singularity, but also guide the ray back from the interior to the exterior region at the correct location and direction. While it seems difficult to achieve directly such a profile in the physical space, the corresponding lower

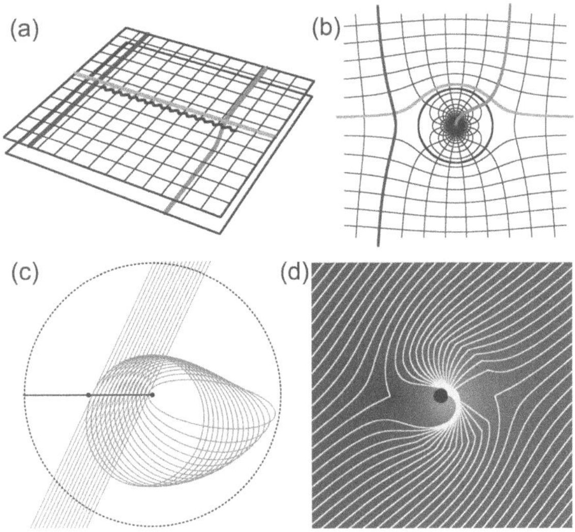

Figure 2.1 The invisibility device from optical conformal mapping. (a) Two Riemann sheets are connected through the branch cut indicated by the wavy black line. Light in this virtual space propagates along straight lines but will enter the other layer once crossing the branch cut, like the ray shown in red. (b) In the physical space, the black circle corresponds to the branch cut. The exterior space is mapped onto the upper sheet in (a) and the interior is mapped onto the lower sheet. Light entering the circle is absorbed by the dielectric singularity at the origin, which in (a) corresponds to infinity of the lower sheet. (c) An index profile analogous to the Kepler potential confines light to the interior of the circle shown by the dashed line. If locating the profile properly on the lower sheet in (a), rays (in blue) that cross the branch cut (shown by the black segment between the two points) will be guided back to their original routes by crossing the cut again after finishing the trip along the elliptic orbits (in red). Therefore, no light can reach the exterior of the circle. Anything put there is invisible. (d) Light propagation in the cloak, where the black core represents the invisible region and the brightness of the background indicates the index profile. (a, b, d) From Ref. [1]. Reprinted with permission from AAAS. (c) From Ref. [32]. © IOP Publishing & Deutsche Physikalische Gesellschaft. CC BY-NC-SA.

Riemann sheet is empty, sufficient to set a nontrivial raw medium compelling light to propagate in closed trajectories. Following these trajectories, light on the lower sheet crosses the branch cut again after finishing one loop and rejoins the upper layer at the same location and direction as it once left there. Meanwhile, the orbits of light should be limited within a finite boundary regardless of the angle of incidence. For the space beyond this borderline, light cannot reach there and as a consequence, anything placed in that area is invisible. Furthermore, noting that the lower Riemann sheet is mapped inside-out by Eq. 2.8 [32], the original dielectric singularity is neatly excluded by the hidden area.

Several index profiles capable of guiding light in closed trajectories have been studied, such as the harmonic-oscillator profile

$$n_1 = \sqrt{1 - r^2/r_0^2} \tag{2.9}$$

related to a Luneburg lens [43], the Kepler profile

$$n_2 = \sqrt{r_0/r - 1} \tag{2.10}$$

corresponding to an Eaton lens [44], and the Maxwell's fish eye (MFE) [43]

$$n_3 = \frac{n_0}{1 + (r/r_0)^2}. \tag{2.11}$$

In all the equations, r_0 and n_0 are constants that control the scale of the indices on the lower sheet. Figure 2.1c illustrates how light is guided with the help of the Kepler profile. The harmonic-oscillator profile works in a similar way, yet the MFE occupies the entire first interior sheet and demands additional layers to hide an object. In Fig. 2.1c, rays on the empty exterior sheet are visualized by the parallel blue lines. Once passing through the branch cut, shown by the black segment between the two points, rays on the lower sheet travel along elliptic orbits but cannot overstep the dashed circle. Finally, they go across the cut for a second time after finishing a loop and continue to propagate along the original blue lines as if nothing has happened. Figure 2.1d shows the ray traces in the physical space, in which the index ranges from 0 to 36 [1]. Nevertheless, the upper limit can be optimized to a much more feasible value by, for instance, using mirrored MFE [45].

Given the above isotropic invisibility cloak, the analytic design process employs abstract operations of light as well as advanced mathematical theorems. Sometimes a rigorous solution does not even exist. To make the strategy more practicable, later advances suggested the use of QCM [46–49]. The QCM design is performed usually via numerical techniques. Before proceeding we sketch its basic idea with the canonical TO formulae.

In Eq. 2.6, the elimination of the off-diagonal elements is not necessarily from Cauchy–Riemann conditions. Alternatively, satisfying the Beltrami equations [35]

$$f\frac{\partial u}{\partial x} = \frac{\partial v}{\partial y}, \frac{1}{f}\frac{\partial v}{\partial x} = -\frac{\partial u}{\partial y} \tag{2.12}$$

also gives a diagonal tensor. The coefficient f is the anisotropic factor dependent on x and y, which characterizes the ratio of the first two elements lying in the xy plane, or specifically, the ratio of the in-plane refractive indices along different basis vectors [50]. When f is unity, Eq. 2.12 reduces to the Cauchy–Riemann conditions and hence, the in-plane index becomes isotropic as defined in Eq. 2.7. Nonetheless, if the anisotropy is minimized to a sufficiently small level, one can drop it and approximate the profile as an isotropic medium. In the design process, when this condition is achieved, the corresponding coordinate frames are close to be orthogonal everywhere. This is how the term "quasi-conformal mapping" originates.

In order to reach the optimal solution, there are a variety of grid generation techniques available [51], such as minimizing the Modified–Liao function [46], solving the inverse Laplace's equation with sliding boundaries [52], etc. These approaches lead to similar profiles in spite of the different functionals used in the optimization. We omit detailed discussion on this point and introduce instead two applications of QCM.

The first case is a carpet cloak [46, 52–57]. Unlike previous cloaks to hide an isolated object, the carpet cloak provides a different camouflage strategy that renders an uneven surface flat. Without loss of generality, we consider a bump, which in principle could have an arbitrary curvature, on a flat, reflective surface. Optically, the bump ruins the mirror-like response of the plane to external incidence and scatters light into different directions. Therefore, cloaking the bump means to recover the reflection property of the surface. From TO point of view, the upper half plane has its bottom boundary lifted up by the bump, sustaining a local squeeze of volume. Under a conformal mapping connecting the two geometries, this squeeze causes a distortion of coordinate frames

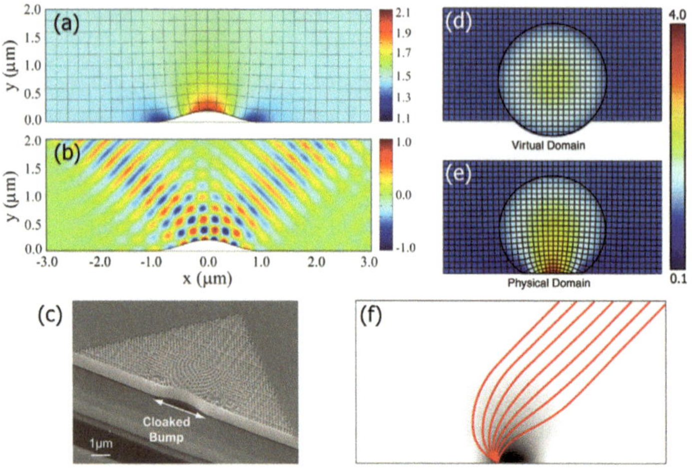

Figure 2.2 (a–c) The carpet cloak and (d–f) flattened Luneburg lens from QCM. (a) The transformed grid and profile of permittivity in the carpet cloak. (b) A Gaussian beam launched at 45° from the left upon the carpet cloak (outlined by the box) is reflected at the same angle to the right. (a, b) Reproduced from Ref. [5] with permission from The Royal Society of Chemistry. (c) Scanning electron microscope (SEM) image of an optical carpet cloak made of silicon. Reprinted by permission from Macmillan Publishers Ltd.: [*Nature Materials*] (Ref. [59]), copyright (2009). (d) A Luneburg lens in vacuum is compressed into (e) a slab of transformation medium. The grids and color maps visualize the space distortion and refractive indices, respectively. (d, e) Reprinted from Ref. [48]. With permission of Springer. (f) Light is focused onto the flattened image plane with the new medium in (e). Reprinted by permission from Macmillan Publishers Ltd.: [*Nature Materials*] (Ref. [47]), copyright (2009).

which averages the deformation into the entire half plane as a new index distribution. In practical designs, the device has to be limited to a finite area [46, 53], where only QCMs can be found by allowing the transformed grids sliding along the edges. Figures 2.2a and 2.2b show a QCM solution and its cloaking effect on a bump. It should be stressed that the grid lines are almost orthogonal, resulting in an isotropic medium with a good approximation. Because of the achievable index ranges, experimental demonstrations of carpet cloaks have been performed at almost all frequency bands, covering from microwave to visible light [58–66]. An SEM image of an optical carpet cloak made of dielectrics is shown in Fig. 2.2c [59], in which the density of air holes in a Si slab is varied in order to realize the required refractive index profile (Fig. 2.2a).

Another interesting device is the flattened Luneburg lens [47, 67]. A Luneburg lens is a spherically symmetric GRIN lens known for the capability to focus incoming plane waves from all directions equally well. However, for a long time, its spherical focal locus restricts the usage in practice. QCM resolves this issue in 2D by reshaping the index profile into another geometry with a planar facet. As shown in Fig. 2.2d, a Luneburg lens is pressed into an air slab that serves as the background medium. This treatment is also applicable for imitating concave reflectors with flat dielectric slabs [68]. The transformed profile in Fig. 2.2e contains a high-index area near the bottom due to the large local deformation. Nevertheless, the anisotropy is minimized to the lowest extent. Figure 2.2f illustrates the ray traces (red lines) in a flattened lens, where the brightness indicates the index values. Realizations of this device have been carried out at microwave and infrared [47, 69, 70], yet for higher frequencies are still awaiting the advances of metamaterial fabrication techniques.

2.3 Transformation Optical Design with an Analogy Strategy

From the implementation point of view, conformal and quasi-conformal maps would be a preferred class of transformations for optical design. The anisotropy of the resulting materials is minimized to the lowest extent and thus the requirement on complex building blocks is relaxed. Although exhibiting this unparalleled advantage, surprisingly, devices derived from conformal mapping are still limited to a few prototypes in a long time [1, 46, 47, 53, 71], which is in sharp contrast to the boom in anisotropic designs. The reason of this unbalanced development is, conformal maps need to preserve local angles everywhere between the line elements. This is a strong extra constraint over general transformations. In TO design, while a general transformation is only required to fit appropriate boundary conditions with little concerns about the interior distortion of coordinate lines, a conformal map has to deal with both issues simultaneously. Rigorous analytic solutions exist in only several simple geometries or the unbounded, entire complex plane. As a consequence, the usage of conformal mapping in optical design is uncommon.

The barrier to purely mathematical treatment demands alternative answers. An analogy approach, inspired by the geometrical similarity of field line patterns in different physical systems, is one of the candidates providing approximate solutions. In this section, we review recent attempts in this direction. Of all the examples presented below, some are designed based on analogue problems in other physical systems, and some can be interpreted by drawing proper analogies, although originally they may be derived with a completely different approach. We note that there is another branch of analogy work aiming at extending TO to other subjects of physics [20–24, 72–75], which is based on the formal similarity of the governing equations. This topic will not be covered in this chapter. Alternatively, readers can refer to a recent review that offers a timely and comprehensive summary on its progress [76].

2.3.1 *Analogies with Fluid Mechanics*

By helping transfer concepts and methods from a mature field to a new one, analogy plays an important role in the study of many branches of physics. A famous example in electromagnetism is the analogy between lines of force (field lines) and fluid flow developed by Maxwell [77]. To some extent, this thinking is even valid for TO and other derived transformation physics, which treat field lines as if they are attached to the coordinate frames and give rules for how these lines can be manipulated by an equivalent material [78]. Taking advantage of the light-fluid analogy, the principle of cloaking can be interpreted with an intuitive picture, that light is guided to wrap around an object like water flows around a stone [2, 28]. Because conformal mapping is widely used in solving 2D flow problems, we begin by examining the possible optical counterparts of the elementary flows, with the expectation of benefiting from both physical intuition and mathematical convenience.

2.3.1.1 Optical sinks

According to the potential theory of fluid mechanics, a 2D irrotational flow is the linear superposition of a series of basic flows, including uniform flow, source/sink, doublet (dipole) and vortex, etc [42, 79]. Each of these elementary flows can be expressed by a simple analytic function $w = w(z)$ on the complex plane, where its real part illustrates the streamlines, imaginary part depicts the velocity potential lines, and modulus of the derivative $|dw/dz|$ gives the magnitude of flow velocity. For instance, a sink is described by

$$w_s = -\log z \tag{2.13}$$

and a vortex follows the form

$$w_v = \pm \mathrm{i} \cdot \log z. \tag{2.14}$$

The $+/-$ sign in Eq. 2.14 denotes the vortex is rotating clockwise/counterclockwise. Note that the streamlines and potential lines form a pattern of orthogonal grids in both flows but switch their roles, as shown in Fig. 2.3a.

It is interesting to revisit these analytic functions from the perspective of TO. Suppose they each carry out certain functionalities. The corresponding refractive indices are then given by $n = |dw/dz|$. Noticeably, the derivatives of Eqs. 2.13 and 2.14 are different only in a factor of the imaginary unit, meaning that they actually lead to the

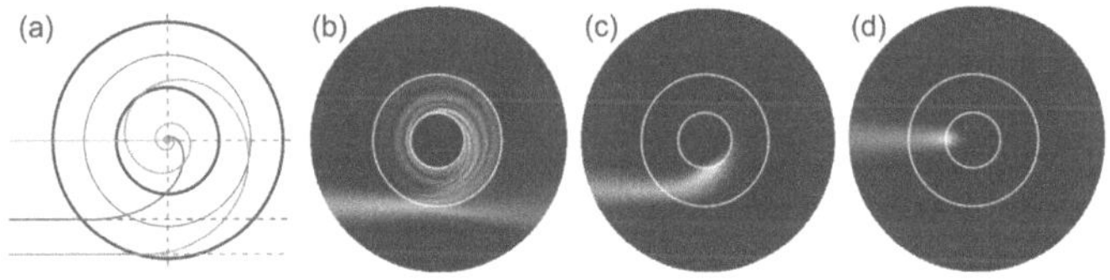

Figure 2.3 Sinks and optical black holes. (a) Two orthogonal sets of equivalue contours of the function log(z) characterize the streamlines/potential lines of a sink/vortex (dashed black lines, in the radial direction) and a vortex/sink (solid black lines, in the azimuthal direction), respectively. Meanwhile, $w = \log(z)$ transforms straight light rays incident from different positions in the virtual space into curved ones (red, blue, and green curves) ended at the origin in the real space. (b–d) An optical black hole wraps incoming light beams and captures them into the central region like a sink absorbs streams. (b–d) Reprinted with permission from Ref. [84]. Copyright The Optical Society.

same index profile $n = 1/|z|$. Then how does light propagate in such a medium? There are two ways to think of this problem. On one hand, following the general procedure of TO, one may expect that light rays travel straight through the virtual space. Under the mapping of Eq. 2.13 or 2.14, the path is transformed into a curved trajectory in the real space. By illustrating the images of several such testing rays in different directions, the functionality of the medium can be determined. On the other hand, however, the analogy between light and fluid can help us arrive at the same answer with much less mathematical consideration. In Fig. 2.3a, if replacing the physical meanings of the orthogonal lines with light rays and wave fronts, the chart is still valid for light propagating in an isotropic medium. Because a sink absorbs fluid isotropically from all directions, light carrying only radial momentum will be attracted straight toward the central singularity in the analogue medium. Similar to fluid in a vortex running in circles, light with purely tangential momentum will propagate in the same manner. Therefore for an arbitrary incidence, it should follow both trends and finally spiral into the center, rendering the profile $n = 1/|z|$ acting as an optical black hole [80–83] or absorber [84], as shown in Fig. 2.3b–d. The profile was originally derived by making optical-mechanical analogy, where either the Hamiltonian or the Lagrangian has to be employed. By comparison, the geometrically intuitive analogy translated from the flow of sinks might be the shortest pathway.

In addition to sinks, sources that emit fluid are also useful building blocks when constructing analogue problems. Systems containing sinks and/or sources, such as doublet ($w = 1/z$), quadrupole ($w = 1/z^2$) [84], Rankine half-body ($w = z \pm \log(z)$) [85] and Rankine body (e.g., $w = z + 1/z$) [1], can be used to create complex profiles for absorbing or guiding light. Although the field line patterns become abstract in the presence of multipoles, the key point is that all streamlines terminate at a finite area where sinks/sources are located.

2.3.1.2 Airfoil carpet cloak

Although Pendry et al. described cloaking in their pioneering work with an intuitive flow picture [2], field lines in an anisotropic cloak do not coincide with the geometry of streamlines. In their proposal, light is bent only inside the cloak shell instead of being distorted in the entire space like a real fluid flow does. Nevertheless, Leonhardt considered another case in his cornerstone paper, that is, an isotropic cloak with refractive index

$$n = \left|1 - a^2/z^2\right| \tag{2.15}$$

given by the Joukowsky transform (Eq. 2.8). The profile requires additional dielectric insertions to act as an omnidirectional cloak but is intrinsically invisible along x axis in the limit of geometric optics [86]. This unidirectional invisibility is usually interpreted in such a way that the cylinder $|z| = a$ is mapped onto a conducting sheet between $w = \pm 2a$ along the direction of incidence. But again, we can check it from the viewpoint of fluid mechanics. Noting that Eq. 2.8 describes the superposition of a uniform horizontal stream and a doublet, i.e., a special case of Rankine body, the field pattern should illustrate the chart of a flow around an object outlined by the zero streamline. In fact, Joukowsky transform presents the rigorous analytic solution of irrotational ideal flows past a circular cylinder or a semicylinder on a wall [42], as shown by the blue frames in Fig. 2.4a. If we replace streamlines (along the horizontal direction), potential lines (along the vertical direction), the bump and wall respectively with light rays, wave fronts and a reflective surface without changing any detail of the grids, the chart retains fitness and we can easily understand why the bump is invisible for incidence along x axis. The only difference is, the bending of field lines is not caused by the immersed object but rather a curved space, or equivalently, a change of medium. So far we have drawn the light-fluid analogy for a cloak by considering the similarity between field lines, which holds for only one direction. However, because of its wave nature, light in the transformation medium can exceed this limitation. As the Joukowski transform is a conformal map, the reflection angles at oblique incidence are preserved when the flat surface is reshaped into a curved one. In Fig. 2.4a, if we take into account the trajectories shown in color, the pattern actually represents the ray traces in a carpet cloak at different incidence angles [31, 53].

A carpet cloak constructed from an isotropic medium suffers from one intrinsic problem. Since the resulting profile fills up the

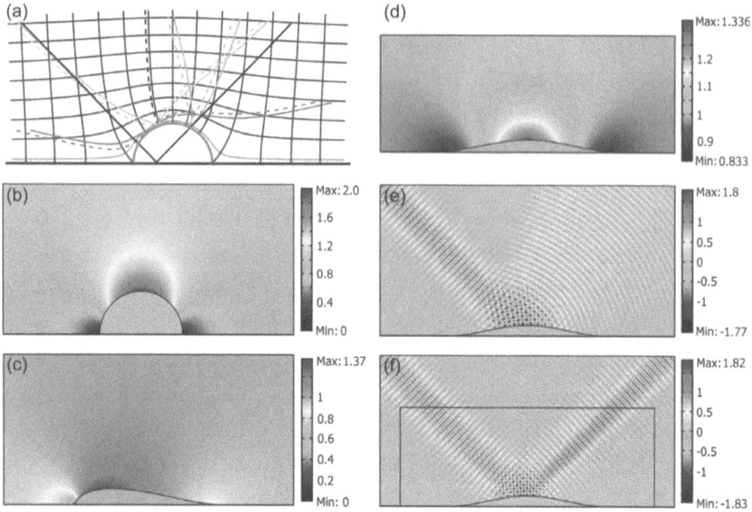

Figure 2.4 Airfoil carpet cloak. (a) The blue frames from the Joukowsky transform illustrate a uniform horizontal flow past a semicylinder. From the TO point of view, this pattern traces the light rays and wave fronts of a horizontal incidence in a carpet cloak. (b) The refractive index and flow velocity are equal in quantity, given by the derivative of the transformation. The dielectric singularities analogous to stagnation points appear at sharp corners. (c) An inverse Joukowsky transform maps the semicylinder into a semi-airfoil. The singularity on the right is removed by the smooth transition of the trailing edge. (d) Flipping over the rear of an airfoil creates a complete carpet cloak. The resulting index does not contain any singularities. Full-wave simulations of a bare bump (e) and a cloaked bump (f) illuminated by a Gaussian beam demonstrate the cloaking effect. The cloak is outlined by the black solid rectangle. From Ref. [53]. © IOP Publishing & Deutsche Physikalische Gesellschaft. CC BY-NC-SA.

entire space from the object to infinity, when it is truncated at a finite distance for implementation, a lateral shift of the reflected beam arises, making the system detectable [87]. The shift diminishes when increasing the size of the cloak or decreasing the abruptness of the bump. Thus in practical designs, the bump is always assumed to be compact and smooth. Obviously, the semicylinder carpet cloak does not satisfy these requirements. This will not only cause large imperfection in terms of the reflection shift but will also introduce fabrication challenges. Figure 2.4b depicts the index derived from Eq. 2.15, where two regions of low-index-containing dielectric singularities appear at the corners of the bump. To tackle these problems, we can still apply the light-fluid analogy approach, but we need to dive deeper from the geometrical similarity of field lines to some quantitative analogies. Fortunately, a pair of quantities which are both equal in value and intuitive in physical meaning can be found from the analogue systems. In TO, the refractive index under conformal mapping follows $n = |dw/dz| \cdot n_w$, with n_w being unity for an empty virtual space; while in fluid mechanics, the same formula $v = |dw/dz|$ gives the magnitude of flow velocity. Although the physical meanings are totally different, their quantitative equivalence allows us to predict and manipulate the indices of optical media with known solutions of fluid mechanics. From now on, a pattern similar to what is shown in Fig. 2.4b possesses dual meanings—it may refer to either an index profile or a flow velocity distribution, depending on the context.

When flowing around a semicylinder, the bottommost streamline has no way to veer but has to stop at sharp corners, forming two stagnation points therein. The adjacent stream, represented by the green curve in Fig. 2.4a, also slows down sharply near the corners queuing to bypass the bump. This analogue picture clearly reveals the origin of the low-index areas in Fig. 2.4b in the language of fluid dynamics, which might not be deduced such straightforwardly within optics itself. The removal of singularities becomes easy with reference to the recipes used in the analogue fluid problem. Here shown is a design analogous to an airfoil following a two-step transformation. First, we perform on Eq. 2.8 an inverse Joukowsky transform

$$z' = \frac{w' \pm \sqrt{w'^2 - a'^2}}{2}, \tag{2.16}$$

where z' and w' are respective coordinates before and after the transformation, and a' is a constant smaller than a. Equation 2.16 further maps the semicylinder into a semi-airfoil with zero angle of attack, whose outline is perpendicular to the ground at the leading edge on the left, but has a tangential tail at the rear part on the right. According to the Kutta condition, flow velocity must vanish at sharp corners while keep continuous on smooth surfaces. This contrast is clearly reflected in the color map in Fig. 2.4c.

The second step is to make both edges smooth to avoid any singularity. The exact solutions of this problem are elusive. Nevertheless, a simple approximation can be achieved by flipping over the rear of the airfoil and sewing up the two halves into a symmetric bump. Since an airfoil is almost flat near the top region, the operation impacts slightly on the smoothness of the coordinate lines. In Fig. 2.4d, we see that the final index exhibits a feasible range and its distribution is very close to those obtained with numerical techniques [46, 52]. The cloaking effect is examined by several challenging tests [53], where the wavelength is set at about half of the bump height in order to recognize any phase distortion caused by the object. As expected, the airfoil carpet cloak works well in keeping the shape of the reflected beams, while the shift is controlled to an acceptable level; see Figs. 2.4e and 2.4f for comparison.

2.3.1.3 Magnus carpet cloak

Starting from the semicylinder carpet cloak, the recipe for achieving feasible materials is not unique. Besides transforming the cylinder into smooth bumps, the additive property of elementary flows enables one to tune the field line patterns and optimize the indices intentionally with a suitable choice of streams. For example, inspired by the Magnus effect in fluid mechanics, we add a vortex term into the Joukowsky transform and then get

$$w = z + \frac{a^2}{z} + \mathrm{i}\Gamma \cdot \log(z), \tag{2.17}$$

where Γ is the strength of the vortex. The streamline pattern from Eq. 2.17 is sketched in Fig. 2.5a. One can see that compared with Fig. 2.4a, the streamlines in the upper half-plane are now bent more significantly when coming closer to the cylinder and intersect with the ground (denoted by the black solid lines). In proximity of the cylinder rotating clockwise, fluid at the original stagnation points receives kinetic energy from the vortex and begins to flow along the surface. Meanwhile, new stagnation points emerge somewhere on the lower half of the cylinder, as the motion of the original flow and the vortex cancel out. Consequently, the dielectric singularities, which are located at the intersections of the zero streamline (shown

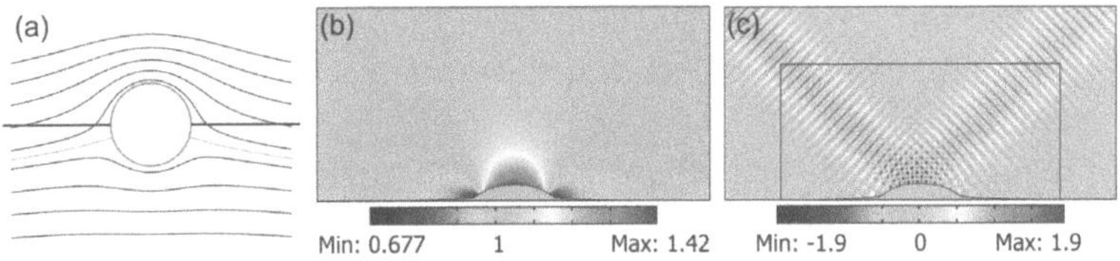

Figure 2.5 Magnus carpet cloak. (a) The presence of a vortex changes the velocity distribution around the circular cylinder. Stagnation points, denoted by the intersections of the green curves and the red circle, are pushed into the lower half-plane. Thus in the analogue carpet cloak, dielectric singularities are below the ground and excluded from the device. (b) The index profile of an elliptical Magnus carpet cloak. (c) Electric field pattern of a Gaussian beam incident on the cloak at 45° from the left. From Ref. [53]. © IOP Publishing & Deutsche Physikalische Gesellschaft. CC BY-NC-SA.

in green) and the cylinder, are depressed below the ground, outside the region of our concern.

Because a vortex has its magnitude decreasing with larger distance from the center, we expect that the distortion of field lines near the cylinder would be large, giving rise to unfeasible material parameters. To improve this, a flattened geometry is helpful. In Figs. 2.5b and 2.5c, we simulate an elliptical Magnus carpet cloak, which is obtained by transforming the circular cylinder in Fig. 2.5a into an ellipse. This treatment is conformal and is commonly used in fluid mechanics dealing with geometries other than circles. The outline of the bump is chosen to be a streamline nearly tangential to the ground when they intersect. Although the index range is larger than that of the airfoil design, which is the cost to fit a more prominent bump, the cloaking capacity is satisfactory. Moreover, one can choose another conformal streamline above the current outline for both a larger cloaking area and a smaller index range.

2.3.2 *Analogies with Electrostatics*

Conformal mapping is a common tool to help solve problems described by the Laplace's equation. Related subjects are not limited to fluid mechanics but also include electrostatics, heat transfer and classical mechanics, etc. This diversity leads to the fact that there are a huge number of field line patterns stemmed from very vast applications. Analogy also contributes to the study of the systems governed by Laplace's equation. For example, the point sources in fluid mechanics, electrostatics and steady heat transfer problems are analogous to each other, and their solutions all share the same expression. The versatility of the method of images is promised consequently, giving convenience to solving a class of problems. Although it is form-invariant under conformal mapping, Helmholtz equation that characterizes wave propagation was seldom treated in the same way before TO was developed. In spite of the different governing equations, it would be constructive for us to revisit the solutions known from nonoptical subjects. Below we consider analogies with electrostatics, which is well studied, and is more intuitive in connection with the concepts in optics.

2.3.2.1 Charge lenses

We begin by designing a collimator in 3D space, which consists of a point source and a lens with a flat aperture. The point source in optics has many counterparts in other branches of physics to generate flux. In order to illustrate field lines perpendicular to a certain surface, the most intuitive analogue might be the electrostatic fields suspended at a conductor. When setting a charge near a conducting plane, the electric field lines between the charge and the plane construct a pattern that fulfils all the features of the light rays inside a collimating lens. The orthogonal surfaces, namely the equipotential surfaces Φ, are bent as well to be coincident with wave fronts (denoted as S). Noting that conformal mapping is unable to serve in three dimensions, we have to use the eikonal equation $n = |\nabla S|$ to attain an isotropic design [43, 88]. However, equipotential surfaces cannot be simply defined as wave fronts. Because $\nabla\Phi$ is divergent at the charge, such assignment will cause a dielectric singularity at the position of the source. To avoid this trouble while keep taking advantage of the analogy between light rays and electrostatic field lines, we redefine the function of wave fronts with a scaling transformation [28]

$$S = \arctan(1/\Phi). \tag{2.18}$$

In this way, S satisfies the physical meaning of phase by being zero at the source and accumulating to a finite value at the plane.

Figure 2.6a illustrates a cross-sectional view of the resulting index. The boundary is chosen to fit an analogue electric field line parallel to the aperture at the apex. As can be seen, unfeasible indices only appear on the two sides, which can be mitigated by choosing a different truncating boundary or by immersing the lens into a background with index larger than 1. The full-wave simulation of 3D collimating is difficult to conduct due to the lack of isotropic sources. Nevertheless, the performance of the lens can still be examined by checking its capacity as a receiver. Figure 2.6b shows the focusing effect of the 3D collimating lens. A Gaussian beam launched vertically from bottom to top converges precisely on the apex, and the flat wave fronts in vacuum are well preserved with only slight reflection caused by the impedance mismatch at the aperture.

The charge lens in two dimensions is available from conformal mapping, which makes the derivation simpler [89–91]. Known from electrostatics [34], a Möbius transformation

$$w = \mathrm{i}\frac{z-\mathrm{i}}{z+\mathrm{i}} \tag{2.19}$$

maps radial field lines into curved ones that terminate at a charge and a conducting plane. Light rays in the lens reproduce this pattern, as shown in Fig. 2.6c.

While a conducting surface is equivalent to an image charge at a certain position, the medium derived from field lines retains the capacity to retrieve both amplitude and phase information of the analogue point source, giving rise to a subwavelength image at the position of the analogue image charge. Figure 2.6d illustrates a snapshot of the field pattern in this dipole lens [90]. We denote that the two peaks are not opposite in sign like their analogue charges. Rather, the image has a lag phase determined by the choice of transformation. Another interesting application we can obtain from electrostatics is to play the tricks of number of point sources [92, 93]. In vacuum two spherical waves interfere and form complex ripples in the space. By making an analogy with electrostatic field lines from point charges of the same sign, it is possible to realize waves that do not interference. Seen from the far field (Fig. 2.6e), the multiple point sources radiate like one.

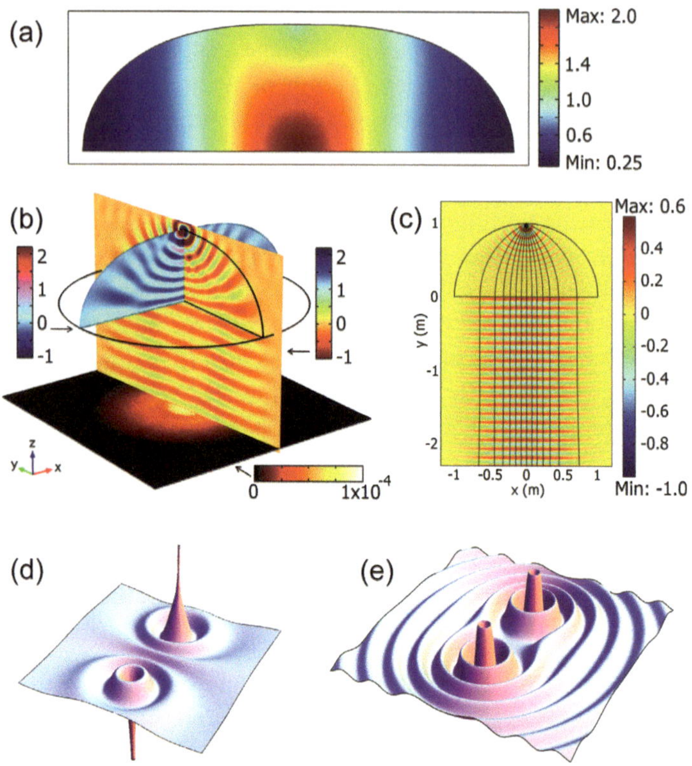

Figure 2.6 Charge lenses. (a) The cross-sectional view of refractive index of the 3D collimator. (b) A Gaussian beam incident from bottom is focused on the apex. Different color maps are used for clarity. (a, b) From Ref. [53]. © IOP Publishing & Deutsche Physikalische Gesellschaft. CC BY-NC-SA. (c) The 2D collimating lens converts a cylindrical wave into a highly oriented beam. Reprinted with permission from Ref. [89]. Copyright The Optical Society. (d) A dipole lens capable of subwavelength imaging is analogous to a charge and its mirror image, and (e) a medium in which two point sources do not interfere is in reference to a pair of charges in the same sign. (d) Reprinted with permission from Ref. [90]. Copyright The Optical Society. (e) Reprinted from Ref. [93] by permission of Taylor & Francis Ltd, http://www.tandfonline.com.

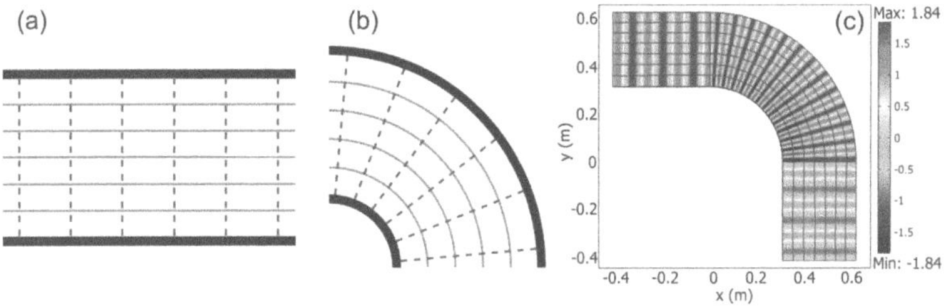

Figure 2.7 The field lines in a planar capacitor (a) and a cylindrical capacitor (b) can be used to describe the wave propagation in waveguides. (c) Magnetic field distribution inside a 90° waveguide bend. The field is launched from the left, and black curves show the trajectories of power flow.

2.3.2.2 Capacitor waveguide bend

To calculate the capacitance of a cylindrical capacitor, the classical method is to transform it conformally into a planar capacitor and translate the known solution of the simple configuration [34]. Figures 2.7(a) and (b) show respectively a part of the planar and cylindrical capacitors, where the colored solid lines illustrate equipotential lines and the dashed lines represent electric field lines. Denoting the coordinates in these two systems as $z = x + \mathrm{i}y$ and $w = u + \mathrm{i}v$, the transformation is normally expressed as

$$z = e^{\mathrm{i} \cdot w/A}, \tag{2.20}$$

where A is an arbitrary scaling factor. We have seen this formula before. It is the inverse form of Eq. 2.13 or 2.14 that gives rise to $n = A/|z|$ in the real space. Fixing v as a constant, obviously, the end of vector z will draw a circle of radius $|z|$ when u moves for a distance of $2\pi A$. Therefore, Eq. 2.20 transforms lines parallel to u axis on the w plane into concentric circles on the z plane, coincident with the geometry of the capacitors in Figs. 2.7a and 2.7b. One can also prove that lines parallel to the v axis are mapped into radial half-lines. Again, the field line pattern from electrostatics can be interpreted in a different manner by replacing equipotential lines with light rays and electric field lines with wave fronts. Figure 2.7c shows the magnetic field pattern inside a bend of 90°. We can see clearly that how power flow lines curve as the equipotential lines do in Fig. 2.7b, while wave fronts are perpendicular to the boundaries, aligned in the transverse direction.

2.4 Transformation Plasmonics

Surface plasmon polaritons are EM waves at a metal surface that are coupled to the collective electron density oscillations [94, 95]. Owing to the unique features of tight field confinement and strong field enhancement, SPPs are among the most promising candidates for a wide variety of applications such as subwavelength imaging [96], nano optical circuitry [97], and biosensing [98], etc. To engineer the propagation of SPPs for a specific functionality, the most general strategy is patterning the metallic films with subwavelength structures [99, 100]. However, the patterning introduces unavoidable scattering loss through which the energy bound at metal–dielectric interfaces is partially transferred into free space as propagating waves [94, 101]. This undesired effect, in addition to the intrinsic Ohmic losses of metals, further shortens the propagation length of SPPs and limits their utility. TO is established on the operation of Maxwell's equations, which give an exact description of EM waves on all length scales. Hence in principle, transformation media are capable of molding the flow of SPPs without extra loss caused by geometric changes. Particularly, devices designed from conformal mapping guarantee isotropic responses to the fields polarized perpendicular to the cross section and parallel to the interface. This implies that we can avoid the demand of fabricating anisotropic materials on the nanometric scale. In this section, we review the theoretical and experimental advances in transformation plasmonics.

2.4.1 *Transformation Optics for SPPs*

As demonstrated in Section 2.2, under a conformal mapping, the permittivity and permeability of the transformed medium are identical, uniaxial tensors. The two elements in the transformation plane retain unity, and only the third component varies in the way assigned by the square of Eq. 2.5. SPPs are transverse magnetic (TM)-polarized EM waves bound at metal–dielectric interfaces. The strict control of SPPs hence relies on a proper magnetic response out of the cross-sectional plane. In practice, however, the lack of optical materials with magnetic properties prevents it being feasible.

Fortunately in many applications, the size of the transformation domain is sufficiently large compared with SPPs wavelengths, and the distortion of coordinate grids is not severe. These conveniences allow us to characterize a plasmonic device with the effective mode index of SPPs, which can be carried out by considering merely the dielectric property of materials [102, 103].

A subsequent concern arises from the field distribution of SPPs. As bound surface waves, SPPs have the maximal intensity at the surface along which they propagate, and the fields decay exponentially with distance into both sides of the interface. When approaching the surface plasmon resonance frequency, the energy carried by each part of the fields, either in metal or in the dielectric, tends to be half the total energy. This implies another impracticable challenge that metal also needs to satisfy the transformed index profile. On the other hand, in the frequency range apart from the resonance, which is of more interest in practice because of the relatively low loss and long propagation length of SPPs, the majority of total energy resides in the dielectric side. For example, in the case of an air–silver interface, more than 95% of the total energy of SPPs is in air for wavelength larger than 600 nm. As a result, rational implementation of plasmonic devices can involve only transformed dielectrics, which are isotropic and nonmagnetic, but leave the metallic part unchanged.

2.4.1.1 Carpet cloak for SPPs

Based on the above discussion, some designs in the preceding sections can be reconsidered in the context of plasmonics. A representative example is the carpet cloak. The functionality of this device renders it particularly suitable for the purpose of scattering elimination on a patterned surface. Figure 2.8a shows the schematic of SPP propagation through a carpet cloak, where a ridge on a metallic slab is covered by a transformation medium indicated by the transparent cladding. The SPP waves, whose magnetic component is polarized parallel to the ridge line (y axis), propagate on the plane with a typical intensity profile decaying away from the surface. Once encountering the cloak, SPPs are guided smoothly to climb over the ridge without being scattered or deflected, as denoted

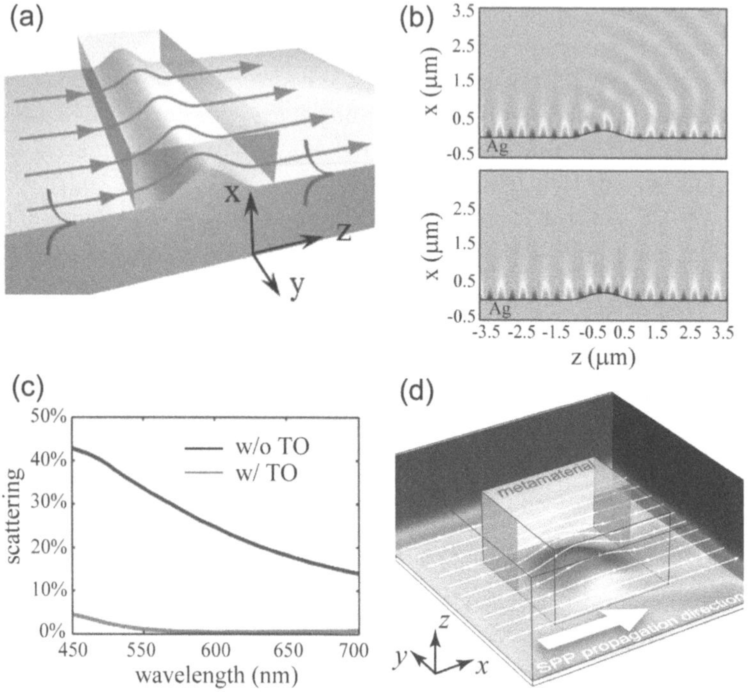

Figure 2.8 (a) Schematic of the SPP propagation past a ridge covered by a carpet cloak. (b) Top panel: a bare bump on the air–silver interface scatters partial SPP waves into free space. Bottom panel: SPPs are well confined on the uneven surface, if the dielectric half-space is filled with a transformation medium obtained by approaches introduced in preceding sections. (c) Comparison of scattering loss of SPP energy at different wavelengths without (blue curve) and with (red curve) the cloak. (a–c) Reprinted with permission from Ref. [103]. Copyright 2010 American Chemical Society. (d) Power flow (shown in color map) and streamlines of SPPs in a 3D anisotropic carpet cloak. Reprinted with permission from Ref. [102]. Copyright 2010 American Chemical Society.

by the arrowed streamlines in red. If we neglect the intrinsic loss of the metal, all EM energy is transported by the surface waves. Thus the field intensity profiles before and after the ridge are identical.

Further numerical studies are performed to investigate a specific design quantitatively. Without loss of generality, let us consider a semi-infinite silver slab that has a single sinusoidal bump on the air–silver interface. The index profile of the cloak is available from QCM. Figure 2.8b shows the 2D simulation results of the same incident SPP propagating past the bump without and with the cloak, respectively. The SPPs are launched from the left boundary at 633 nm wavelength. It can be clearly seen, in the top panel, that there is an appreciable amount of energy (about 26%) scattered into the upper right quadrant, even the bump is smooth and shallow. Also evident on the right side of the bump is the reduced intensity of field, which confirms the loss of energy from another perspective. In contrast, nearly no scattered field is observable in the bottom panel, when the transformation medium is applied to the dielectric space.

Analogous to what has been done to evaluate the performance of a carpet cloak for propagating waves [54], the energy carried in each field component is calculated based on the simulation results. First of all, in accord with the field profiles in Fig. 2.8b, the SPP energy contained in the metal part is only about 3.05% of the total energy. This fact supports the previous assumption that the transformation in metal can be ignored. Furthermore, by integrating in the dielectric space the energy carried in the evanescent surface wave and in the scattered propagating wave respectively, the scattering loss of SPPs can be determined. Figure 2.8c compares the results of the above two configurations over a wavelength range from 450 to 850 nm. Within the entire spectrum, the cloak effectually suppresses at least 90% of scattering from the bump. In particular, for wavelengths larger than 600 nm, the cloaking effect is strikingly satisfactory by showing almost zero scattering. While for shorter wavelengths, the deviation of the effective index from the original full tensors is no longer negligible. Therefore, the scattering loss increases with decreasing wavelength, which reaches a maximal of 4.5% at 450 nm. A similar trend is observed for the scattering from a bare bump. The reason is, without presence of the transformation medium, the bump is more pronounced to SPPs of shorter wavelengths. The scattering loss varies from 14% to 43%, which is fairly larger than its counterpart from the Ohmic loss of silver (~4% at 633 nm).

It should be stressed that 2D anisotropic carpet cloaks for SPPs can also avoid magnetic response under proper approximations. In this case, only the in-plane elements of the dielectric tensor take

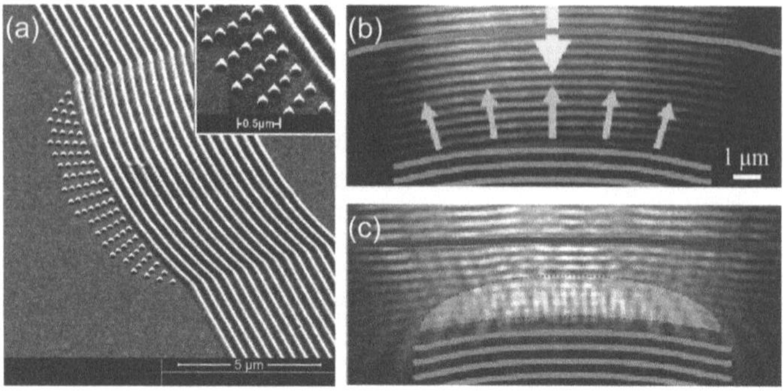

Figure 2.9 (a) SEM image of the crescent heterogeneous carpet cloak deduced from QCM. (b, c) Leakage radiation images show the beating patterns of SPPs impinging on a curved Bragg-type reflector (red solid lines) without (b) and with (c) the cloak (purple dotted lines), respectively. The wavelength is 800 nm. Reprinted with permission from Ref. [106]. Copyright The Optical Society.

effect, and it is possible to construct such cloaks with some naturally occurring crystals or composites [104]. For the concealment of 3D bumps, isotropic cloaks or even nonmagnetic cloaks are not available [105]. For example, a design by lifting up the coordinate grids leads to biaxial permittivity and permeability [102]. In order to guide SPPs from all directions correctly, the full implementation of the parameters is required. A simulation of SPP propagation in the 3D carpet cloak is shown in Fig. 2.8d, where the streamlines clearly indicate the detour of the power flow of SPPs over the bump.

Renger et al. experimentally demonstrated a QCM-based plasmonic carpet cloak [106]. The experimental configuration is shown in Fig. 2.9a. TiO_2 nanopillars are patterned on a gold surface to realize the required effective refractive index profile of the crescent-moon-like cloak. A curved Bragg-type reflector (15 gold lines) is used as the object to be hidden behind the carpet. Leakage radiation microscopy is employed to map the spatial distribution of SPPs at the air–gold interface. For a bare curved Bragg mirror, the reflected SPPs exhibit a convex wave front (Fig. 2.9b), following the curvature of the reflector. In contrast, the TiO_2 nanopillars recover a fringe pattern in the reflected SPPs with a nearly straight wave front, resembling the reflection from a flat Bragg mirror (Fig. 2.9c). Data analysis further quantifies that the wave front curvature is reduced by a factor of 3.7 in the presence of the TiO_2 carpet.

2.4.1.2 Plasmonic waveguiding devices

The transport of energy along curved surfaces is another application of wide interest in wave physics and engineering. As to the guiding of SPPs, the amount of energy leaked to free space depends strongly on the curvature of the surface, besides the surface roughness and discontinuities. When the curvature radius exceeds a critical value, the tangential wave vector of the SPP mode supported by the curved surface is smaller than the photon momentum in free space. Accordingly, the radial component of the wave vector becomes real, meaning that the corresponding EM fields cannot be confined at the surface but is leaked to free space as a radiation loss [107]. In Fig. 2.10a we show such an example by wrapping an air–silver interface into a 180° bend. As can be seen, almost all energy leaks to free space soon after entering the bending region. To let most energy of SPPs make a U-turn, appropriate dielectric claddings are needed.

TO has been applied to the creation of arbitrary waveguide bends [36, 52, 89]. Among various proposals, the isotropic bends are particularly attractive and have been realized on the microwave [108] and photonics [109] platforms. Taking the simplest index

$$n = A/r, \tag{2.21}$$

which is derived from the capacitor analogue in the last section, the energy will flow along the contour lines of r as if it travels on a flat geometry in the virtual space. A major advantage of this design is, the index contains only one dependent variable, the radius. This allows an arbitrary bending angle and an arbitrary choice of the factor A to minimize the impedance mismatch at the ends. Figure 2.10b illustrates the magnetic field of SPPs in a transformed bend. The radius of the inner boundary is 2 μm as in Fig. 2.10a and $A = 2.3$ μm. With the designed material, almost all SPPs propagate smoothly around the bend and reach the bottom interface, suffering little radiation loss. Again, the transmission of SPPs through the bend with and without the transformation medium is calculated over a

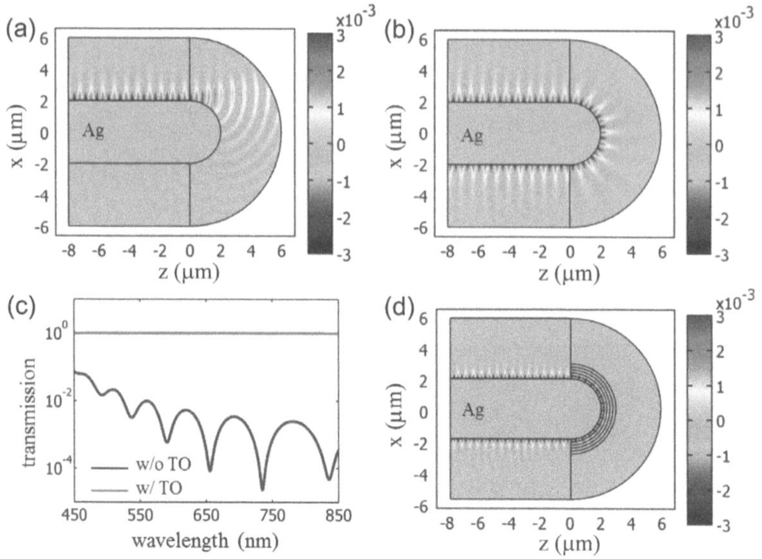

Figure 2.10 Plasmonic waveguide bend. (a) An SPP propagating on a U-shaped air–silver interface has most of its energy leak to free space. (b) Field distribution indicates nearly perfect transmission of SPPs after the transformation medium is applied. (c) Comparison of the transmitted SPP energy at different wavelengths without (blue) and with (red) the TO-based bend. (d) SPP propagation in a realistic design of the bend in (b), constructed by a multilayer structure. All fields shown are the magnetic components, and the wavelength is 780 nm. Reprinted with permission from Ref. [103]. Copyright 2010 American Chemical Society.

broad wavelength range. The results are compared in Fig. 2.10c. By neglecting Ohmic loss (about 1.7%) in the simulation, almost perfect transmission can be achieved in the transformed bend over the entire spectrum. In comparison, the transmission in the empty bend is orders of magnitude lower and decreases with increasing wavelength.

The feasible range and radius-dependent-only distribution of the refractive index largely relieves the fabrication difficulties. For instance, multilayers with proper dielectric constants can play effectively the role of the GRIN medium in Eq. 2.21. Figure 2.10d shows the magnetic component of an SPP propagating through a simplified bend made of a multilayer. The dielectric background of the straight waveguides is silicon dioxide instead of air, in order to shorten the penetration depth of SPPs. The refractive index of the bend is thus amplified by the same factor of 1.45. Benefiting from this change, the new bend can be implemented by only five 200 nm thick layers with refractive indices linearly changing from 1.595 to 1.045. Even if the original continuous profile is now highly discretized, most energy of SPPs can still reach the bottom interface with the guide of the annular dielectric stack.

Not limited to carpet cloaks and waveguide bends, TO has inspired various planar devices for plasmonic applications [110–112]. For example, many functional elements derived from the so-called embedded coordinate transformations [113] can be carried out on plasmonic platforms, although anisotropy is inevitable oftentimes. Representative designs include the waveguide adapters and couplers, which enable adiabatic coupling of SPPs between slabs with different thickness [104]. The required magnetic response is reduced to unity normally by scaling the dielectric tensor [17]. However, the materials are intrinsically nonmagnetic when an area-preserving map is applicable [114]. The implementation of the anisotropic permittivity profiles can then follow the same strategy used for macroscopic optical cloaks [115, 116]. For other devices such as beam shifters and splitters that detour light in the lateral dimension, realizations are still difficult due to the requirement of magnetic responses.

2.4.2 *GRIN Plasmonic Lenses*

The spatially varying material parameters produce one of the major challenges in the implementation of transformation media, especially at optical frequencies. As the dimension of the index variation decreases to the order of a few tens of nanometers, it is difficult to realize the required index profile by structuring the dielectrics or mixing nanoparticles [103]. Fortunately, the bound property of SPPs enables one to define the effective mode index, which can be adjusted by changing the thickness of a cladding dielectric film on the metallic slab [117]. In such dielectric/dielectric/metal structures, the dispersion relation of

SPPs is given by a transcendental equation

$$\tanh(k_2\varepsilon_2 d) = -\frac{k_1k_2 + k_2k_m}{k_2^2 + k_1k_m} \tag{2.22}$$

with

$$k_{1(2,m)} = \frac{\sqrt{\beta^2 - \varepsilon_{1(2,m)}\omega^2/c^2}}{\varepsilon_{1(2,m)}}, \tag{2.23}$$

where the subscripts 1, 2 and m correspond to the top dielectric background, the sandwiched dielectric cladding and the bottom metallic slab, respectively. The thickness of the cladding layer is d, and SPP wave vector is denoted as β. According to the above equations, an effective control of the mode index, defined as $n_{\text{eff}} = \beta/k_0 = \beta c/\omega$, can be achieved by gradually changing the thickness d within a moderate extent.

As a proof-of-principle demonstration, a plasmonic Luneburg lens is fabricated on a poly(methyl methacrylate) (PMMA) film on top of a gold surface. The refractive index of a Luneburg lens fulfills the expression

$$n = \sqrt{2 - (r/R)^2}. \tag{2.24}$$

Here, r is the radius measured from the center and R characterizes the extension of the lens. In the planar version, the index distribution reduces from being spherically symmetric to rotationally symmetric. The functionality of such a lens is to focus plane-wave-like SPPs at the rim of the lens for arbitrary incident angles. Figure 2.11a shows the 3D simulation result of an SPP propagating through a Luneburg lens in the positive z direction. One can clearly see that the launched SPP wave is focused precisely at the edge of the circle on the far side. By solving Eqs. 2.22 through 2.24, the index profile can be mapped to a height profile of the dielectric cladding. Figure 2.11b shows the SEM image of a fabricated Luneburg lens, which is a PMMA cone with a diameter of 13 μm in base and a maximum height of about 180 nm. In the fabrication, gray-scale electron beam lithography is used to tailor the height profile of PMMA with high accuracy.

The performance of the plasmonic Luneburg lens is verified by two different optical characterization techniques. Figure 2.11c provides the intensity image obtained by leakage radiation microscopy, which records the SPP beam propagating on the entire metal surface.

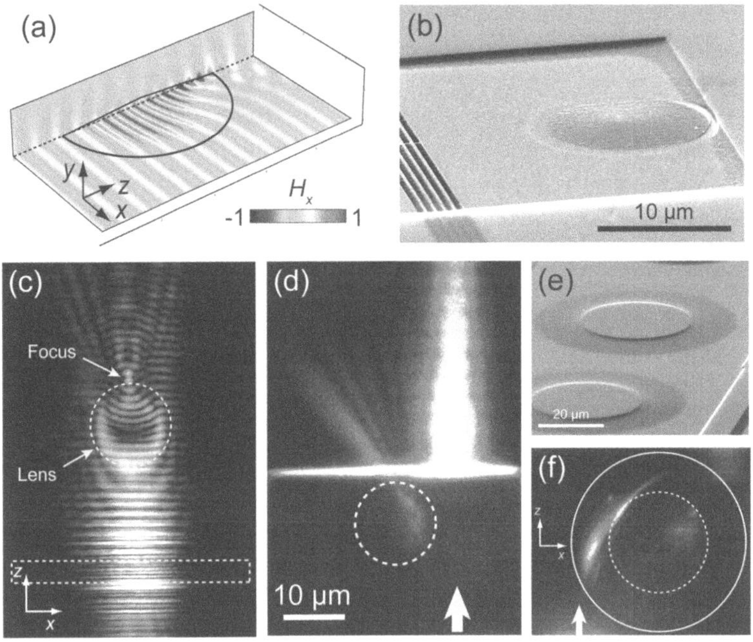

Figure 2.11 Plasmonic Luneburg lens (a–d) and Eaton lens (e, f). (a) Normalized magnetic field H_x for an SPP propagating through a planar Luneburg lens. (b) SEM image of a plasmonic Luneburg lens made of PMMA on a gold film. The grating on the left is made of air grooves for launching the SPPs. (c) Leakage radiation image for SPPs at 800 nm passing through a Luneburg lens. (d) Fluorescence image of the focusing of an SPP Gaussian beam launched from the bottom. The launching position is transposed to the right of the lens center. (e) SEM image of a plasmonic Eaton lens. (f) Fluorescence image of the bending of SPPs. The incident wave propagating along positive z direction is bent rightward after entering the lens region. Yellow arrows in (d) and (f) indicate the launching location and direction of SPPs. Reprinted with permission from Ref. [117]. Copyright © 2011 by Nature Publishing Group.

Briefly, the lens is fabricated on a thin film of gold (50 nm) on top of a 150 μm thick glass substrate. The thin film allows coupling of SPPs from the top interface to the bottom glass substrate, where a coherent leakage radiation is generated for imaging. In Fig. 2.11c, the fringes show clearly the phase evolution of the SPP from launching to focusing. The pattern is in good agreement with the simulation result in Fig. 2.11a despite the intensity at the center is dim, which is because the leakage radiation signal decreases when the mode index is large.

The second imaging approach is the fluorescence microscopy. In this case, dye molecules are incorporated into the PMMA layer. To demonstrate not only the focusing but also the propagation of SPPs after passing the lens region, the PMMA behind the lens is not removed. Figure 2.11d shows the fluorescence image of an SPP propagating through the sample. Although the launching location is shifted in the lateral direction for a considerable distance, the beam is still focused at the correct point on the perimeter. Moreover, the beam path of the SPP after the focal point is also recorded successfully in the expected direction. The high intensity area on the right is the result of the direct illumination of the incident SPP beam and the large number of dyes in the thick PMMA background.

Besides the Luneburg lens, a plasmonic Eaton lens is demonstrated using the same technique. Traditional Eaton lenses are also spherically symmetric devices [44], whereas they can be reduced to a planar version working in 2D. The refractive indices are described by implicit equations as a function of radius. Depending on the choice of index profiles, light rays passing through an Eaton lens will undergo different bending angles [6]. Figures 2.11e shows the SEM picture of a plasmonic Eaton lens designed to bend light for 90°. However, in the experiment, the bending angle is expected to be smaller, because the refractive index diverges at the center and is truncated to an achievable value. The fluorescence image of SPPs propagating through the lens is shown in Fig. 2.11f, where the turning of beam trajectory is clearly indicated by the fluorescence intensity. As many planar devices have a moderate range of refractive index, we foresee that more functional plasmonic elements can be realized and eventually, an integrated plasmonic circuit is achievable [118].

2.4.3 *Transformation Optics for LSPs*

Lastly, we briefly introduce the applications of TO for localized surface plasmons and light harvesting [29].

In the investigation of LSPs, the usual procedure is to determine the eigenfrequencies and in turn the potentials and fields of the system in the quasi-static limit by solving Laplace's equation [95]. On the other hand, it is well known that Laplace's equation is form-invariant under conformal mappings. Hence, the electrostatic potentials in two systems connected by a conformal mapping hold a one-to-one correspondence and the material in each domain is unchanged [119]. The consequent benefit is that the study of many complex structures, either irregular or singular in shape, can be performed in the counterpart systems belonging to the canonic catalogue of simple geometries [120, 121]. In other words, LSPs can be studied via the adiabatic transformation of SPP modes.

Figures 2.12a and 2.12c sketch two examples in 2D to illustrate this TO-based strategy. With the excitation of a dipole source located near the surface, the infinite metal structures, e.g., a metallic film or two semi-infinite metallic slabs separated by a dielectric film, support SPPs propagating along the surfaces over a broad spectrum. By applying a conformal transformation, here a complex inversion

$$z' = \frac{g^2}{z} \tag{2.25}$$

with g an arbitrary scaling factor, the infinite planar structures (shown in the top panels in z coordinates) are transformed into deep-subwavelength particles (bottom panels with z'), and the dipole source is converted to a uniform electric field. Since the properties of SPPs on flat interfaces have been studied thoroughly, the solutions in the transformed systems can be deduced in a remarkably simple manner. Shown in the top panels of Figs. 2.12a and 2.12c, in lossless systems, SPPs excited by the dipole will propagate along the slabs to infinity with constant magnitude. After the transformation, however, the SPP modes in finite structures couple to external fields, while the broad continuous spectra are preserved. In such systems, the mode profiles are wrapped consistently with the curved surfaces. During the propagation of SPPs toward the structure singularity with continuously shrinking wavelengths, the group velocity decreases in proportion and eventually vanishes. As a result, SPP waves accumulate toward the singular point but can never reach it, leading to a divergence of energy density. In practical systems, finite loss exists but a

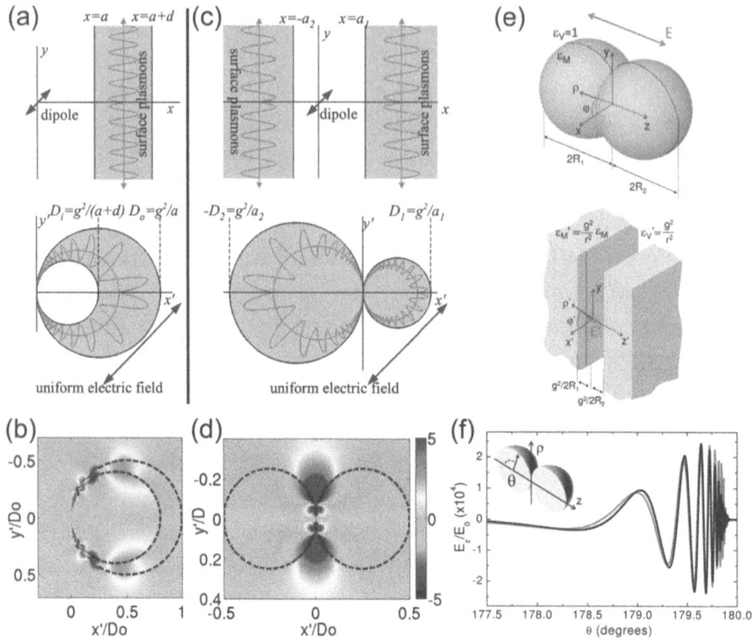

Figure 2.12 TO for LSPs. (a, c) Under a 2D complex inversion, a thin film of metal is transformed into a crescent (a), and two semi-infinite metallic slabs sandwiched with a thin film of dielectric are transformed into two kissing cylinders (c). In both cases, a line dipole exciting SPPs on the extended surfaces is converted into a uniform electric filed illuminating the nanoparticles. In the quasi-static approximation, the material properties are unchanged, and SPP modes are compressed toward the singular structures with preserved, broadband spectra. (b, d) The amplitude of the x' component of the electric field normalized by the x'-polarized incident plane wave in the crescent (b) and in the kissing cylinders (d). The color scale is truncated to ±5 for clarity. Simulations are performed for silver at $\omega = 0.9\omega_{\text{sp}}$. (a–d) Reprinted with permission from Ref. [119]. Copyright 2010 American Chemical Society. (e) In 3D, an inversion transforms two kissing spheres into two semi-infinite slabs. Moreover, the planar system is filled with materials having spatially dependent dielectric functions. (f) The electric field amplitude enhancement near the touching point of a silver dimer ($r = 20$ nm) at 500 THz. (e, f) Reprinted (figure) with permission from Ref. [128]. Copyright (2010) by the American Physical Society.

balance between energy accumulation and dissipation can always be established. Therefore, the above TO-based approach provides both physical insight and easy access to the design and analysis of broadband light-harvesting devices. Figures 2.12b and 2.12d, which correspond to the schematics in (a) and (c), show the normalized electric field in the crescent and in the kissing cylinders, respectively. From the field patterns, one can observe that the modes are excited predominantly at the broad parts of the structures and are squeezed toward the singularities with greatly enhanced amplitude.

The complex inversion transforms finite circular surfaces into infinite planar surfaces and vice versa. Other analytic functions, such as the complex exponentials and their compositions with inversion and translation, are capable of converting periodic structures into isolated objects. For periodic metallic films that couple to a dipole array, an exponential function transforms the system into a metallic wedge or a V-shaped groove coupling to a single dipole source [122]. By applying an additional Möbius transformation with proper coefficients, the wedge or groove is then mapped into a crescent or a pair of overlapping nanowires [122, 123]. The latter can be considered as a rough surface in the extreme case when one of the cylinders has a significantly larger dimension [124]. Separated nanowires [125, 126] and metallic rings [127] have also been investigated using the same technique. As can be expected, more complicated plasmonic structures will become accessible with the large family of conformal mappings [120].

The framework can be generalized to the study of 3D geometries. For instance, the 3D inversion connects a dimer of two kissing nanospheres to a pair of semi-infinite slabs, as shown in Fig. 2.12e. However, this extension is not a conformal mapping, so the transformed equation is no longer Laplace's equation and the refractive index of the transformed medium becomes spatially varying [128]. Such limitations restrict the use of 3D transformations to explore LSPs in more complex structures [129]. Despite these differences, the underlying physics is very close to that in the 2D problems, enabling the quantitative analysis of light-harvesting efficiency [130], nonlocal effects [131, 132] and robustness of blunt [133, 134]. Figure 2.12f displays the normalized electric field amplitude near the contact point of two identical kissing nanospheres, from which the remarkable field enhancement and superfocusing capabilities of the structure are clearly revealed. Lately, Pendry et al. have further extended the scope of TO beyond a description of the spectroscopy to the van der Waals interactions in 3D plasmonic systems [135, 136]. Undoubtedly, this powerful tool will keep showing its superior ability in computing EM phenomena at extreme length scales [29, 135].

2.5 Conformal Mapping in Anisotropic Devices

Stemmed from complex analysis, conformal mapping is considered an elegant technique for solving a variety of problems in 2D. As in TO, it leads to isotropic materials by which magnetic resonating structures are avoided and only electric resonant elements or even nondispersive dielectrics can work. A major concern is, nevertheless, that most real devices in practice are 3D. Only under a few circumstances the third dimension is not important [59, 106, 109] or is translationally invariant [62, 103], which allows treating the problems as in the 2D cases with reduced polarization. On the other hand, given by the Liouville's theorem, 3D conformal transformations are hardly available in Euclidean spaces [86], meaning that the elimination of anisotropy in 3D is of intrinsic difficulty. Therefore, the role of conformal mapping in constructing 3D anisotropic devices becomes a topic worth of special attention. This section introduces several representative examples, covering applications from lensing, imaging to cloaking.

2.5.1 *Devices from Stacked 2D Profiles*

Most 3D TO devices involving conformal mapping can be considered as the extensions of 2D profiles to some degree. For instance, if one imagines a 2D profile being an infinitesimally thin film with a refractive index varying in the plane, stacking a number of these sheets (infinite) or disks (finite) into a pile will cover a 3D volume. The functionality of the 3D profile could be either similar to or distinct from its 2D constituents, depending on how the layers are arranged. Although the new profile is normally

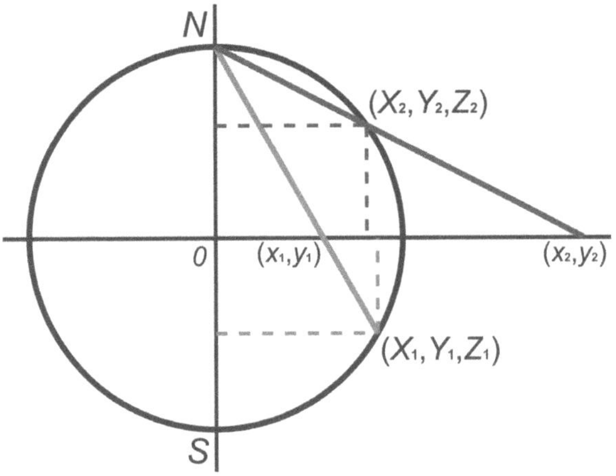

Figure 2.13 Stereographic projection in cross section. It maps each point (X, Y, Z) on the surface of the sphere to a corresponding point (x, y) on the plane and vice versa. The Southern Hemisphere is projected onto the disk bounded by the equator, while the northern half is mapped beyond.

anisotropic, it may still improve on other aspects, such as feasibility and compactness, compared with devices directly obtained from arbitrary 3D transformations.

2.5.1.1 Stereographic projection

We begin by introducing a powerful tool of geometry, the stereographic projection. As an exceptional example, it is a conformal transformation valid in spaces of dimension even greater than two [6]. For the convenience of discussion, here we consider only the 3D case, where a spherical surface is mapped onto a plane through its equator.[a] Figure 2.13 shows in cross section how the transformation works. For any point on the surface of the sphere, a straight line connecting the point and the North Pole N intersects with the plane at a unique point, and the reverse is true for any points on the plane. Therefore, a one-to-one correspondence is established between the points from the two geometries. Denoting the coordinates of points on the sphere by (X, Y, Z) and the plane by (x, y), the projection

[a]Some authors define the projection by mapping the sphere onto a plane tangent to it at the South Pole.

takes the form:

$$x = \frac{X}{1 - Z/R}, y = \frac{Y}{1 - Z/R}, \tag{2.26}$$

where $R = \sqrt{X^2 + Y^2 + Z^2}$ is the radius of the sphere.

Invented by Ptolemy for cartographic use, the stereographic projection is a powerful tool in the study of geometry and many of its applications. An interesting case from geometrical optics is the Maxwell's fish eye [43], a solution of the absolute instruments capable of imaging a 3D domain stigmatically [88, 137]. The derivation of MFE via stereographic projection is extremely simple and elegant. Comparing the line element on a spherical surface of radius R and its image after projection

$$ds = \sqrt{dX^2 + dY^2 + dZ^2} = \frac{2}{1 + (x^2 + y^2)/R^2}\sqrt{dx^2 + dy^2}, \tag{2.27}$$

the curvature of the sphere appears as a spatially dependent factor on the plane. The factor defines a change of measure of length, and for light it is equivalent to experience a medium with the refractive index

$$n = n(x, y) = \frac{2}{1 + (x^2 + y^2)/R^2}. \tag{2.28}$$

Hence the behaviors of rays in MFE can be studied conveniently by just considering their stereographic images, i.e., geodesic lines on a uniform sphere. Related investigations on this profile have recently extended to wave optics in the context of TO [138] and aroused broad controversy on whether perfect imaging can be achieved by introducing a mirror on the equator [139]. Although the debate has not been fully settled and it turns out that unlimited resolution by an MFE lens is unlikely to be realistic, the derived systems with subwavelength imaging capability were shown possible [140–142]. Further studies on this topic are still ongoing [143].

Another successful application of stereographic projection in 3D optical design is the non-Euclidean cloak [33]. In this work, the projection shows its superior ability on the translation of abstract operations on spheres and the implementation of the joint geometry of a sphere and a plane. Since the topic is out of scope of this section, we would like to refer interested readers to Refs. [33, 144, 145] and the supporting material therein for more details.

2.5.1.2 Collimating lenses and superantennas

A stereographic projection maps the surface of a sphere onto a plane and vice versa. Here we first take the consideration in the forward direction, i.e., projecting spherical surfaces onto planes.

As a starting point of optical design, it is natural to use spherical surfaces to visualize the wave fronts emanating from an isotropic light source. Subsequently, how to reshape this divergent wave into a collimated beam is the main focus of the design of collimating lenses and antennas. TO enables almost arbitrary manipulation of light trajectories and wave fronts, providing potential approaches to solve this problem [27]. Nevertheless, the main challenge in design arises just from the arbitrariness. To perform a certain functionality by TO, there are, in principle, an infinite number of transformations that can be applied as long as the requisite boundary condition is satisfied. Among this huge amount of candidates, making the choice that leads to a device with both feasible materials and a compact size would be rather difficult. In TO-based antennas, near-zero and/or very large values in the constitutive tensors often appear, resulting from the drastic changes of wave front curvature near the source or boundary discontinuities [146].

Fortunately, the stereographic projection offers an alternative way to think about this problem. Let us consider a sphere with a point source located at the center, which constitutes the transformation domain in the virtual space. Because the projection itself describes the mapping from curved wave fronts into the desired, flat ones, one may expect performing the transformation layer by layer, shrinking from boundary to source, to create the collimating lens. While it is reasonable to do so on the boundary, we cannot map all the concentric layers in this volume into flat sheets. When the spherical wave fronts shrink to the source, their curvatures tend to infinity; hence no media with finite parameters will suffice to flatten them. To remove such singularities, an unusual style is employed to represent the transformation domain as shown in Fig. 2.14a. The volume is now covered by a set of tangent Riemann spheres, which are fixed at a common North Pole N. Supposing we perform the stereographic projection on the Southern Hemisphere of each layer, a corresponding set of circular disks will pile up

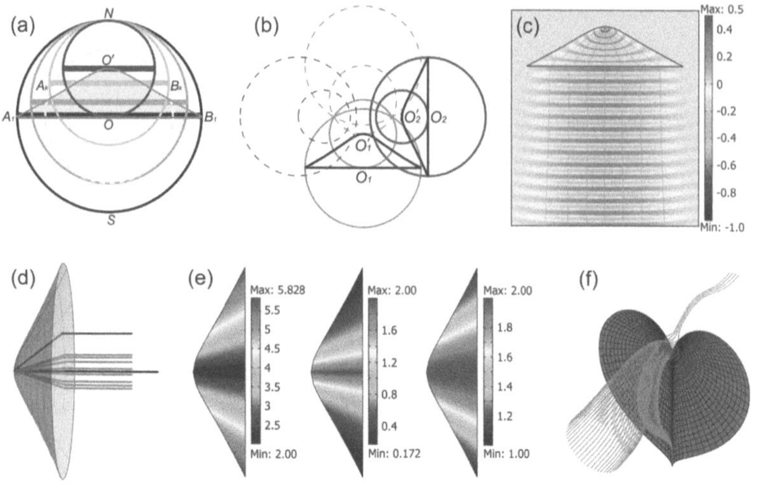

Figure 2.14 Light collimating and focusing by stacked transformation media. (a) Schematic of the transformation which projects tangent Riemann spheres onto parallel disks. The disks pile up and build the 3D lens (gray region). (b) The bispherical representation (in blue) of the transformation domain leads to a thinner lens, while the toroidal representation (in red) causes a thicker lens. (c, d) Light collimating is verified by 2D full-wave simulation (c) and 3D ray tracing (d), respectively. (e) Profiles of the constitutive tensor expressed in eigenvalues. (a–e) From Ref. [146]. © IOP Publishing & Deutsche Physikalische Gesellschaft. CC BY-NC-SA. (f) A superantenna focuses light into a needle-sharp beam and emits it at the South Pole. From Ref. [147]. © IOP Publishing & Deutsche Physikalische Gesellschaft. CC BY-NC-SA.

continuously in descending order of size like the Tower of Hanoi. One possible transformation that describes the above geometry is given by

$$x = \frac{X}{R_0 - Z} R, \; y = \frac{Y}{R_0 - Z} R, \; z = R_0 - R, \tag{2.29}$$

with

$$R = \frac{X^2 + Y^2 + (R_0 - Z)^2}{2(R_0 - Z)}. \tag{2.30}$$

Here R and R_0 denote respectively the radii of the inner Riemann spheres and the outer boundary, and only the volume below the equator A_1B_1 $(Z \leq 0)$ is used for lens construction.

We emphasize that the mapping is not an assignment of wave fronts except on the boundary, since the Riemann spheres do not coincide with equiphase surfaces, which are concentric layers. However, this geometry shifts the singularity from the origin to the North Pole, which is beyond the transformation domain actually involved. The curvature of each wave front layer is greater than that of the corresponding Riemann sphere, like, e.g., the yellow dashed line and green solid line in Fig. 2.14a, meaning that wave fronts in the physical space are no longer concentric but become more and more warped from the flat aperture to the point source. This gradual and smooth transition promises a feasible range of the resulting material parameters. In addition, the compactness of the device also benefits from the stacking strategy. From Fig. 2.14a, one can easily deduce that the thickness of the lens is limited to one quarter of its width, and this ratio can be further reduced if bispherical coordinates are adopted to represent the transformation domain, as shown in Fig. 2.14b.

Using Eqs. 2.29 and 2.30, we further derive the tensor of permittivity and permeability as in the following form

$$\overline{\overline{\varepsilon}} = \overline{\overline{\mu}} = 8 \cdot [1 + \frac{x^2 + y^2}{(R_0 - z)^2}]^{-1}$$
$$\times \begin{pmatrix} \frac{1}{4} + \frac{x^2}{(R_0 - z)^2} & \frac{xy}{(R_0 - z)^2} & -\frac{1}{2}\frac{x}{R_0 - z} \\ \frac{xy}{(R_0 - z)^2} & \frac{1}{4} + \frac{y^2}{(R_0 - z)^2} & -\frac{1}{2}\frac{y}{R_0 - z} \\ -\frac{1}{2}\frac{x}{R_0 - z} & -\frac{1}{2}\frac{y}{R_0 - z} & \frac{1}{4} \end{pmatrix}. \quad (2.31)$$

Noticeably, the material is anisotropic. The reason is that although the stereographic projection maps each Riemann sphere onto a flat sheet conformally, its extension is not conformal when performed layer by layer to cover a 3D volume. As a consequence, light rays in the collimating lens do not follow the direction of wave vectors as in isotropic media. While a hemispherical wave front emanating from the source is converted into a flat disk continuously, light rays act differently. As shown in Fig. 2.14c, inside the lens, light rays travel smoothly and intersect with the aperture at different angles. Once passing through the interface and entering the free space, they are confined abruptly to a uniform direction along the wave vector, which is always perpendicular to wave fronts. Seen in 3D view, rays within a light cone experience refraction at the interface and then emerge as a beam with high directivity; see Fig. 2.14d.

The constitutive tensor in Eq. 2.31 is expressed in a complicated form. To check the feasibility of the material for implementation, information from eigenvalues may be more desirable. Denoting the eigenbasis by (u, v, w), the tensor is diagonal with three elements being

$$\varepsilon_u = \mu_u = \frac{4}{1+\Delta}\left(\frac{1}{2} + \Delta + \sqrt{\Delta^2 + \Delta}\right), \tag{2.32}$$

$$\varepsilon_v = \mu_v = \frac{4}{1+\Delta}\left(\frac{1}{2} + \Delta - \sqrt{\Delta^2 + \Delta}\right), \tag{2.33}$$

$$\varepsilon_w = \mu_w = \frac{2}{1+\Delta}, \tag{2.34}$$

where $\Delta = (x^2 + y^2)/(R_0 - z)^2$. The third eigenvalue is particularly worthy of discussion. For a given height $z = z_0$, namely a certain sheet parallel to the aperture, the spatial dependence of ε_w and μ_w follows exactly the profile of MFE. This result is not surprising since we build up the 3D lens by stacking 2D disks, each of which is obtained from a stereographic projection. Although for each infinitesimally thin layer, the MFE deals only with light travelling on it, when piled up, the volume contributes to the control of light in the third dimension despite not being able to maintain isotropy. Figure 2.14e illustrates the three eigenvalues in the cross-sectional view. We see that ε_w and μ_w indeed range from 1 to 2 and vary as an MFE does on each horizontal layer, where the minimum value is tailored by the lens's outline. Meanwhile, the other two components are finely restricted within achievable limits.

The inverse mapping of stereographic projection is mainly considered when designing imaging systems, as we shall discuss in the next subsection. However, there are still some attempts on guiding light with this powerful tool [147]. Figure 2.14f shows the design of a "superantenna" consisting of concentric spherical layers. Mapped from the flat Riemann sheet used in 2D conformal cloak (see Figs. 2.1a and 2.1b in Section 2.2), each layer carries correspondingly a circular branch cut which traps light in a certain region in the physical space (or another Riemann sheet in the virtual space). Therefore, stacking these layers into a foliage geometry creates an anisotropic sphere containing a 3D heart-shaped branch cut. Any light that enters this region will be focused to the South Pole and emitted there within a needle-sharp beam. In contrast, light hitting the rest part of the sphere can propagate through as if the device is transparent. Although this extraordinary control of light requires zero and infinite indices at the South Pole and may not be a practical blueprint, it is still an interesting exploration on constructing 3D devices with 2D conformal transformations.

2.5.1.3 Geodesic waveguides for subwavelength imaging

In this section we will discuss the inverse mapping of stereographic projection. Because this transformation connects a uniform sphere to an MFE on the flat plane, many of its applications revolve around the imaging capability.

The subwavelength imaging property of a 2D MFE has been analytically demonstrated [90, 138, 148] and experimentally observed at the microwave regime [140]. Translating the profile along the surface normal for a small distance or stacking MFE layers to a finite thickness creates a 3D slab. Following the similar procedure in the 2D work, one can prove that the slab works still like an MFE does when the guided field is polarized along the slab thickness [149]. In other words, it is a planar waveguide filled with MFE profiles parallel to the boundaries. Under the inverse mapping of stereographic projection, the image of the planar waveguide is a spherical shell, as shown in Fig. 2.15a. The electric field inside is transformed consequently to be polarized along the radial direction, while the dielectric parameters of the inclusion can be chosen to be isotropic and inversely proportional to the radius. This result is straightforward if interpreting the transformation as a stack of 2D projections, where both the in-plane material parameters and out-of-plane polarization are preserved. Since TO is valid for EM fields at all length scales, researchers expect that the subwavelength imaging feature of the MFE waveguide will also be conserved after the transformation. Recently, superresolution up to $\lambda/105$ in

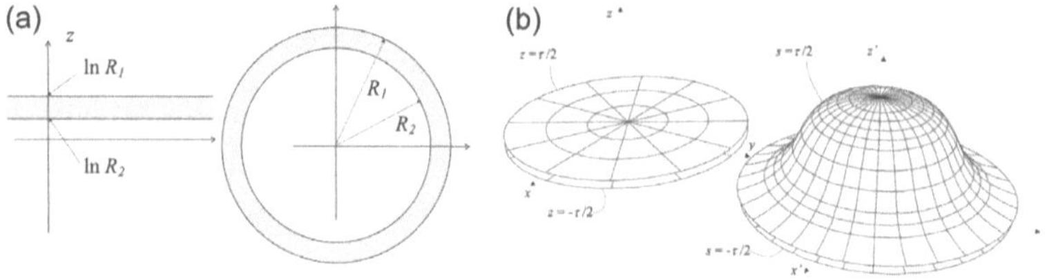

Figure 2.15 Geodesic waveguides for subwavelength imaging. (a) A planar waveguide filled with MFE layers is transformed into a spherical waveguide. The ability of subwavelength imaging is preserved during the transformation. (b) A planar waveguide filled with an inhomogeneous medium is transformed into a curved geodesic waveguide with rotational symmetry. From Ref. [149]. © IOP Publishing & Deutsche Physikalische Gesellschaft. CC BY-NC-SA.

the spherical waveguide has been theoretically and experimentally demonstrated in a narrow band of microwave frequencies [142]. The extension of bandwidth is still challenging and relies on deeper understandings of some fundamental problems on image formation.

The above idea can be generalized to more complicated geometries with irregular cross sections and rotational symmetry. Figure 2.15b illustrates such a schematic design. In this case, the stereographic projection is unable to serve because of the nonspherical geometry of the waveguide. Instead, other specific conformal maps are required to fit the boundary contours. Similar to the previous case, the transformed fields are polarized perpendicular to the surface of the geodesic waveguide. Also, the equivalue surfaces of the material parameters, owing to the layer-by-layer projection, are still conformal and parallel to the curved boundaries.

2.5.2 *Devices of Azimuthal Invariance*

As mentioned above, 3D conformal transformations are hardly available in Euclidean spaces, meaning that the construction of 3D isotropic TO devices would be of intrinsic difficulty. Not only stacking conformal transformations layer by layer gives rise to anisotropic profiles, translating such a map uniformly along the third dimension leads to anisotropy as well. Moreover, in order to deal with waves of arbitrary polarization, a strict implementation would require equal electric and magnetic responses [150]. This is normally unrealistic. Under some circumstances, the requirements can be relaxed, such as cloaking a ridge to incidence polarized along the ridge line, which is equivalent to a 2D problem and has been resolved by conformal mapping/QCM. In this translationally invariant, depolarized situation, a dielectric implementation can be applied [103]. However, when coming into three dimensions, the challenge remains.

One possible alleviant comes from conformal mapping [105]. Noting that translating a 2D profile uniformly along the third dimension is a special case of stacking, analogously, one can choose to perform a rotational translation to build a 3D device of azimuthal invariance. Figures 2.16a and 2.16b compare the schematics of carpet cloaking by these two operations. The benefit of rotating

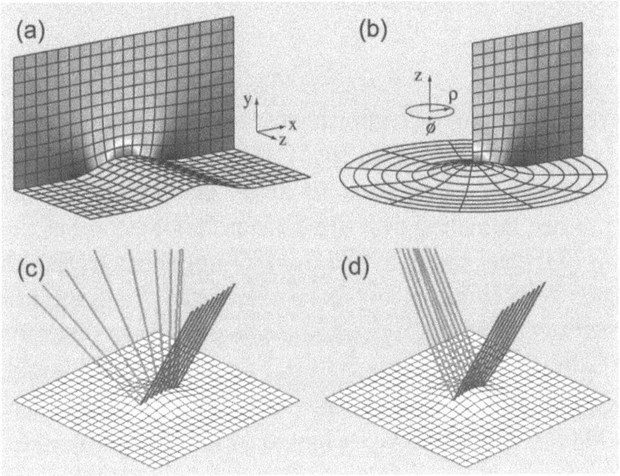

Figure 2.16 The 3D carpet cloak created by rotating a 2D profile. (a) A translationally invariant carpet cloak is either isotropic but valid only for one polarization, or anisotropic and with equal electric and magnetic responses in order to work for arbitrary incidence. (b) An azimuthally invariant carpet cloak is intrinsically anisotropic, but the anisotropy can be relaxed in the eikonal limit. (c, d) Comparison of light scattering by a 3D conducting bump without (c) and with (d) the simplified anisotropic carpet cloak. The incident and reflected rays are plotted in blue and red, respectively. Reprinted (figure) with permission from Ref. [105]. Copyright (2010) by the American Physical Society.

2D maps is not to achieve compact devices (as in the case of the collimating lens) or isotropic media (as in the geodesic waveguides). In fact, the resulting materials are intrinsically anisotropic. The transformed constitutive tensor in cylindrical coordinates can be expressed by

$$\varepsilon_\rho{}' = \mu_\rho{}' \approx \varepsilon_z{}' = \mu_z{}' \approx \beta,\ \varepsilon_\phi{}' = \mu_\phi{}' = \alpha/\beta. \tag{2.35}$$

Here α is the reciprocal of the Jacobian determinant of the transformation, which is a combination of rotation and the in-plane conformal mapping/QCM, and β is the ratio of the scale factor in the virtual space to that in the physical space. The tensor is approximately uniaxial. Compared with another design obtained from vertical compression of the volume [102], it is more convenient for implementation. In addition, the requirement on anisotropy in Eq. 2.35 can be further relaxed under the Eikonal approximation. The substitute tensors are

$$\varepsilon_\rho = \varepsilon_z = \beta^2,\ \varepsilon_\phi = \alpha, \tag{2.36}$$

$$\mu_\rho = \mu_z = 1,\ \mu_\phi = \alpha/\beta^2, \tag{2.37}$$

which require magnetic response only in one direction. The comparison of ray tracing analysis in Figs. 2.16c and 2.16d demonstrates the performance of the simplified 3D carpet cloak. In both simulations, light rays are launched in a tilted plane not containing the axis of rotation and thus carry momentum components in the azimuthal direction. Nevertheless, in sharp contrast to a bare bump, the reflected rays from the simplified cloak show only slight distortion, resulting from the reduced material parameters. It should also be stressed that anisotropy cannot be completely eliminated in devices of this kind. A neat proof is presented by Landy et al. in Ref. [105] by considering the local dispersion relation for the system.

The approach of rotating 2D maps into 3D profiles works for a class of systems of azimuthal invariance. This allows alternative access to the design of many 3D devices. For instance, to create a Luneburg lens with a flat focal plane, using 3D conformal mesh generation technique would be complicated though possible. In contrast, rotating the profile of a 2D flattened Luneburg lens from QCM is much simpler and introduces only slight anisotropy. In short, conformal mapping helps not only to reduce the complex material parameters, but also to ease the TO design process in 3D.

2.6 Outlook

In this chapter, we have reviewed the fundamentals and applications of conformal mapping in TO. Since the material properties could be isotropic and nonmagnetic after conformal mapping, this technique has become a preferred subset of transformations in the design and analysis of GRIN media, on-chip photonic circuits, and plasmonic light-guiding and light-harvesting devices. The promise of simplification of the resulting materials renders the blueprints of TO devices practical and leads to potential interactions with other nontransformation designs, such as the cloaks based on scattering cancellation [38, 40] or the bifunctional metamaterials working in the static/steady-state limit [151–155]. As the explorations of transformation physics keep being "pushed forward," these advantages become more significant. For example, parity-time (PT)-symmetric metamaterials with balanced loss and gain have been used for unidirectional light transport [156]. By applying transformations to the complex spatial coordinates, imaginary parts can be introduced into the resulting parameters, giving rise to precise loss-gain distributions for the desired PT potential [157–159]. Apparently, the number of the independent material parameters doubles in the PT transformation media, and conformal mapping would serve as the most favorable tool to alleviate this complexity. In Section 2.5 we introduce the idea of treating a 2D profile as an infinitesimally thin film. Recent progress in graphene plasmonics has shined light on this direction [160]. With a properly arranged spatial distribution of bias, a conductivity pattern, which leads to the required effective index profile for SPP waves, can be achieved on the graphene sheet [161]. This one-atom-thick platform is flexible in fitting curved surfaces [162–164] and has the potential to pave a different way for the construction of anisotropic metamaterials with multilayer structures [165].

References

1. Leonhardt, U. (2006). Optical conformal mapping, *Science*, **312**, pp. 1777–1780.

2. Pendry, J. B., Schurig, D., and Smith, D. R. (2006). Controlling electromagnetic fields, *Science*, **312**, pp. 1780–1782.
3. Greenleaf, A., Kurylev, Y., Lassas, M., and Uhlmann, G. (2009). Cloaking devices, electromagnetic wormholes, and transformation optics, *SIAM Review*, **51**, pp. 3–33.
4. Chen, H., Chan, C. T., and Sheng, P. (2010). Transformation optics and metamaterials, *Nat. Mater.*, **9**, pp. 387–396.
5. Liu, Y., and Zhang, X. (2012). Recent advances in transformation optics, *Nanoscale*, **4**, pp. 5277–5292.
6. Leonhardt, U., and Philbin, T. (2012) *Geometry and Light: the Science of Invisibility* (Courier Dover Publications).
7. Thompson, R. T., and Cummer, S. A. (2012) *Advances in Imaging and Electron Physics*, ed. Peter W. H., Chapter 5 "*Transformation Optics*," (Elsevier), pp. 195–295.
8. Post, E. J. (1997) *Formal Structure of Electromagnetics: General Covariance and Electromagnetics* (Courier Dover Publications).
9. Jackson, J. D. (1975) *Classical Electrodynamics*, 2nd Ed. (Wiley).
10. Kong, J. A. (1990) *Electromagnetic Wave Theory*, 2nd Sub Ed. (Wiley).
11. Ward, A. J., and Pendry, J. B. (1996). Refraction and geometry in Maxwell's equations, *J. Mod. Opt.*, **43**, pp. 773–793.
12. Pendry, J. B. (2000). Negative refraction makes a perfect lens, *Phys. Rev. Lett.*, **85**, pp. 3966–3969.
13. Pendry, J. B., and Ramakrishna, S. A. (2003). Focusing light using negative refraction, *J. Phys. Cond. Matter*, **15**, pp. 6345–6364.
14. Greenleaf, A., Lassas, M., and Uhlmann, G. (2003). Anisotropic conductivities that cannot be detected by EIT, *Physiol. Meas.*, **24**, pp. 413–419.
15. Greenleaf, A., Lassas, M., and Uhlmann, G. (2003). On nonuniqueness for Calderon's inverse problem, *Math. Res. Lett.*, **10**, pp. 685–693.
16. Schurig, D., Pendry, J. B., and Smith, D. R. (2006). Calculation of material properties and ray tracing in transformation media, *Opt. Express*, **14**, pp. 9794–9804.
17. Cummer, S. A., Popa, B.-I., Schurig, D., Smith, D. R., and Pendry, J. (2006). Full-wave simulations of electromagnetic cloaking structures, *Phys. Rev. E*, **74**, 036621.
18. Zhao, Y., Argyropoulos, C., and Hao, Y. (2008). Full-wave finite-difference time-domain simulation of electromagnetic cloaking structures, *Opt. Express*, **16**, pp. 6717–6730.

19. Schurig, D., Mock, J. J., Justice, B. J., Cummer, S. A., Pendry, J. B., Starr, A. F., and Smith, D. R. (2006). Metamaterial electromagnetic cloak at microwave frequencies, *Science*, **314**, pp. 977–980.

20. Chen, H., and Chan, C. T. (2007). Acoustic cloaking in three dimensions using acoustic metamaterials, *Appl. Phys. Lett.*, **91**, 183518.

21. Cummer, S. A., Popa, B.-I., Schurig, D., Smith, D. R., Pendry, J., Rahm, M., and Starr, A. (2008). Scattering theory derivation of a 3D acoustic cloaking shell, *Phys. Rev. Lett.*, **100**, 024301.

22. Farhat, M., Guenneau, S., Enoch, S., and Movchan, A. B. (2009). Cloaking bending waves propagating in thin elastic plates, *Phys. Rev. B*, **79**, 033102.

23. Zhang, S., Genov, D. A., Sun, C., and Zhang, X. (2008). Cloaking of matter waves, *Phys. Rev. Lett.*, **100**, 123002.

24. Urzhumov, Y. A., and Smith, D. R. (2011). Fluid flow control with transformation media, *Phys. Rev. Lett.*, **107**, 074501.

25. Guenneau, S., Amra, C., and Veynante, D. (2012). Transformation thermodynamics: cloaking and concentrating heat flux, *Opt. Express*, **20**, pp. 8207–8218.

26. Guenneau, S., and Puvirajesinghe, T. M. (2013). Fick's second law transformed: one path to cloaking in mass diffusion, *J. R. Soc. Interface*, **10**, 20130106.

27. Kundtz, N. B., Smith, D. R., and Pendry, J. B. (2011). Electromagnetic design with transformation optics, *Proc. IEEE*, **99**, pp. 1622–1633.

28. Alexander, V. K., and Vladimir, M. S. (2011). Transformation optics and metamaterials, *Phys. Usp.*, **54**, pp. 53–63.

29. Pendry, J. B., Aubry, A., Smith, D. R., and Maier, S. A. (2012). Transformation optics and subwavelength control of light, *Science*, **337**, pp. 549–552.

30. Chen, H., and Chan, C. T. (2010). Acoustic cloaking and transformation acoustics, *J. Phys. D*, **43**, 113001.

31. Ma, Y., Liu, Y., Lan, L., Wu, T., Jiang, W., Ong, C. K., and He, S. (2013). First experimental demonstration of an isotropic electromagnetic cloak with strict conformal mapping, *Sci. Rep.*, **3**, 2182.

32. Leonhardt, U. (2006). Notes on conformal invisibility devices, *New J. Phys.*, **8**, 118.

33. Leonhardt, U., and Tyc, T. (2009). Broadband invisibility by non-Euclidean cloaking, *Science*, **323**, pp. 110–112.

34. Ablowitz, M. J., and Fokas, A. S. (2003) *Complex Variables: Introduction and Applications*, 2nd Ed. (Cambridge University Press).

35. Smith, D. R., Urzhumov, Y., Kundtz, N. B., and Landy, N. I. (2010). Enhancing imaging systems using transformation optics, *Opt. Express*, **18**, pp. 21238–21251.
36. Liu, D., Gabrielli, L. H., Lipson, M., and Johnson, S. G. (2013). Transformation inverse design, *Opt. Express*, **21**, pp. 14223–14243.
37. Alù, A., and Engheta, N. (2005). Achieving transparency with plasmonic and metamaterial coatings, *Phys. Rev. E*, **72**, 016623.
38. Alù, A., and Engheta, N. (2008). Plasmonic and metamaterial cloaking: physical mechanisms and potentials, *J. Opt. A*, **10**, 093002.
39. Alù, A. (2009). Mantle cloak: invisibility induced by a surface, *Phys. Rev. B*, **80**, 245115.
40. Chen, P.-Y., Soric, J., and Alù, A. (2012). Invisibility and cloaking based on scattering cancellation, *Adv. Mater.*, **24**, pp. OP281-OP304.
41. Xu, L., and Chen, H. (2013). Transformation optics with artificial Riemann sheets, *New J. Phys.*, **15**, 113013.
42. Batchelor, G. K. (2000) *An Introduction to Fluid Dynamics* (Cambridge University Press).
43. Luneberg, R. (1964) *Mathematical Theory of Optics* (University of California Press).
44. Eaton, J. (1952). On spherically symmetric lenses, *Trans. IRE Antennas Propag.*, **4**, pp. 66–71.
45. Wu, Q., Xu, Y., Li, H., and Chen, H. (2013). Cloaking and imaging at the same time, *Europhys. Lett.*, **101**, 34004.
46. Li, J., and Pendry, J. B. (2008). Hiding under the carpet: a new strategy for cloaking, *Phys. Rev. Lett.*, **101**, 203901.
47. Kundtz, N., and Smith, D. R. (2010). Extreme-angle broadband metamaterial lens, *Nat. Mater.*, **9**, pp. 129–132.
48. Landy, N., Urzhumov, Y., and Smith, D. (2014) *Transformation Electromagnetics and Metamaterials*, eds. Werner D. H., and Kwon D.-H., Chapter 1 "Quasi-Conformal Approaches for Two and Three-Dimensional Transformation Optical Media" (Springer, London), pp. 1–32.
49. Zentgraf, T., Valentine, J., Tapia, N., Li, J., and Zhang, X. (2010). An optical "Janus" device for integrated photonics, *Adv. Mater.*, **22**, pp. 2561–2564.
50. Rahm, M., Schurig, D., Roberts, D. A., Cummer, S. A., Smith, D. R., and Pendry, J. B. (2008). Design of electromagnetic cloaks and concentrators using form-invariant coordinate transformations of Maxwell's equations, *Photon. Nanostruct. Fundam. Appl.*, **6**, pp. 87–95.

51. Knupp, P. M., and Steinberg, S. (1993) *Fundamentals of Grid Generation*, Vol. 1 (CRC press, Boca Raton).
52. Chang, Z., Zhou, X., Hu, J., and Hu, G. (2010). Design method for quasi-isotropic transformation materials based on inverse Laplace's equation with sliding boundaries, *Opt. Express*, **18**, pp. 6089–6096.
53. Yao, K., Chen, H., Liu, Y., and Jiang, X. (2014). An analogy strategy for transformation optics, *New J. Phys.*, **16**, 063008.
54. Kallos, E., Argyropoulos, C., and Hao, Y. (2009). Ground-plane quasicloaking for free space, *Phys. Rev. A*, **79**, 063825.
55. Zhang, P., Lobet, M., and He, S. (2010). Carpet cloaking on a dielectric half-space, *Opt. Express*, **18**, pp. 18158–18163.
56. Schmied, R., Halimeh, J. C., and Wegener, M. (2010). Conformal carpet and grating cloaks, *Opt. Express*, **18**, pp. 24361–24367.
57. Xi, S., Chen, H., Wu, B.-I., and Kong, J.-A. (2009). One-directional perfect cloak created with homogeneous material, *IEEE Microw. Wirel. Compon. Lett.*, **19**, pp. 131–133.
58. Liu, R., Ji, C., Mock, J. J., Chin, J. Y., Cui, T. J., and Smith, D. R. (2009). Broadband ground-plane cloak, *Science*, **323**, pp. 366–369.
59. Valentine, J., Li, J., Zentgraf, T., Bartal, G., and Zhang, X. (2009). An optical cloak made of dielectrics, *Nat. Mater.*, **8**, pp. 568–571.
60. Gabrielli, L. H., Cardenas, J., Poitras, C. B., and Lipson, M. (2009). Silicon nanostructure cloak operating at optical frequencies, *Nat. Photon.*, **3**, pp. 461–463.
61. Zhou, F., Bao, Y., Cao, W., Stuart, C. T., Gu, J., Zhang, W., and Sun, C. (2011). Hiding a realistic object using a broadband Terahertz invisibility cloak, *Sci. Rep.*, **1**, 78.
62. Ergin, T., Stenger, N., Brenner, P., Pendry, J. B., and Wegener, M. (2010). Three-dimensional invisibility cloak at optical wavelengths, *Science*, **328**, pp. 337–339.
63. Fischer, J., Ergin, T., and Wegener, M. (2011). Three-dimensional polarization-independent visible-frequency carpet invisibility cloak, *Opt. Lett.*, **36**, pp. 2059–2061.
64. Gharghi, M., Gladden, C., Zentgraf, T., Liu, Y., Yin, X., Valentine, J., and Zhang, X. (2011). A carpet cloak for visible light, *Nano Lett.*, **11**, pp. 2825–2828.
65. Shin, D., Urzhumov, Y., Jung, Y., Kang, G., Baek, S., Choi, M., Park, H., Kim, K., and Smith, D. R. (2012). Broadband electromagnetic cloaking with smart metamaterials, *Nat. Commun.*, **3**, 1213.

66. Bückmann, T., Thiel, M., Kadic, M., Schittny, R., and Wegener, M. (2014). An elasto-mechanical unfeelability cloak made of pentamode metamaterials, *Nat. Commun.*, **5**, 4130.
67. Schurig, D. (2008). An aberration-free lens with zero F-number, *New J. Phys.*, **10**, 115034.
68. Tang, W., Argyropoulos, C., Kallos, E., Song, W., and Hao, Y. (2010). Discrete coordinate transformation for designing all-dielectric flat antennas, *IEEE Trans. Antennas Propag.*, **58**, pp. 3795–3804.
69. Ma, H. F., and Cui, T. J. (2010). Three-dimensional broadband and broad-angle transformation-optics lens, *Nat. Commun.*, **1**, 124.
70. Hunt, J., Tyler, T., Dhar, S., Tsai, Y.-J., Bowen, P., Larouche, S., Jokerst, N. M., and Smith, D. R. (2012). Planar, flattened Luneburg lens at infrared wavelengths, *Opt. Express*, **20**, pp. 1706–1713.
71. Schmiele, M., Varma, V. S., Rockstuhl, C., and Lederer, F. (2010). Designing optical elements from isotropic materials by using transformation optics, *Phys. Rev. A*, **81**, 033837.
72. Milton, G. W., Briane, M., and Willis, J. R. (2006). On cloaking for elasticity and physical equations with a transformation invariant form, *New J. Phys.*, **8**, 248.
73. Li, J. Y., Gao, Y., and Huang, J. P. (2010). A bifunctional cloak using transformation media, *J. Appl. Phys.*, **108**, 074504.
74. Schittny, R., Kadic, M., Guenneau, S., and Wegener, M. (2013). Experiments on transformation thermodynamics: molding the flow of heat, *Phys. Rev. Lett.*, **110**, 195901.
75. García-Meca, C., Carloni, S., Barceló, C., Jannes, G., Sánchez-Dehesa, J., and Martínez, A. (2013). Analogue transformations in physics and their application to acoustics, *Sci. Rep.*, **3**, 2009.
76. Kadic, M., Bückmann, T., Schittny, R., and Wegener, M. (2013). Metamaterials beyond electromagnetism, *Rep. Prog. Phys.*, **76**, 126501.
77. Maxwell, J. C. (1964) *The Scientific Papers of James Clerk Maxwell*, ed. Niven, W. D. (Dover).
78. Bergamin, L., Alitalo, P., and Tretyakov, S. A. (2011). Nonlinear transformation optics and engineering of the Kerr effect, *Phys. Rev. B*, **84**, 205103.
79. Prandtl, L. (2010) *Prandtl-Essentials of Fluid Mechanics*, 3rd Ed., ed. Oertel, H. (Springer, New York).
80. Narimanov, E. E., and Kildishev, A. V. (2009). Optical black hole: broadband omnidirectional light absorber, *Appl. Phys. Lett.*, **95**, 041106.

81. Genov, D. A., Zhang, S., and Zhang, X. (2009). Mimicking celestial mechanics in metamaterials, *Nat. Phys.*, **5**, pp. 687–692.

82. Sheng, C., Liu, H., Wang, Y., Zhu, S. N., and Genov, D. A. (2013). Trapping light by mimicking gravitational lensing, *Nat. Photon.*, **7**, pp. 902–906.

83. San-Román-Alerigi, D. P., Slimane, A. B., Ng, T. K., Alsunaidi, M., and Ooi, B. S. (2013). A possible approach on optical analogues of gravitational attractors, *Opt. Express*, **21**, pp. 8298–8310.

84. Kildishev, A. V., Prokopeva, L. J., and Narimanov, E. E. (2010). Cylinder light concentrator and absorber: theoretical description, *Opt. Express*, **18**, pp. 16646–16662.

85. Jiang, X., Yao, K., Wu, Q., Xu, Y., and Chen, H. (2012). Conformal transformations to achieve unidirectional behavior of light, *New J. Phys.*, **14**, 053023.

86. Urzhumov, Y., Landy, N., and Smith, D. R. (2012). Isotropic-medium three-dimensional cloaks for acoustic and electromagnetic waves, *J. Appl. Phys.*, **111**, 053105.

87. Zhang, B., Chan, T., and Wu, B.-I. (2010). Lateral shift makes a ground-plane cloak detectable, *Phys. Rev. Lett.*, **104**, 233903.

88. Born, M., and Wolf, E. (1999). *Principles of Optics: Electromagnetic Theory of Propagation, Interference and Diffraction of Light*, 7th Ed. (Cambridge University Press).

89. Yao, K., and Jiang, X. (2011). Designing feasible optical devices via conformal mapping, *J. Opt. Soc. Am. B*, **28**, pp. 1037–1042.

90. Benítez, P., Miñano, J. C., and González, J. C. (2010). Perfect focusing of scalar wave fields in three dimensions, *Opt. Express*, **18**, pp. 7650–7663.

91. Gu, C., Yao, K., Lu, W., Lai, Y., Chen, H., Hou, B., and Jiang, X. (2012). Experimental realization of a broadband conformal mapping lens for directional emission, *Appl. Phys. Lett.*, **100**, 261907.

92. Chen, H., Xu, Y., Li, H., and Tyc, T. (2013). Playing the tricks of numbers of light sources, *New J. Phys.*, **15**, 093034.

93. Philbin, T. G. (2014). Making geometrical optics exact, *J. Mod. Opt.*, **61**, pp. 552–557.

94. Raether, H. (1988) *Surface Plasmons on Smooth and Rough Surfaces and on Gratings* (Springer, Berlin, Heidelberg).

95. Zayats, A. V., Smolyaninov, I. I., and Maradudin, A. A. (2005). Nano-optics of surface plasmon polaritons, *Phys. Rep.*, **408**, pp. 131–314.

96. Kawata, S., Inouye, Y., and Verma, P. (2009). Plasmonics for near-field nano-imaging and superlensing, *Nat. Photon.*, **3**, pp. 388–394.
97. Schuller, J. A., Barnard, E. S., Cai, W., Jun, Y. C., White, J. S., and Brongersma, M. L. (2010). Plasmonics for extreme light concentration and manipulation, *Nat. Mater.*, **9**, pp. 193–204.
98. Yao, K., and Liu, Y. (2014). Plasmonic metamaterials, *Nanotechnol. Rev.*, **3**, pp. 177–210.
99. Barnes, W. L., Dereux, A., and Ebbesen, T. W. (2003). Surface plasmon subwavelength optics, *Nature*, **424**, pp. 824–830.
100. Gramotnev, D. K., and Bozhevolnyi, S. I. (2010). Plasmonics beyond the diffraction limit, *Nat. Photon.*, **4**, pp. 83–91.
101. Elser, J., and Podolskiy, V. A. (2008). Scattering-free plasmonic optics with anisotropic metamaterials, *Phys. Rev. Lett.*, **100**, 066402.
102. Huidobro, P. A., Nesterov, M. L., Martín-Moreno, L., and García-Vidal, F. J. (2010). Transformation optics for plasmonics, *Nano Lett.*, **10**, pp. 1985–1990.
103. Liu, Y., Zentgraf, T., Bartal, G., and Zhang, X. (2010). Transformational plasmon optics, *Nano Lett.*, **10**, pp. 1991–1997.
104. Zhang, J., Xiao, S., Wubs, M., and Mortensen, N. A. (2011). Surface plasmon wave adapter designed with transformation optics, *ACS Nano*, **5**, pp. 4359–4364.
105. Landy, N. I., Kundtz, N., and Smith, D. R. (2010). Designing three-dimensional transformation optical media using quasiconformal coordinate transformations, *Phys. Rev. Lett.*, **105**, 193902.
106. Renger, J., Kadic, M., Dupont, G., Aćimović, S. S., Guenneau, S., Quidant, R., and Enoch, S. (2010). Hidden progress: broadband plasmonic invisibility, *Opt. Express*, **18**, pp. 15757–15768.
107. Hasegawa, K., Nöckel, J. U., and Deutsch, M. (2004). Surface plasmon polariton propagation around bends at a metal–dielectric interface, *Appl. Phys. Lett.*, **84**, pp. 1835–1837.
108. Mei, Z. L., and Cui, T. J. (2009). Experimental realization of a broadband bend structure using gradient index metamaterials, *Opt. Express*, **17**, pp. 18354–18363.
109. Gabrielli, L. H., Liu, D., Johnson, S. G., and Lipson, M. (2012). On-chip transformation optics for multimode waveguide bends, *Nat. Commun.*, **3**, 1217.
110. Huidobro, P. A., Nesterov, M. L., Martín-Moreno, L., and García-Vidal, F. J. (2011). Moulding the flow of surface plasmons using conformal and quasiconformal mappings, *New J. Phys.*, **13**, 033011.

111. Kong, X.-T., Li, Z.-B., and Tian, J.-G. (2013). Mode converter in metal-insulator-metal plasmonic waveguide designed by transformation optics, *Opt. Express*, **21**, pp. 9437–9446.

112. García-Meca, C., Ortuño, R., Martí, J., and Martínez, A. (2012). Exciting surface plasmons with transformation media, *Plasmonics*, **7**, pp. 701–707.

113. Rahm, M., Cummer, S. A., Schurig, D., Pendry, J. B., and Smith, D. R. (2008). Optical design of reflectionless complex media by finite embedded coordinate transformations, *Phys. Rev. Lett.*, **100**, 063903.

114. Xu, H., Wang, X., Yu, T., Sun, H., and Zhang, B. (2012). Radiation-suppressed plasmonic open resonators designed by nonmagnetic transformation optics, *Sci. Rep.*, **2**, 784.

115. Chen, X., Luo, Y., Zhang, J., Jiang, K., Pendry, J. B., and Zhang, S. (2011). Macroscopic invisibility cloaking of visible light, *Nat. Commun.*, **2**, 176.

116. Zhang, B., Luo, Y., Liu, X., and Barbastathis, G. (2011). Macroscopic invisibility cloak for visible light, *Phys. Rev. Lett.*, **106**, 033901.

117. Zentgraf, T., Liu, Y., Mikkelsen, M. H., Valentine, J., and Zhang, X. (2011). Plasmonic Luneburg and Eaton lenses, *Nat. Nanotechnol.*, **6**, pp. 151–155.

118. Wu, Q., Turpin, J. P., and Werner, D. H. (2012). Integrated photonic systems based on transformation optics enabled gradient index devices, *Light Sci. Appl.*, **1**, e38.

119. Aubry, A., Lei, D. Y., Fernández-Domínguez, A. I., Sonnefraud, Y., Maier, S. A., and Pendry, J. B. (2010). Plasmonic light-harvesting devices over the whole visible spectrum, *Nano Lett.*, **10**, pp. 2574–2579.

120. Aubry, A., and Pendry, J. B. (2013) *Active Plasmonics and Tuneable Plasmonic Metamaterials*, eds. Zayats, A. V., and Maier, S. A., Chapter 4 "Transformation Optics for Plasmonics," (John Wiley & Sons), pp. 105–152.

121. Zeng, Y., Liu, J., and Werner, D. H. (2011). General properties of two-dimensional conformal transformations in electrostatics, *Opt. Express*, **19**, pp. 20035–20047.

122. Luo, Y., Pendry, J. B., and Aubry, A. (2010). Surface plasmons and singularities, *Nano Lett.*, **10**, pp. 4186–4191.

123. Lei, D. Y., Aubry, A., Luo, Y., Maier, S. A., and Pendry, J. B. (2010). Plasmonic interaction between overlapping nanowires, *ACS Nano*, **5**, pp. 597–607.

124. Luo, Y., Aubry, A., and Pendry, J. B. (2011). Electromagnetic contribution to surface-enhanced Raman scattering from rough metal surfaces: a transformation optics approach, *Phys. Rev. B*, **83**, 155422.

125. Aubry, A., Lei, D. Y., Maier, S. A., and Pendry, J. B. (2010). Interaction between plasmonic nanoparticles revisited with transformation optics, *Phys. Rev. Lett.*, **105**, 233901.

126. Aubry, A., Lei, D. Y., Maier, S. A., and Pendry, J. B. (2011). Plasmonic hybridization between nanowires and a metallic surface: a transformation optics approach, *ACS Nano*, **5**, pp. 3293–3308.

127. Zhang, J., and Zayats, A. (2013). Multiple Fano resonances in single-layer nonconcentric core-shell nanostructures, *Opt. Express*, **21**, pp. 8426–8436.

128. Fernández-Domínguez, A. I., Maier, S. A., and Pendry, J. B. (2010). Collection and concentration of light by touching spheres: a transformation optics approach, *Phys. Rev. Lett.*, **105**, 266807.

129. Kraft, M., Pendry, J. B., Maier, S. A., and Luo, Y. (2014). Transformation optics and hidden symmetries, *Phys. Rev. B*, **89**, 245125.

130. Fernández-Domínguez, A. I., Luo, Y., Wiener, A., Pendry, J. B., and Maier, S. A. (2012). Theory of three-dimensional nanocrescent light harvesters, *Nano Lett.*, **12**, pp. 5946–5953.

131. Luo, Y., Fernandez-Dominguez, A. I., Wiener, A., Maier, S. A., and Pendry, J. B. (2013). Surface plasmons and nonlocality: a simple model, *Phys. Rev. Lett.*, **111**, 093901.

132. Fernández-Domínguez, A. I., Wiener, A., García-Vidal, F. J., Maier, S. A., and Pendry, J. B. (2012). Transformation-optics description of nonlocal effects in plasmonic nanostructures, *Phys. Rev. Lett.*, **108**, 106802.

133. Luo, Y., Lei, D. Y., Maier, S. A., and Pendry, J. B. (2012). Broadband light harvesting nanostructures robust to edge bluntness, *Phys. Rev. Lett.*, **108**, 023901.

134. Luo, Y., Lei, D. Y., Maier, S. A., and Pendry, J. B. (2012). Transformation-optics description of plasmonic nanostructures containing blunt edges/corners: from symmetric to asymmetric edge rounding, *ACS Nano*, **6**, pp. 6492–6506.

135. Pendry, J. B., Fernandez-Dominguez, A. I., Luo, Y., and Zhao, R. (2013). Capturing photons with transformation optics, *Nat. Phys.*, **9**, pp. 518–522.

136. Zhao, R., Luo, Y., Fernández-Domínguez, A. I., and Pendry, J. B. (2013). Description of van der Waals interactions using transformation optics, *Phys. Rev. Lett.*, **111**, 033602.

137. Tyc, T., Herzánová, L., Šarbort, M., and Bering, K. (2011). Absolute instruments and perfect imaging in geometrical optics, *New J. Phys.*, **13**, 115004.

138. Leonhardt, U. (2009). Perfect imaging without negative refraction, *New J. Phys.*, **11**, 093040.

139. Tyc, T., and Zhang, X. (2011). Perfect lenses in focus, *Nature*, **480**, pp. 42–43.

140. Ma, Y. G., Sahebdivan, S., Ong, C. K., Tyc, T., and Leonhardt, U. (2011). Evidence for subwavelength imaging with positive refraction, *New J. Phys.*, **13**, 033016.

141. Ma, Y. G., Sahebdivan, S., Ong, C. K., Tyc, T., and Leonhardt, U. (2012). Subwavelength imaging with materials of in-principle arbitrarily low index contrast, *New J. Phys.*, **14**, 025001.

142. Miñano, J. C., Sánchez-Dehesa, J., González, J. C., Benítez, P., Grabovičkić, D., Carbonell, J., and Ahmadpanahi, H. (2014). Experimental evidence of super-resolution better than $\lambda/105$ with positive refraction, *New J. Phys.*, **16**, 033015.

143. Tyc, T., and Danner, A. (2014). Resolution of Maxwell's fisheye with an optimal active drain, *New J. Phys.*, **16**, 063001.

144. Perczel, J., Tyc, T., and Leonhardt, U. (2011). Invisibility cloaking without superluminal propagation, *New J. Phys.*, **13**, 083007.

145. Tyc, T., Chen, H., Chan, C. T., and Leonhardt, U. (2010). Non-Euclidean cloaking for light waves, *IEEE J. Sel. Top. Quantum Electron.*, **16**, pp. 418–426.

146. Yao, K., Jiang, X., and Chen, H. (2012). Collimating lenses from non-Euclidean transformation optics, *New J. Phys.*, **14**, 023011.

147. Leonhardt, U., and Tyc, T. (2008). Superantenna made of transformation media, *New J. Phys.*, **10**, 115026.

148. Leonhardt, U., and Philbin, T. G. (2010). Perfect imaging with positive refraction in three dimensions, *Phys. Rev. A*, **81**, 011804.

149. Miñano, J. C., Benítez, P., and González, J. C. (2010). Perfect imaging with geodesic waveguides, *New J. Phys.*, **12**, 123023.

150. Liu, G., Li, C., Zhang, C., Sun, Z., and Fang, G. (2013). Experimental verification of field concentrator by full tensor transmission-line metamaterials, *Phys. Rev. B*, **87**, 155125.

151. Moccia, M., Castaldi, G., Savo, S., Sato, Y., and Galdi, V. (2014). Independent manipulation of heat and electrical current via bifunctional metamaterials, *Phys. Rev. X*, **4**, 021025.

152. Gömöry, F., Solovyov, M., Šouc, J., Navau, C., Prat-Camps, J., and Sanchez, A. (2012). Experimental realization of a magnetic cloak, *Science*, **335**, pp. 1466–1468.

153. Xu, H., Shi, X., Gao, F., Sun, H., and Zhang, B. (2014). Ultrathin three-dimensional thermal cloak, *Phys. Rev. Lett.*, **112**, 054301.

154. Han, T., Bai, X., Gao, D., Thong, J. T. L., Li, B., and Qiu, C.-W. (2014). Experimental demonstration of a bilayer thermal cloak, *Phys. Rev. Lett.*, **112**, 054302.

155. Schittny, R., Kadic, M., Bückmann, T., and Wegener, M. (2014). Invisibility cloaking in a diffusive light scattering medium, *Science*, **345**, pp. 427–429.

156. Lin, Z., Ramezani, H., Eichelkraut, T., Kottos, T., Cao, H., and Christodoulides, D. N. (2011). Unidirectional invisibility induced by *PT*-symmetric periodic structures, *Phys. Rev. Lett.*, **106**, 213901.

157. Castaldi, G., Savoia, S., Galdi, V., Alù, A., and Engheta, N. (2013). *PT* metamaterials via complex-coordinate transformation optics, *Phys. Rev. Lett.*, **110**, 173901.

158. Popa, B.-I., and Cummer, S. A. (2011). Complex coordinates in transformation optics, *Phys. Rev. A*, **84**, 063837.

159. Zharova, N. A., Zharov, A. A., and Zharov, A. A., Jr. (2014). Conformal transformations in design of the coatings with gain-loss permittivity, *Phys. Rev. A*, **89**, 043801.

160. Grigorenko, A. N., Polini, M., and Novoselov, K. S. (2012). Graphene plasmonics, *Nat. Photon.*, **6**, pp. 749–758.

161. Vakil, A., and Engheta, N. (2011). Transformation optics using graphene, *Science*, **332**, pp. 1291–1294.

162. Lu, W. B., Zhu, W., Xu, H. J., Ni, Z. H., Dong, Z. G., and Cui, T. J. (2013). Flexible transformation plasmonics using graphene, *Opt. Express*, **21**, pp. 10475–10482.

163. Mitchell-Thomas, R. C., McManus, T. M., Quevedo-Teruel, O., Horsley, S. A. R., and Hao, Y. (2013). Perfect surface wave cloaks, *Phys. Rev. Lett.*, **111**, 213901.

164. Chen, P.-Y., and Alù, A. (2011). Atomically thin surface cloak using graphene monolayers, *ACS Nano*, **5**, pp. 5855–5863.

165. Wang, J., Xu, Y., Chen, H., and Zhang, B. (2012). Ultraviolet dielectric hyperlens with layered graphene and boron nitride, *J. Mater. Chem.*, **22**, pp. 15863–15868.

Chapter 3

Quasiconformal Transformation Media and Their Electrostatic Analogy

Jensen Li,[a] Fu Liu,[a] Zheng Chang,[b] and Gengkai Hu[c]
[a]*School of Physics and Astronomy, University of Birmingham, Birmingham B15 2TT, United Kingdom*
[b]*CNMM and AML, Department of Engineering Mechanics, Tsinghua University, Beijing 100084, P. R. China*
[c]*School of Aerospace Engineering, Beijing Institute of Technology, Beijing 100081, P. R. China*
j.li@bham.ac.uk, hugeng@bit.edu.cn

Transformation optics generally needs gradient anisotropic metamaterials with extreme material parameters for realization. For applications in which a larger-frequency bandwidth is desired, one possible route is to seek coordinate maps that can minimize the generated index range and anisotropy. In this chapter, we demonstrate quasiconformal map (QCM) as a useful tool for minimizing anisotropy, while the index range can be further minimized by avoiding sharp corners in boundaries. We will give examples with both analytic and numerical approaches and show the corresponding map generation can be simplified by taking an electrostatic analogy. We will also discuss its extension to acoustic and elastic waves and investigate an elastodynamic carpet cloak. These considerations are useful in a wide class of applications such

Transformation Wave Physics: Electromagnetics, Elastodynamics, and Thermodynamics
Edited by Mohamed Farhat, Pai-Yen Chen, Sebastien Guenneau, and Stefan Enoch

ISBN 978-981-4669-95-5 (Hardcover), 978-981-4669-96-2 (eBook)
www.panstanford.com

as invisibility cloaks and optimized integrated optical devices at the optical wavelengths.

3.1 Introduction

Transformation optics (TO) is an emerging technique in manipulating light by drawing an equivalence between two different configurations using a coordinate transformation [1–3]. It enables a powerful and intuitive approach to design optical devices, with prominent examples including invisibility cloaks [2–18], concentrators/optical black holes [19–21], field rotators [22] and optical illusion devices [23]. TO usually starts from a simple virtual configuration, which is a vacuum in the case of designing an invisibility cloak. A coordinate transformation on the virtual system induces a corresponding material transformation, based on the form invariance property of the Maxwell's equations. Generally speaking, the induced material parameters become anisotropic and their values can span a large range of values, which are to be realized using metamaterials with resonating constitutive parameters. For applications like invisibility cloaks, this can limit the working-frequency bandwidth of the resultant devices. On the other hand, the coordinate transformation can be quite arbitrary as long as the transformation gives rise to the same functionality. There is certainly a huge room for us to choose an optimal coordinate transformation in giving rise to small anisotropy and small range of indices, which can facilitate easier fabrication and experimental realizations. One of such attempts is to use the so-called quasiconformal maps (QCMs). In this chapter, we focus on obtaining such coordinate maps to minimize anisotropy. This will be beneficial to TO applications in which allowed anisotropy is limited, for example, in the optical wavelengths, to design a range of transformation optical devices, including waveguide benders, couplers and surface-conformal antennas [24–33]. Furthermore, by choosing appropriate boundaries to avoid sharp corners for certain applications like carpet cloaks and unidirectional cloaks [5–18], index range can be lowered as well. In Section 3.2, we establish the QCM as the optimal choice in minimizing anisotropy and establish the corresponding electrostatic analogy. In Section 3.3, we give an analytical example of QCM in which its properties can be revealed immediately. We develop the numerical technique simplified with the electrostatic analogy in obtaining a QCM with arbitrary shape of device boundaries. Such a numerical approach brings us an advantage that further constraints can be imposed on the device boundary. In this case, the generated map will deviate a bit from the ideal QCM with a larger anisotropy, but the device boundary can be forced to be matched to the outside environment numerically. In Section 3.4, we extend the discussion to other classical waves, such as acoustic and elastic waves. We discuss an elastodynamic carpet cloak. We then summarize our findings and give further outlook on its applications.

3.2 Transformation Optics with Anisotropy Minimization

3.2.1 *Minimizing Anisotropy*

TO involves a coordinate transformation, which induces a corresponding transformation of material parameters, the electric permittivity and the magnetic permeability, as a result of the form-invariant property of Maxwell's equations. To discuss the anisotropy of the induced material parameters for minimization using an appropriate map and to seek further simplification, we would like to rewrite some of the formulas in the TO framework [5]. To begin our discussion, we only consider two dimensional transformations for simplicity. The geometric nature of a coordinate transformation between coordinates (x', y') and coordinates (x, y) are nicely captured by the Jacobian matrix Λ and the metric g:

$$\Lambda = \begin{pmatrix} \partial_{x'}x & \partial_{y'}x \\ \partial_{x'}y & \partial_{y'}y \end{pmatrix} \text{ and } g = \Lambda^{\mathrm{T}}\Lambda. \tag{3.1}$$

Next, we use $(x', y') \xrightarrow{\Lambda} (x, y)$ to abbreviate the above transformation and use the primed notation to represent different symbols defined in the (x', y') space. Suppose we are working on the E_z polarization for in-plane wave propagation. The permittivity along the z direction, a scalar profile, and the in-plane permeability (2×2 matrix) are transformed according to TO [2] by

$$\epsilon(x, y) = \frac{\epsilon'(x, y)}{\sqrt{\det g}} = \epsilon'(x, y)\left|\nabla x' \times \nabla y'\right|,$$

$$\mu(x, y) = \frac{\Lambda \mu'(x, y) \Lambda^T}{\sqrt{\det g}} \text{ with } \frac{\mu_T}{\mu_L} + \frac{\mu_L}{\mu_T} + 2 = \frac{Tr(\mu)^2}{\det \mu}, \quad (3.2)$$

where μ_T and μ_L are the two principal values (eigenvalues) of matrix μ. The quantity $Tr(\mu)^2/\det\mu$, approaching a value of 4 for isotropic μ as a minimum, thus measures the degree of anisotropy of μ if we are only dealing with positive-definite permeability. If μ' is isotropic, the anisotropy is solely a property of the coordinate transformation so that

$$\text{isotropic } \mu' : \frac{Tr(\mu)^2}{\det \mu} = \frac{Tr(g)^2}{\det g}. \quad (3.3)$$

On the other hand, if Λ is a conformal map, the extent of anisotropy $(Tr(\mu)^2/\det\mu)$ is preserved so that

$$\text{conformal } \Lambda : \quad \mu = \left(\hat{x}' \; \hat{y}'\right) \mu' \left(\hat{x}' \; \hat{y}'\right)^T \text{ with } \frac{Tr(\mu)^2}{\det \mu} = \frac{Tr(\mu')^2}{\det \mu'}, \quad (3.4)$$

where $\hat{x}'$ and $\hat{y}'$ are the unit vectors along the orthogonal x' and y' coordinate lines in the physical space. As a conformal map preserves anisotropy, we assign a set of isothermal coordinates (u, v), comformal-mapped from (x, y), to describe the boundaries of the (x, y) space, which can now be specified by $0 \leq u \leq M$ and $0 \leq v \leq 1$ where M is called the conformal module of the (x, y) space. M is fixed once the boundary and four corners are fixed. Similarly, we use (u', v') as isothermal coordinates to describe the (x', y') space by $0 \leq u' \leq M'$ and $0 \leq v' \leq 1$. Such a setting is schematically represented by Fig. 3.1. Both spaces are having shapes of generally curved quadrilaterals.

We have decomposed the transformation $(x', y') \xrightarrow{\Lambda} (x, y)$ into $(x', y') \xrightarrow{\Lambda_1} (u', v') \xrightarrow{\Lambda_2} (u, v) \xrightarrow{\Lambda_3} (x, y)$ with metric $g_i = \Lambda_i^T \Lambda_i$ and the final Jacobian as $\Lambda = \Lambda_3 \Lambda_2 \Lambda_1$. As both Λ_1 and Λ_3 are conformal and if we start from a medium of isotropic μ' (suitable for a wide range of TO applications), we immediately obtain from the properties listed in Eq. 3.3 and Eq. 3.4 that

$$\frac{Tr(\mu)}{\sqrt{\det \mu}} = \frac{Tr(g_2)}{\sqrt{\det g_2}}. \quad (3.5)$$

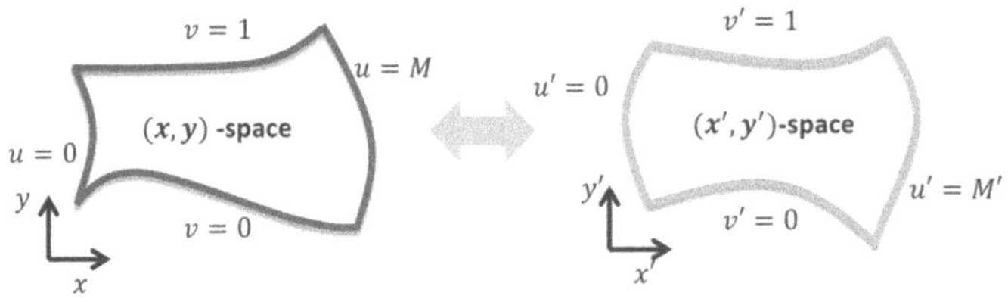

Figure 3.1 The general coordinate transformation maps corners to corners and edges to edges. Sense/orientation is also preserved. $(u, v)/(u'v')$ are the isothermal coordinates naturally describing the boundaries of the two spaces. The optimal map in minimizing anisotropy is a linear scaling between the (u, v) and the $(u'v')$ space.

Therefore, minimizing the anisotropy of the medium in (x, y) space is equivalent to minimizing the anisotropy introduced from Λ_2 or g_2. Next, by employing the generalized mean inequalities $\sqrt{1/M' \int (\partial u/\partial u')^2 du'} \geq 1/M' \int (\partial u/\partial u')\, du' = M/M'$ and $\sqrt{\int (\partial v/\partial v')^2 dv'} \geq \int (\partial v/\partial v')\, dv' = 1$ for the sense-preserving map (i.e., only one-to-one mapping is allowed), we have the averaged anisotropy in the (u, v) space satisfying the following inequality:

$$\frac{1}{M}\int \frac{Tr\,(g_2)}{\sqrt{\det g_2}} dudv = \frac{1}{M}\int Tr\,(g_2)\, du'dv' \geq \frac{M}{M'} + \frac{M'}{M}, \tag{3.6}$$

with equality happens at the following optimal map:

$$\frac{u}{M} = \frac{u'}{M'},\quad v = v', \tag{3.7}$$

which minimizes the averaged anisotropy. Since the maximum of the anisotropy profile must be larger than or equal to its averaged value, we finally obtain

$$\max\left(\frac{Tr\,(\mu)}{\sqrt{\det \mu}}\right) \geq \frac{M}{M'} + \frac{M'}{M}, \tag{3.8}$$

The equality happens at the same optimal map with constant anisotropy $Tr\,(\mu)/\sqrt{\det \mu} = M/M' + M'/M$ and hence the optimal map also minimizes the maximum anisotropy. We note that the same conclusion can be obtained alternatively by considering $\int Tr\,(g_2)/\sqrt{\det g_2} du'dv' = \int Tr\,(g_2^{-1})\, dudv$ instead of Eq. 3.6. This is the approach used in Ref. [5] where the space (x', y') is simply a rectangle, while minimization of the same integral but with squared

Kernel (the so called Modified–Liao functional to avoid grid folding) is applied to numerically generate the optimal map of smallest anisotropy. The optimum map in that case is actually a kind of QCMs of constant real Beltrami coefficient. It is a composition of a conformal map and a linear scaling map. Here, with curved boundary also in the (x', y') space, the optimal map is a composition of two conformal maps and a linear scaling map in the middle. It is called a Teichmüller map, also a kind of QCMs which have a complex Beltrami coefficient of constant modulus and spatially varying argument [34]. Here, we simply call these optimal maps with constant anisotropy as the QCMs for convenience (also in Refs. [5, 34]). We note that the above optimal map is already the map with the smallest possible anisotropy, subject to the only restrictions that the four corners be mapped to other four prescribed corners, while each boundary be mapped to the corresponding one up to a sliding degree of freedom for the coordinates (we called this slipping boundary condition). Further restrictions due to practical considerations, like a completely fixed boundary, increases the achievable minimum value in Eq. 3.8. On the other hand, one can have a better choice of the positions of the corners and the shapes of the boundaries to lower the anisotropy [31, 33]. The advantage of considering such restricted kind of QCMs (Teichmüller maps) is that they are analytically simple and yet numerically general enough to allow designing TO devices with prescribed boundaries. As we shall see, it allows further development and simplification for the map generation.

3.2.2 *Electrostatic Analogy*

From the last section, the QCM, more accurately a Teichmüller map, can be obtained by minimizing a functional, and such a map is then used to design a TO device with prescribed boundaries. In fact, the map/grid generation is very often the first step in numerically solving partial differential equations (PDEs) [35]. One would therefore expect that the design process can become simpler if we use an existing PDE solver to solve a related or analog problem to obtain the required numerical map or equivalently the TO medium profile. It has been shown that a QCM can be obtained by solving two concurrent Laplace equations on the two coordinates [25], which is our starting point here to establish an electrostatic analogy and to seek further simplification in the map generation. Before we go on, it is also worth to mention that there is a well-known electrostatic analogy to ray optics that replaces the constant phase surfaces and the light rays in ray optics by the equipotential surfaces and the electric field lines in electrostatics. From a known electrostatic solution, we can scale the potential to give the same rate change of phase propagation. The magnitude of the analog electric field gives the required isotropic index profile in achieving a set of light rays along the analog electric field lines [36, 37]. Instead of invoking geometric optics for isotropic medium as a limiting case, here we introduce an electrostatic analogy to TO in the map generation level. As we shall see, the unknown scaling factor to the analog electrostatic potential or electric field is actually the ratio of conformal modules of the physical and virtual space. Such a formal analogy also gives us a clear way to use general electrostatic solutions from Poisson's equation as seed solutions to generate QCMs.

The QCM we want to generate is a linear scaling between the two sets of isothermal coordinates (u, v) and (u', v'), whose geometrical properties will be inherited by any new set of "analog" coordinates (ξ, η) with definition:

$$\xi = \alpha u + \xi_0 = \alpha' u' + \xi_0, \quad \eta = \beta v + \eta_0 = \beta' v' + \eta_0, \tag{3.9}$$

where α, α', β, β', ξ_0, η_0 are all arbitrary constants satisfying $\alpha'/\alpha = M/M'$ and $\beta' = \beta$. As $(u, v)/\ (u', v')$ is linked to $(x, y)\ /\ (x', y')$ by a conformal map, we have

$$\begin{aligned} \nabla\left(\xi/\alpha\right) &= \nabla\left(\eta/\beta\right) \times \hat{z}, \\ \nabla'\left(\xi/\alpha'\right) &= \nabla'(\eta/\beta') \times \hat{z}. \end{aligned} \tag{3.10}$$

Moreover, since the target map, the $\xi - \eta$ grid on both (x, y) and (x', y') spaces, to be generated is orthogonal, one can apply the Dirichlet–Neumann boundary condition (Dirichlet boundary condition for ξ and Neumann boundary condition for η or vice versa) on a particular boundary [25]. By taking divergence on Eq. 3.10, we obtain our electrostatic analog problem with analog potential

$V = \xi$:

$$\nabla^2 V = 0,$$
$$\nabla'^2 V = 0, \tag{3.11}$$

so that from its solutions on both (x, y) and (x', y') space, the induced TO medium can be obtained by applying Eq. 3.2 (between (x, y) and (ξ, η) and between (x', y') and (ξ, η)), that is,

$$\epsilon = \epsilon' \gamma \frac{|\nabla V|^2}{|\nabla' V|^2}, \tag{3.12}$$

where γ is the ratio of the conformal moduli (M/M') and can be obtained by integrating Eq. 3.10:

$$\gamma = \frac{\int_{A'}^{B'} \nabla' V \times \hat{z} \cdot d\mathbf{l}'}{\int_{A}^{B} \nabla V \times \hat{z} \cdot d\mathbf{l}}. \tag{3.13}$$

The integrations can be taken along any paths linking any two points A and B in the (x, y) space and along the corresponding transformed path in the (x', y') space. We can simply choose a boundary path of constant V that the two end points in both spaces are already known. The TO medium can be written directly using the solution V on both spaces. Alternatively, we can choose η as the analog potential V and $\gamma = M'/M$ so that Eqs. 3.11–3.13 are still valid. The electrostatic analogy actually simplifies the designing process of TO devices that we only need to solve electrostatic problems on one coordinate/analog potential (either ξ or η). Such a simplification is possible since we have already assumed the $\xi - \eta$ grid on both the virtual (x', y') and the physical (x, y) spaces are orthogonal.

Figure 3.2a shows the electrostatic analog using a capacitor with a fixed external voltage h, also taken as the distance, across the top and bottom plates. The analog electrostatic potential is chosen as η. The top and bottom plates are with the Dirichlet boundary condition, while the left and right plates are with the Neumann boundary condition. The virtual space is often chosen to have a trivial solution: a rectangle $w \times h$ in this case with $V = y'$. Then, a similar electrostatic problem is solved in the physical space with curved boundaries so that the induced permittivity in the physical space is $\epsilon = \epsilon' \gamma |\nabla V|^2$. This result returns to the electrostatic

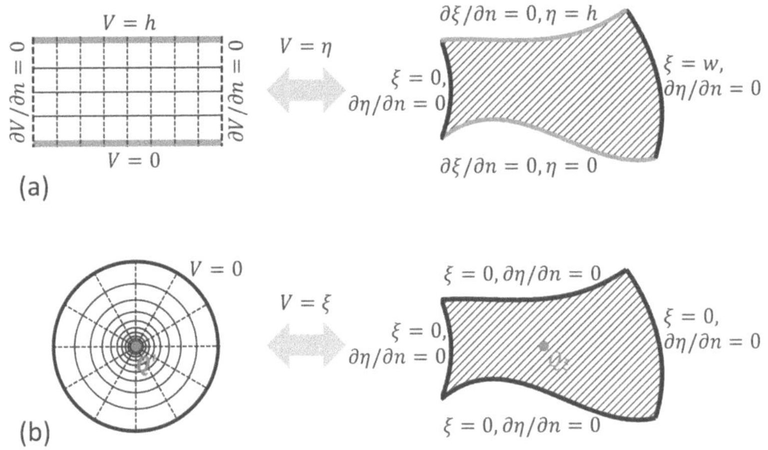

Figure 3.2 Electrostatic configurations as starting points for QCM generation. (a) A capacitor with an external voltage with η chosen as the analog potential. (b) A point charge in a grounded cavity with ξ chosen as the analog potential. Left panels show the trivial solution in the virtual space (x', y'). Right panels show the corresponding physical space (x, y). The orthogonal coordinate (ξ in (a) or η in (b)) is also shown in the physical space for clarity, although it is not necessary.

analogy to ray optics except the additional proportionality constant γ. If we are designing TO devices with prescribed boundaries, the conformal modules of the physical and virtual spaces are generally different.

On the other hand, we can embed point charges (more accurately line charges in 2D) by solving the Poisson's equation on both spaces:

$$\nabla^2 V = -\rho, \;\; \nabla'^2 V = -\rho'. \tag{3.14}$$

By integrating closed loops enclosing all charges in Eq. 3.13, we obtain

$$\gamma = \frac{\oint \nabla' V \cdot \hat{\mathbf{z}} \times d\mathbf{l}'}{\oint \nabla V \cdot \hat{\mathbf{z}} \times d\mathbf{l}} = \frac{Q'_{\text{enc}}}{Q_{\text{enc}}}. \tag{3.15}$$

It gives us an additional flexibility to control the conformal module ratio by setting the ratio between the enclosed line charges. Figure 3.2b shows the electrostatic analog of a point charge in a grounded cavity. We can choose a unit circular disk with a point charge at the center in the virtual space. Poisson's equation in the virtual space has a trivial solution $V = \ln r'$ where $r' = \sqrt{x'^2 + y'^2}$. Then, a similar electrostatic problem is solved in the physical space with curved boundaries so that the induced permittivity in the physical space is $\epsilon = \epsilon' \gamma r'^2 |\nabla V|^2$, in which the factor r'^2 regularizes the singularity at the point charge to have infinite potential. The infinity values of potential at the point charges allow the floating of conformal modules (as a scaled infinity is still infinity). By enforcing $Q_{\text{enc}} = Q'_{\text{enc}}$, it returns to a conformal map whose existence is guaranteed by the Riemann mapping theorem. We have considered a simple shape in the virtual space to have a trivial analytic solution as a seed to numerically generate the QCM. On the other hand, the above method can be applied more generally so that we can numerically solve electrostatic solution in the virtual space with a more complicated shape as well.

3.3 Examples of Quasiconformal Transformation Media

3.3.1 *An Analytic Example*

Actually, we can also use the same technique in the last section to generate analytic QCMs by comparing analytic electrostatic solutions. As a simple example, we design a unidirectional cloak of inner and outer circular boundary of radii a and b. In the physical space, let us consider an electrostatic problem: a perfect electric conducting cylinder (with radius a) under a constant applied static electric field of unit amplitude in the x direction. In the virtual space, we consider the same problem without the cylinder. Then, the electrostatic potential, given by ξ here, is

$$\xi = -x + a^2 x/r^2 = -\left(b^2 - a^2\right) x'/b^2, \tag{3.16}$$

where $r = \sqrt{x^2 + y^2}$. The scaling factor of the potential is chosen so that $\xi = \xi'$ at the outer boundary of the cloak ($r = b$ or $r' = b$) to have continuity between the cloak and the vacuum outside. The ratio of conformal modules γ is obtained by choosing the two end points of path integrations as $A = A' = (0, -b)$ and $B = B' = (0, b)$ on the outer cloak boundary in Eq. 3.13:

$$\gamma = \frac{b^2 - a^2}{b^2 + a^2}. \tag{3.17}$$

The TO medium (transformed from $\epsilon' = 1$) is then obtained by Eq. 3.12:

$$\epsilon = \frac{b^4}{b^4 - a^4}\left(1 + \frac{a^4 + 2a^2\left(x^2 - y^2\right)}{r^4}\right), \tag{3.18}$$

and the permeability is γ along the direction given by $\nabla\xi$ and $1/\gamma$ along the orthogonal direction. Although not necessary now from the analogy approach, by using Eq. 3.10, we can obtain the orthogonal coordinate to ξ as

$$\eta = -y - a^2x/r^2 = -\left(b^2 + a^2\right)y'/b^2. \tag{3.19}$$

The scaling factors are chosen so that $\xi + i\eta$ is a conformal map from $x + iy$ for convenience of discussion and $\eta = \eta'$ at the same outer boundary of the cloak ($r = b$ or $r' = b$). This is actually the conformal map $w = z - a^2/z$ previously considered as an optical conformal map [3] giving the part of permittivity $1 + (a^4 + 2a^2\left(x^2 - y^2\right))/r^4$ now with a small modification by a factor $b^4/(b^4 - a^4)$ and the anisotropy of permeability profile due to the prescribed boundary. Therefore, with the size of cloak larger and larger, this QCM approaches the conformal map.

Figure 3.3a shows the permittivity profile of the cloak with inner/outer radius of 1/4. The virtual coordinates lines with constant x' or constant y' is also shown. It acts as a unidirectional cloak with full-wave simulations (COMSOL Multiphysics) shown in Fig. 3.3b for an incident plane wave from the bottom without any scattering. Because the anisotropy is minimized in the map generation/the electrostatic analogy, the anisotropy is often neglected

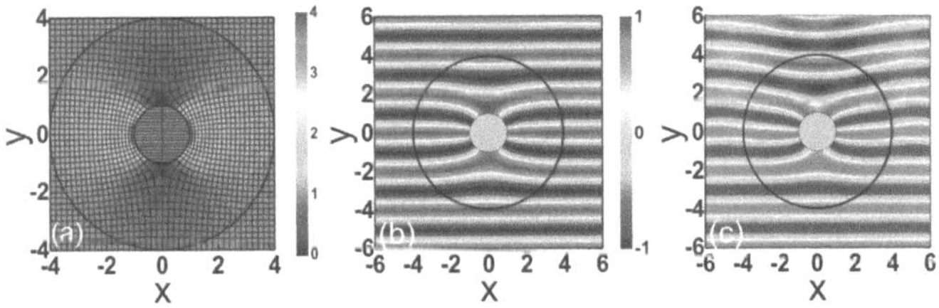

Figure 3.3 A unidirectional cloak (from radii $a = 1$ to 4) with the analytic QCM. (a) Permittivity profile with the virtual x', y' coordinate lines plotted in the physical space. (b) Full-wave simulation with incident plane wave from the bottom. (c) The same full-wave simulation with anisotropy neglected.

as an approximation so that only isotropic dielectric medium is needed for the ease of fabrication. The corresponding full-wave simulation is shown in Fig. 3.3c. In this case, the approximated cloak can largely cancel the original scattering from a bare central cylinder. However, due to this approximation, there is unavoidably aberration [13] introduced as phase distortion when the wave leaves the cloak in the upper part of the simulation, shown in Fig. 3.3c. The aberration in optical path length, due to the approximation, is $\Delta_{\mathrm{ab}} = 2b\left(1 - \sqrt{\gamma}\right) \cong .5$. With a larger cloak size (b) with the object size (a) fixed, Δ_{ab} decreases as $2a^2/b$, a value smaller than the size of the object when $b > 2a$ and approaches zero when the cloak size becomes much larger than the object size.

3.3.2 *Quasiconformal Map with Arbitrary Shape of Device Boundaries*

While the analytic approach in the QCM generation can reveal different aspects of QCMs in TO, it is desirable to have further control on the map/TO medium by taking advantage of our numerical capability in solving an electrostatic problem (Poisson's equation). In particular, although the considered QCM minimizes the anisotropy, the permittivity profile spans from a value of 0 to 4, with a very large ratio between the maximum to minimum value, which is not a straightforward task to tackle analytically. This is due to the existence of sharp corners of the map at $(0, \pm a)$ shown in Fig. 3.3a. Next, we will take the approach in Fig. 3.2 to numerically generate the TO medium using the electrostatic analogy approach.

As the first numerical example, we revisit the quasiconformal carpet cloak. In this case, a simple choice of the virtual space is a rectangle (of dimension $w \times h$, $w = 4$ and $h = 1.5$), while the physical space is the same rectangle with the bottom boundary curved upward as a bump (of height 0.2), as shown in Figs. 3.4a and 3.4b [5]. It is the bump we would like to hide so that if we regard the bottom boundary is a perfectly conducting surface as a mirror, the carpet cloak is to cancel the scattering of this bump.

Here, we follow the capacitor analogy in Fig. 3.2a. A fixed potential of 1.5 units is applied across the top and the bottom boundary, with Neumann boundary condition on the left and the

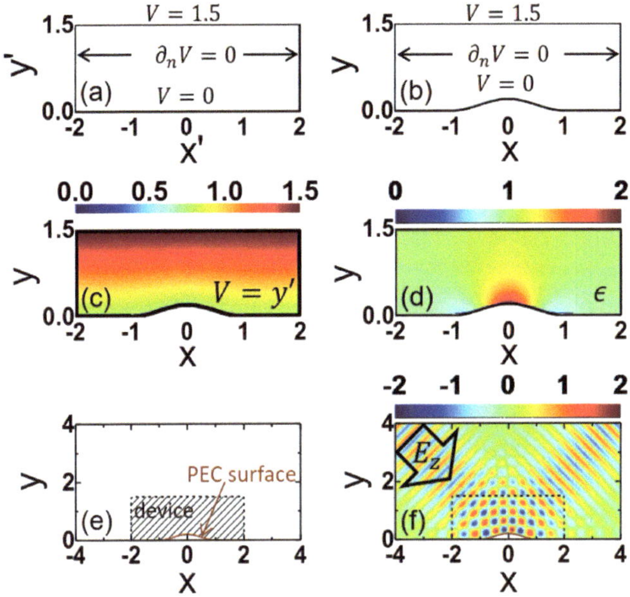

Figure 3.4 A quasiconformal carpet cloak. (a, b) Device shape in the virtual and physical space with the capacitor-type electrostatic problem. A potential of $\Delta V = 1.5$ is applied across the top and bottom plate. (c) The potential profile V (same as the virtual coordinate y') in the physical space. (d) The induced permittivity profile of the cloak. (e, f) Full-wave simulation of E_z profile with an incident beam at 45° to the curved perfect electric conductor (PEC) surface and with the cloak at a free-space wavelength of 0.5.

right boundary. In the virtual space, the trivial solution is simply $V = y'$. In the physical space, the electrostatic problem is solved numerically (using COMSOL Multiphysics) and the potential profile is shown in Fig. 3.4c. The induced permittivity profile of the cloak, ranging from 0.66 to 1.97 can be obtained from the solution V by using Eq. 3.12, plotted in Fig. 3.4d. The constant ratio of conformal module $\gamma \cong .96$ is obtained by using Eq. 3.13 with the path integrations taken along the top boundary of the cloak. When γ is sufficiently near to the value 1, we can neglect the anisotropy as an approximation so that the cloak can be fabricated using isotropic medium. This approximation introduces an aberration as a difference between the apparent height of the cloak and the actual

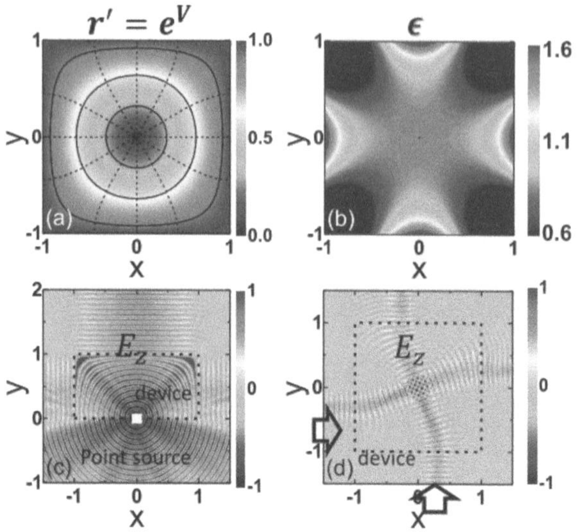

Figure 3.5 Quasiconformal TO medium with electrostatic analogy using a point charge within the cavity. (a) exp(V) profile for a point charge at the center of a square grounded cavity. (b) Induced permittivity in the physical space, with respect to a point charge in a circular cavity of unit radius in the virtual space. (c) A beam collimator using the TO medium with a point source at origin. (d) A beam shifter using the TO medium for both beams incident from the bottom or from the left.

height as $\Delta_{\text{ab}} = h\left(1 - \sqrt{\gamma}\right)$ for the height or $\Delta_{\text{ab}} = w\left(1/\sqrt{\gamma} - 1\right)$ for the width. We can vary the width or height of the cloak and have found numerically that γ varies with w by $1/\gamma \cong 1 + 0.17/w$ and varies with h by $\gamma \cong 1 - 0.061/h$ in this example. Thus, the aberration becomes smaller and approaches zero as the size of the cloak increases. Figures 3.4e and 3.4f show the full-wave simulation under such an approximation for a beam at 45° incident on the curved mirror at a free-space wavelength of 0.5. With the cloak, the scattering of the bump is largely cancelled, compared to the case when there is no cloak.

As the second numerical example, we start from the electrostatic analogy in Fig. 3.2b in considering a point charge in a cavity. We put a unit point charge at the center of a square cavity (of width 2) whose boundary is kept at a constant zero potential. The electrostatic solution in the physical space is shown in Fig. 3.5a by plotting exp(V). In the virtual space, the similar problem with a circular cavity with unit radius is $V = \ln(r')$. The permittivity profile, from Eq. 3.12, is plotted in Fig. 3.5b where we drop permittivity lower than 0.6 as an approximation. Here, we are transforming a unit disk in the virtual space to a square in the physical space. As the shape of the boundary is flattened from a curve, such a medium can be used as a beam collimator or a multiangle beam shifter [26, 31]. Now, we switch from the analogy problem to the TO function at a finite frequency. Figure 3.5c shows the case as a beam collimator when we put a point source (of free-space wavelength 0.08) at the center. We are only using the upper half of the device in this case so that from the simulation result, a collimated beam is formed for the upper part of the radiation from the point source, while the cylindrical wave front is maintained for a point source at the lower part of the simulation for comparison. The wave front within the device follows the analog potential profile. In Fig. 3.5d, the device is used as a beam shifter for a beam shining from the bottom or from the left. The simulation result shows the E_z profile when both beams are incident to the device. The beams follow the analog E-field streamlines, shown as dashed lines in Fig. 3.5a and beam shifting follows.

3.3.3 *From Slipping Boundary to Fixed Boundary*

In the previous examples, the simplicity of implementation in generating the maps comes from the fact that we have assumed a QCM of constant anisotropy so that the intermediate coordinates (ξ, η) have their coordinate lines being orthogonal in both the virtual and the physical space. The slipping boundary used means that there are no restrictions on the mapping of the boundaries. Only the shape is fixed and the corners are fixed. This is the ideal situation, giving us a Teichmüller map which can have analytic properties and electrostatic analogy for assisting its generation. However, for applications which depend strongly on the matching of boundaries, for example, waveguides with fixed input and output facets, further constraints on the boundaries have to be applied [32, 33]. Then, maximum anisotropy should be minimized subject to these additional constraints. The maximum anisotropy of the resulting optimized map will be greater than the ideal case.

As the minimization of the total anisotropy is related to solving a Laplace equation of either virtual coordinate in the physical space [25], one can therefore heuristically solve the Laplace equation for more general restricted type of boundary conditions in generating the maps with small anisotropy. Here, we take this approach for its simplicity in implementation. Let us take the carpet cloak implemented with slipping boundary condition as example. Now, we hold the three boundaries matching to the outside medium (except the bottom boundary) as fixed to be continuous. The resultant map is generally not orthogonal so that we have to solve the two separate Laplace equations for x' and y'. Figure 3.6a shows the result of the virtual coordinate x' by solving $\nabla^2 x' = 0$ with the Neumann boundary condition on the bottom boundary and fixed boundary condition $x' = x$ on the other three boundaries. Figure 3.6b shows the result of the virtual coordinate y' by solving $\nabla^2 y' = 0$ with Dirichlet boundary condition: $y' = 0$ at the bottom boundary and $y' = y$ for the other three. From the resultant map, we can calculate the induced permittivity and anisotropy (still defined by $\gamma + 1/\gamma = Tr(\mu)/\sqrt{\det \mu}$) directly form Eq. 3.2. They are plotted in Figs. 3.6c and 3.6d. While the permittivity profile is found nearly unchanged, γ is now not a constant. The value $\max(\gamma, 1/\gamma)$ ranges from 1 at the two completely fixed corners to around 1.08. Nevertheless, we still neglect the anisotropy and perform the full-wave simulation for a beam incident at 45° to the curved mirror and we found that the cloak performs similar to its counterpart with slipping boundary condition, as shown in Figs. 3.6e and 3.6f.

3.4 Extension to Acoustic and Elastic Waves

Although initially proposed for electromagnetic waves, the principle of transformation optics can be straightforwardly transplanted into other physics branches [38, 39]. In this section, we briefly review the quasiconformal transformation method applied to acoustics and elastodynamics.

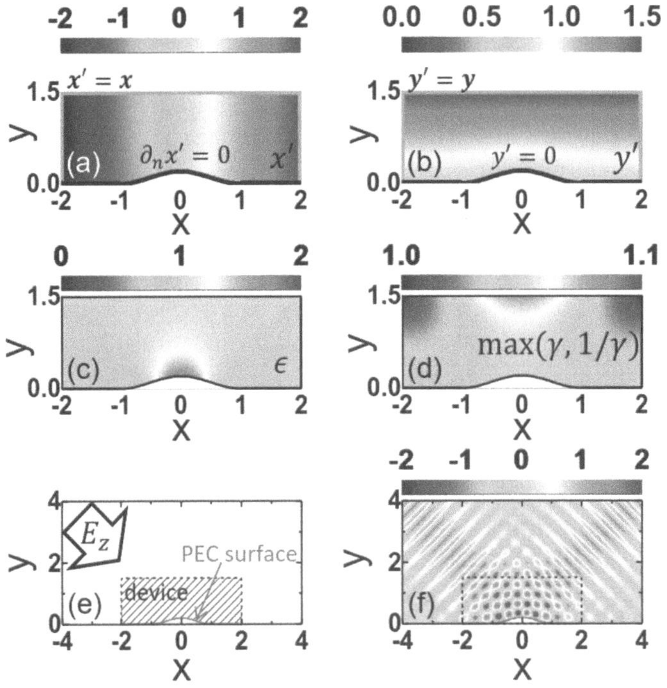

Figure 3.6 A carpet cloak with fixed boundary to vacuum. (a, b) Resultant x' and y' obtained by solving the Laplace equation on x' and y' with Dirichlet–Neumann boundary condition on the bottom boundary, while the other three boundaries are held fixed. (c) Permittivity profile of the cloak. (d) Anisotropy profile. (e, f) Full-wave simulation of E_z profile with an incident beam at 45° to the curved PEC surface and with the cloak at a free-space wavelength of 0.5. The anisotropy is neglected.

3.4.1 *Acoustic Case*

The governing equation of classical acoustic wave can be written as a Helmholtz equation:

$$\nabla \cdot \left(\rho^{-1} \nabla p\right) = \kappa^{-1} \ddot{p}, \tag{3.20}$$

in which p is the sound pressure, ρ is the mass density, and κ is the bulk modulus. The form invariance of Eq. 3.20 under coordinate transformation guarantees the feasibility of the transformation method. In this sense, the acoustic waves can be manipulated with the following material transformation relations [40, 41]:

$$\rho' = \rho\frac{\sqrt{\det g}}{g}, \quad \kappa' = \kappa\sqrt{\det g}. \tag{3.21}$$

As indicated in Eq. 3.21, the anisotropy has been introduced into the transformed mass density. Although not impossible [42–44], this requirement is rather hard to be satisfied for acoustic fluid and dramatically increases the implementation difficulty. Fortunately, similar as the electromagnetic case, if the mapping is (quasi) conformal, the anisotropy can be (approximately) eliminated. While deriving the material transformation formulations, if the Cauchy–Riemann condition, that is,

$$\frac{\partial x}{\partial \xi} = \frac{\partial y}{\partial \eta}, \quad \frac{\partial x}{\partial \eta} = -\frac{\partial y}{\partial \xi}, \tag{3.22}$$

is taken into consideration, the following relations can be obtained via a local-point-of-view transformation method [45]:

$$\rho' = \zeta\rho, \quad \kappa' = \zeta\sqrt{\det g}\kappa. \tag{3.23}$$

Compared with Eq. 3.21, the anisotropy in the mass density has been eliminated. Additionally, a tunable factor, ζ, is introduced to the transformation relations, providing another degree of freedom to adjust the material parameters. The nature of the different material combinations resulted from ζ, can be explained from the perspective of refractive index. Actually, in isotropic acoustic media, the refractive index, that is, $n_a = \sqrt{\kappa/\rho}$, is the only parameter governing the path of sound waves. In this case, with

$$n_a' = \sqrt{\frac{\kappa'}{\rho'}} = \sqrt[4]{\det g}, \tag{3.24}$$

being guaranteed, any combinations of mass density and bulk modulus can be available for the same wave path. With this extra degree of freedom, one can simplify the material parameters by assuming constant mass density ($\zeta = 1$) or constant bulk modulus ($\zeta = (\det g)^{-1/2}$). It is worth to note that, as the Cauchy–Riemann condition leads to the following relations:

$$g = \sqrt{\det g}\mathbf{I}, \tag{3.25}$$

the case of $\zeta = 1$ can be seen as the reduction of Eq. 3.23. On the other hand, the tunable factor ζ also provides us another route to design impedance-matched transformation devices. As the acoustic impedance of the transformed media is

$$z'_a = \sqrt{\rho'\kappa'} = \left(\zeta\,\sqrt[4]{\det g}\right)\sqrt{\rho\kappa} = \left(\zeta\,\sqrt[4]{\det g}\right)z_a, \qquad (3.26)$$

the impedance match condition of $z'_a = z_a$ will naturally leads to $\zeta = (\det g)^{-1/4}$. With this ζ, the designed transformation device can be confidently embedded in the environment media without any boundary mismatch.

With the transformation relations given by Eq. 3.23, all the technique in the aforementioned sections can be applied in the design of acoustic transformation devices. As a numerical example, here we generate an impedance-matched acoustic beam bender, applying the numerical approach provided in Section 3.2. As shown in Figs. 3.7a and 3.7b, a rectangular domain is mapped to a fan-shaped physical domain. The Laplace equation, Eq. 3.11, is calculated together with the sliding boundary conditions. Although the anisotropy factor $\max(\gamma, 1/\gamma)$ is globally 1.17, the conformal approximation is still available. As mentioned above, $\zeta = (\det g)^{-1/4}$ is employed to make sure the impedances of the inlet and outlet boundaries of the beam bender match with the environment media. With the anisotropy being truncated, the calculated $\sqrt{\det g}$ is imported into Eq. 3.23 to obtain the material parameters. In this example, the environment media is set to be water with $\rho = 1000$ $\mathrm{kg \cdot m^{-3}}$ and $\kappa = 2.2$ GPa. Therefore, the distribution of mass density and bulk modulus can be obtained and illustrated, respectively, in Figs. 3.7c and 3.7d. It can be easily calculated that the acoustic impedance of the physical domain is homogeneous 1.482×10^6 $\mathrm{kg \cdot m^{-2} \cdot s^{-1}}$, the same as the impedance of the medium outside. A finite element analysis is performed to show the effectiveness of the designed beam bender, with the simulation setup shown in Fig. 3.7e. With the pressure field demonstrated in Fig. 3.7f, a horizontally incident sound beam is perfectly bent by the designed beam bender, without any reflection on the device boundaries.

3.4.2 *Elastodynamic Case*

Different from fluid media, a solid supports not only compressional wave (P wave) but also shear wave (S wave). This makes the control

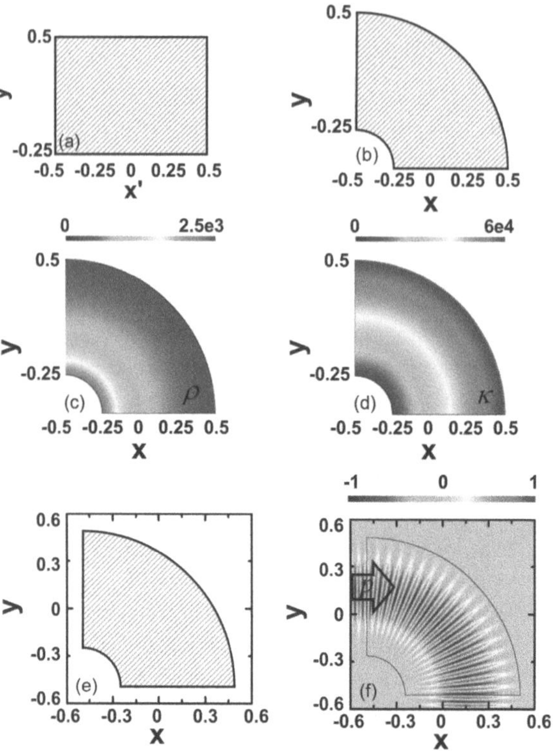

Figure 3.7 An acoustic beam bender. (a, b) Device shape in the virtual and physical space. (c, d) The distribution of mass density and bulk modulus of the beam bender. (e, f) Full-wave simulation of p profile with an acoustic wave beam horizontally incident at the beam bender.

of elastic waves more challenging. Milton et al. [46] showed that the elastodynamic equation, that is, Navier's equation,

$$\nabla \cdot (\mathbf{C} : \nabla \mathbf{u}) = \rho \ddot{u}, \tag{3.27}$$

with **u** being the displacement vector and **C** being the fourth-order elastic tensor, is not form-invariant for a general spatial mapping. Therefore, the transformation method is not directly applicable for elastic waves in a general sense. However, later, Brun et al. [47] reported a precise control of elastic wave with asymmetric transformation relations:

$$\rho' = \rho \frac{1}{\sqrt{\det g}}, \quad C_{i'j'k'l''} = \frac{\Lambda_i^{i'} \delta_j^{j'} \Lambda_k^{k'} \delta_l^{l'}}{\sqrt{\det g}} C_{ijkl}. \tag{3.28}$$

On the other hand, using the above mentioned local-point-of-view transformation method [45], we found that the form invariance of Eq. 3.27 can be approximately preserved with the material relations of

$$\rho' = \rho \frac{g}{\sqrt{\det g}}, \quad C'_{i'j'k'l'} = \frac{\Lambda_i^{i'} \Lambda_j^{j'} \Lambda_k^{k'} \Lambda_l^{l'}}{\sqrt{\det g}} C_{ijkl}, \tag{3.29}$$

in the case where the gradient of the transformed material is small or when the wave frequency is high [48, 49]. The above two sets of transformation relations enable us to design elastodynamic transformation devices by using the transformation method. Similarly as the acoustic case, considering the Cauchy–Riemann condition, the following material transformation relations can be approximately obtained for elastic waves:

$$\rho' = \zeta \rho, \quad \mathbf{C}' = \zeta \sqrt{\det g} \mathbf{C}. \tag{3.30}$$

The relations reveal that, during the transformation, the generalized refractive index tensor $\mathbf{n}_e = (\mathbf{C}/\rho)^{1/2}$ should follow the rule of

$$\mathbf{n}'_e = \sqrt{\mathbf{C}'/\rho'} = \sqrt[4]{\det g} \sqrt{\mathbf{C}/\rho} = \sqrt[4]{\det g} \mathbf{n}_e. \tag{3.31}$$

Equation 3.31 can be considered as a general extension of Eq. 3.24, implying that both the P and S waves can be simultaneously manipulated. However, different from acoustic waves, the governing equation for elastic waves loses its form invariance during transformation, so that Eqs. 3.29 and 3.30 remain as approximated equations. Fortunately, through many numerical examples [50], we found that the isotropic transformation devices designed in accordance with Eq. 3.30 work reasonably well to replace the devices designed by quasiconformal transformation method. Now, we will follow the acoustic case, by tuning the factor ζ to obtain different transformation relations for conformal or quasiconformal mappings. The "constant mass density" or "constant elastic tensor" transformation relations can be expected by choosing $\zeta = 1$ or $\zeta = (\det g)^{-1/2}$, respectively. Also, the impedance matched relations can be obtained to make the generalized impedance tensor

$$\mathbf{z}'_e = \sqrt{\mathbf{C}'\rho'} = \left(\zeta \sqrt[4]{\det g}\right) \sqrt{\mathbf{C}\rho} = \left(\zeta \sqrt[4]{\det g}\right) \mathbf{z}_e, \tag{3.32}$$

equal to $\mathbf{z}_e$, corresponding to the case of $\zeta = (\det g)^{-1/4}$.

In what follows, an elastic carpet cloak is designed. The general principle is the same as the electromagnetic carpet cloak proposed in the previous section. Here, the bump is set to be more irregular to demonstrate the ability of the numerical approach, as illustrated in Fig. 3.8. Sliding boundaries are applied to solve the Laplace equation. In this example, the anisotropy factor $\max(\gamma, 1/\gamma)$ is calculated to be globally 1.07, indicating that the mapping is almost conformal. For the ease of implementation, here we choose the tunable factor $\zeta = (\det g)^{-1/2}$ in order to keep the designed elastic tensor constant. With the background media set to be structural steel with the following material parameters: Young's modulus $E = 200$ GPa, Poisson's ratio $\nu = 0.33$ and mass density $\rho = 7850$ kg·m^{-3}, the distribution of mass density required by the carpet cloak is obtained from Eq. 3.30, and is shown in Fig. 3.8a. In order to validate the designed material parameters, finite element simulations are performed. In this example, the bottom boundary is set to be rigid. Figure 3.8b shows the total displacement field of a harmonic S wave incident on the designed elastodynamic carpet cloak. For comparison, the total displacement fields of the curved boundary without the carpet cloak and a rigid flat boundary

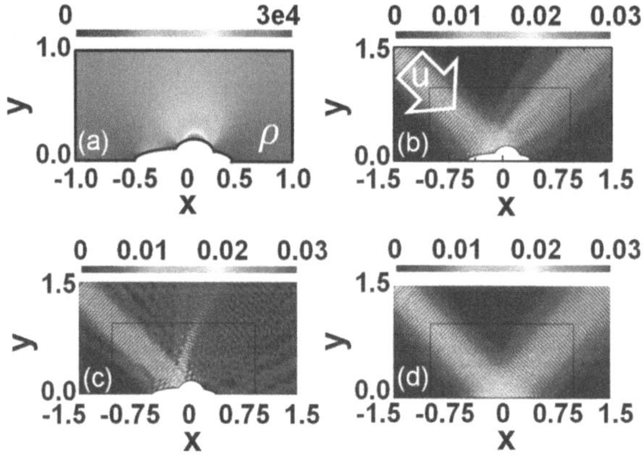

Figure 3.8 An elastodynamic carpet cloak with irregular configuration. (a) Mass density profile of the carpet cloak. (b, c) Total displacement field of an incident S-wave beam with the frequency 40 kHz to the curved rigid boundary with and without the carpet cloak. (d) Total displacement field of an incident P-wave beam to the flat rigid boundary.

are shown in Figs. 3.8c and 3.8d, respectively. Different from the electromagnetic and the acoustic cases, mode conversion takes place at the boundary in the case of oblique incidence. Nevertheless, the designed cloak can restore not only the propagation directions of reflective P and S waves, but also the energy distribution between P and S components, keeping the displacement field looks very similar between Figs. 3.8b and 3.8d.

3.5 Conclusion

In this chapter, we have established QCMs as a useful tool to develop transformation media with minimized anisotropy. By using the method of solving Laplace equations or equivalent electrostatic problems, QCMs can be generated in a straightforward manner. The same method can be applied to problems with further restrictions on the boundaries for practical applications. An electrostatic analogy is also developed to further simplify the generation of QCMs. We have also discussed the extension of these techniques to acoustic and elastic waves.

Acknowledgments

J.L. would like to acknowledge funding support from the European Union's Seventh Framework Programme under Grant Agreement No. 630979. G.K.H. would like to acknowledge funding support for NSFC under No.11472044. Z.C. would like to acknowledge the support from the China Postdoctoral Science Foundation (No. 2014M550054).

References

1. Pendry, J. B., Holden, A. J., Robbins, D. J., and Stewart, W. J. (1999). Magnetism from conductors and enhanced nonlinear phenomena, *IEEE Trans. on Microw. Theory Techn.*, **47**, pp. 2075–2084.
2. Pendry, J. B., Schurig, D., and Smith, D. R. (2006). Controlling electromagnetic fields, *Science*, **312**, pp. 1780–1782.

3. Leonhardt, U. (2006). Optical conformal mapping, *Science*, **312**, pp. 1777–1780.
4. Schurig, D., Mock, J. J., Justice, B. J., Cummer, S. A., Pendry, J. B., Starr, A. F., and Smith, D. R. (2006). Metamaterial electromagnetic cloak at microwave frequencies, *Science*, **314**, pp. 977–980.
5. Li, J., and Pendry, J. B. (2008). Hiding under the carpet: a new strategy for cloaking, *Phys. Rev. Lett.*, **101**, 203901.
6. Jiang, W. X., Cui, T. J., Yang, X. M., Cheng, Q., Liu, R., and Smith, D. R. (2008). Invisibility cloak without singularity, *Appl. Phys. Lett.*, **93**, 194102.
7. Liu, R., Ji, C., Mock, J. J., Chin, J. Y., Cui, T. J., and Smith, D. R. (2009). Broadband ground-plane cloak, *Science*, **323**, pp. 366–369.
8. Leonhardt, U., and Tyc, T. (2009). Broadband invisibility by non-Euclidean cloaking, *Science*, **323**, pp. 110–112.
9. Valentine, J., Li, J., Zentgraf, T., Bartal, G., and Zhang, X. (2009). An optical cloak made of dielectrics, *Nat. Mater.*, **8**, pp. 568–571.
10. Gabrielli, L. H., Cardenas, J., Poitras, C. B., and Lipson, M. (2009). Silicon nanostructure cloak operating at optical frequencies, *Nat. Photonics*, **3**, pp. 461–463.
11. Ergin, T., Stenger, N., Brenner, P., Pendry, J. B., and Wegener, M. (2010). Three-dimensional invisibility cloak at optical wavelengths, *Science*, **328**, pp. 337–339.
12. Ma, H. F., and Cui, T. J. (2010). Three-dimensional broadband ground-plane cloak made of metamaterials, *Nat. Commun.*, **1**, 21.
13. Zhang, B., Chan, T., and Wu, B. I. (2010). Lateral shift makes a ground-plane cloak detectable, *Phys. Rev. Lett.*, **104**, 233903.
14. Zhou, F., Bao, Y., Cao, W., Stuart, C. T., Gu, J., Zhang, W., and Sun, C. (2011). Hiding a realistic object using a broadband terahertz invisibility cloak, *Sci. Rep.*, **1**, 78.
15. Gharghi, M., Gladden, C., Zentgraf, T., Liu, Y., Yin, X., Valentine, J., and Zhang, X. (2011). A carpet cloak for visible light, *Nano lett.*, **11**, pp. 2825–2828.
16. Chen, X., Luo, Y., Zhang, J., Jiang, K., Pendry, J. B., and Zhang, S. (2011). Macroscopic invisibility cloaking of visible light, *Nat. Commun.*, **2**, 176.
17. Zhang, B., Luo, Y., Liu, X., and Barbastathis, G. (2011). Macroscopic invisibility cloak for visible light, *Phys. Rev. Lett.*, **106**, 033901.
18. Landy, N., and Smith, D. R. (2013). A full-parameter unidirectional metamaterial cloak for microwaves, *Nat. Mater.*, **12**, pp. 25–28.

19. Rahm, M., Schurig, D., Roberts, D. A., Cummer, S. A., Smith, D. R., and Pendry, J. B. (2008). Design of electromagnetic cloaks and concentrators using form-invariant coordinate transformations of Maxwell's equations, *Photon. Nanostr. Fundam. Appl.*, **6**, pp. 87–95.
20. Kildishev, A. V., and Shalaev, V. M. (2008). Engineering space for light via transformation optics, *Opt. Lett.*, **33**, pp. 43–45.
21. Genov, D. A., Zhang, S., and Zhang, X. (2009). Mimicking celestial mechanics in metamaterials, *Nat. Phys.*, **5**, pp. 687–692.
22. Chen, H., Hou, B., Chen, S., Ao, X., Wen, W., and Chan, C. T. (2009). Design and experimental realization of a broadband transformation media field rotator at microwave frequencies, *Phys. Rev. Lett.*, **102**, 183903.
23. Lai, Y., Ng, J., Chen, H., Han, D., Xiao, J., Zhang, Z. Q., and Chan, C. T. (2009). Illusion optics: the optical transformation of an object into another object, *Phys. Rev. Lett.*, **102**, 253902.
24. Landy, N. I., and Padilla, W. J. (2009). Guiding light with conformal transformations, *Opt. Express*, **17**, pp. 14872–14879.
25. Chang, Z., Zhou, X., Hu, J., and Hu, G. (2010). Design method for quasi-isotropic transformation materials based on inverse Laplace's equation with sliding boundaries, *Opt. Express*, **18**, pp. 6089–6096.
26. Zentgraf, T., Valentine, J., Tapia, N., Li, J., and Zhang, X. (2010). An optical "Janus" device for integrated photonics, *Advanced Materials*, **22**, pp. 2561–2564.
27. Landy, N. I., Kundtz, N., and Smith, D. R. (2010). Designing three-dimensional transformation optical media using quasiconformal coordinate transformations, *Phys. Rev. Lett.*, **105**, 193902.
28. Mei, Z. L., Bai, J., and Cui, T. J. (2011). Experimental verification of a broadband planar focusing antenna based on transformation optics, *New J. Phys.*, **13**, 063028.
29. Yang, R., Tang, W., and Hao, Y. (2011). A broadband zone plate lens from transformation optics, *Opt. Express*, **19**, pp. 12348–12355.
30. García-Meca, C., Martínez, A., and Leonhardt, U. (2011). Engineering antenna radiation patterns via quasi-conformal mappings, *Opt. Express*, **19**, pp. 23743–23750.
31. Wu, Q., Turpin, J. P., and Werner, D. H. (2012). Integrated photonic systems based on transformation optics enabled gradient index devices, *Light Sci. Appl.*, **1**, e38.

32. Gabrielli, L. H., Liu, D., Johnson, S. G., and Lipson, M. (2012). On-chip transformation optics for multimode waveguide bends, *Nat. Commun.*, **3**, 1217.
33. Liu, D., Gabrielli, L. H., Lipson, M., and Johnson, S. G. (2013). Transformation inverse design, *Opt. Express*, **21**, pp. 14223–14243.
34. Weber, O., Myles, A., and Zorin, D. (2012). Computing extremal quasiconformal maps, *Computer Graphics Forum*, **31**, pp. 1679–1689.
35. Chan, W. M. (2010) *Handbook of grid generation*. Willardson, Thompson, J. F., Soni, B. K., and Weatherill, N. P., Chapter 5 "Hyperbolic Methods for Surface and Field Grid Generation," (CRC press, USA).
36. Mei, Z. L., Bai, J., Niu, T. M., and Cui, T. J. (2010). Design of arbitrarily directional cloaks by solving the Laplace's equation, *J. Appl. Phys.*, **107**, 124502.
37. Yao, K., Chen, H., Liu, Y., and Jiang, X. (2014). An analogy strategy for transformation optics, *New J. Phys.*, **16**, 063008.
38. Guenneau, S., Amra, C., and Veynante, D. (2012). Transformation thermodynamics: cloaking and concentrating heat flux, *Opt. Express*, **20**, pp. 8207–8218.
39. Zhang, S., Genov, D. A., Sun, C., and Zhang, X. (2008). Cloaking of matter waves, *Phys. Rev. Lett.*, **100**, 123002.
40. Chen, H., and Chan, C. T. (2007). Acoustic cloaking in three dimensions using acoustic metamaterials, *Appl. Phys. Lett.*, **91**, 183518.
41. Cummer, S. A., and Schurig, D. (2007). One path to acoustic cloaking, *New J. Phys.*, **9**, 45.
42. Torrent, D., and Sánchez-Dehesa, J. (2010). Anisotropic mass density by radially periodic fluid structures, *Phys. Rev. Lett.*, **105**, 174301.
43. Zhang, S., Xia, C., and Fang, N. (2011). Broadband acoustic cloak for ultrasound waves, *Phys. Rev. Lett.*, **106**, 024301.
44. Popa, B. I., Zigoneanu, L., and Cummer, S. A. (2011). Experimental acoustic ground cloak in air, *Phys. Rev. Lett.*, **106**, 253901.
45. Chang, Z., Hu, J., and Hu, G. K. (2010). Transformation method and wave control, *Acta Mech. Sin.*, **26**, pp. 889–898.
46. Milton, G. W., Briane, M., and Willis, J. R. (2006). On cloaking for elasticity and physical equations with a transformation invariant form, *New J. Phys.*, **8**, 248.
47. Brun, M., Guenneau, S., and Movchan, A. B. (2009). Achieving control of in-plane elastic waves, *Appl. Phys. Lett.*, **94**, 061903.

48. Hu, J., Chang, Z., and Hu, G. (2011). Approximate method for controlling solid elastic waves by transformation media, *Phys. Rev. B*, **84**, 201101.
49. Chang, Z., Liu, X., Hu, G., and Hu, J. (2012). Transformation ray method: Controlling high frequency elastic waves (L), *J. Acoust. Soc. Am.*, **132**, pp. 2942–2945.
50. Chang, Z., Hu, J., Hu, G., Tao, R., and Wang, Y. (2011). Controlling elastic waves with isotropic materials, *Appl. Phys. Lett.*, **98**, 121904.

Chapter 4

Control of Electromagnetic Flux in Inhomogeneous Anisotropic Media

Jie Luo,[a] Yun Lai,[a] and C. T. Chan[b]
[a]*College of Physics, Optoelectronics and Energy and Collaborative Innovation Center of Suzhou Nano Science and Technology, Soochow University Suzhou 215006, P. R. China*
[b]*Department of Physics and Institute for Advanced Study, Hong Kong University of Science and Technology, Clear Water Bay, Hong Kong*
laiyun@suda.edu.cn

Control of electromagnetic (EM) flux and energy is a fundamental scientific issue with great implications in theory and applications. Based on the principles of refraction and reflection, lenses and waveguides composed of dielectrics and metals have become the most successful devices to guide and control EM waves and fluxes. Recently, the rapid development in the field of artificial EM structures such as photonic crystals, plasmonic structures, and metamaterials has inspired new ways to control EM waves and fluxes beyond traditional methods, leading to intriguing devices such as superlens, hyperlens, invisibility cloaks, illusion devices, and tunneling waveguides. In this chapter, we introduce a mechanism to control the spatial distribution of either the electric or the magnetic field instead of both fields. We demonstrate that EM fluxes can be controlled in an almost arbitrary manner in wavelength and

Transformation Wave Physics: Electromagnetics, Elastodynamics, and Thermodynamics
Edited by Mohamed Farhat, Pai-Yen Chen, Sebastien Guenneau, and Stefan Enoch

ISBN 978-981-4669-95-5 (Hardcover), 978-981-4669-96-2 (eBook)
www.panstanford.com

subwavelength scales. Interestingly, the principle of flux control in our method does not rely on surface waves as in the field of plasmonics, but relies on the evanescent waves induced by the strong anisotropy and designed inhomogeneity of media. Along with the ability of controlling EM flux, such inhomogeneous anisotropic media exhibit surprisingly robust high transmittance. Combining with transformation optics, such high transmittance property can be utilized to build waveguide devices with almost arbitrary shapes and bending angles. As our method only needs to control the electric field instead of the magnetic field, it also provides an efficient approach for flux control with nonmagnetic dielectric media.

4.1 Introduction

In the past several decades, the rapid development in the fields of photonic crystals [1–3], plasmonics [4–6], and metamaterials [7–9] has inspired a tremendous amount of research on artificial electromagnetic (EM) media beyond natural materials, and their plausible applications beyond traditional electromagnetics and optics. These artificial EM media have the ability of controlling EM flux in unprecedented ways. Photonic crystals [1–3] offer the possibility of forbidding and guiding the propagation of EM waves within certain frequency ranges due to photonic band gaps and unusual dispersions. By utilizing surface plasmons on metal surfaces, subwavelength controlling of EM waves has been achieved, leading to the flourishing field of plasmonics [4–6] with promising applications such as nanoscale plasmonic chips [5] and nanolasing [10]. Another significant advance is the discovery of metamaterials [7–9, 11–25], which are artificial materials designed to exhibit effective permittivity and permeability values nonexistent in nature. The metamaterial technology endows independent control of the effective permittivity and permeability with both positive and negative values, anisotropic parameters, and gradient index. These unusual constitutive parameters give rise to various intriguing phenomena including negative refraction [11, 12], superlensing [13, 14], and hyperlensing [15–17], cloaking [20–22], and illusion optics [24, 25]. In particular, the theory of transformation optics [20, 22] provides a unique way to use metamaterials to control the spatial distribution of electric and magnetic fields, and therefore the EM flux.

In all previous paradigms of wave control, including refraction, reflection, photonic crystals, plasmonics and transformation optics, both the electric and magnetic fields are controlled simultaneously, so as to control the EM flux. Controlling EM waves and flux are often considered to be the same. However, there is actually a difference between them. From the formula of Poynting vector $\mathbf{S} = \mathbf{E} \times \mathbf{H}$, we can see that EM flux **S** can be controlled by manipulating either the electric field **E** or the magnetic field **H**, not necessarily both fields at the same time. This simple idea stimulated the research works we are going to summarize in this chapter. As we shall see next, EM flux control can indeed be achieved by controlling only the electric field **E** or the magnetic field **H**.

In this chapter we will review our work on controlling EM flux by controlling either electric or magnetic fields. We demonstrate almost arbitrary control of EM flux in both wavelength and subwavelength scales in strongly anisotropic inhomogeneous media. Robust high transmission is achievable in such media due to the impedance averaging effect. Based on such unique property, intriguing applications such as waveguide devices with almost arbitrary shapes and bending angles can be constructed. Our work represents a principle to confine and guide EM fluxes in an almost arbitrary way, which is totally different from those in reflections, photonic crystals, plasmonics and transformation optics.

4.2 Inhomogeneous Anisotropic Zero-Index Media

As mentioned in the abstract, the flux controlling ability of our method is bestowed by strong anisotropy and designed inhomogeneity. To obtain the strong anisotropy, one way is to tune one component of the permittivity/permeability to be near zero.

Zero-index media are a special kind of materials exhibiting near-zero refractive index. Zero-index media can be classified into single-zero media with permittivity near zero, that is, epsilon-near-zero (ENZ) media; permeability near zero, that is, mu-near-zero

(MNZ) media; and double-zero media, with both permittivity and permeability near zero. In the past decade, zero-index media have been extensively investigated, and many interesting applications have been proposed including tailoring wave front shapes [26–33], squeezing and tunneling EM waves in narrow waveguides [34–36], and holes [37, 38], constructing displacement-current wires and nanocircuit boards [39–41], obtaining subwavelength imaging [42–45], manipulating transmission, scattering, and source radiation by engineering defects [46–54], obtaining perfect absorbers [55–59], etc.

Anisotropic zero-index media is a special class of zero-index media. In the directions perpendicular to the zero component of permittivity/permeability, the wavelength is infinity. But in the direction of the zero component, the wavelength is finite. Interestingly, for metamaterials, anisotropic zero-index media are often easier to achieve than isotropic zero-index media. It is because the typical resonators inside the unit cell of the metamaterials, such as split rings and cut wires, are themselves anisotropic [51, 52, 60–63].

In this chapter, we will consider transverse magnetic (TM)-polarized incident waves with the magnetic field in the z direction and electric fields in the xy plane. In this case, we need anisotropic epsilon-near-zero (AENZ) media, in which one component of permittivity is nearly zero. Similar phenomena can also be obtained in the inhomogeneous anisotropic mu-near-zero (AMNZ) media for transverse electric (TE)-polarized waves.

Now we consider TM-polarized waves propagating in the x direction. The relevant parameters of the inhomogeneous AENZ media are characterized by:

$$\varepsilon_y = f(x, y) >> \varepsilon_x \to 0^+, \quad \mu = 1 \tag{4.1}$$

where ε_x and ε_y are the x and y components of the relative permittivity tensor, μ is the relative permeability, and $f(x, y)$ is the inhomogeneity profile for the y component of the permittivity tensor. Note that the wavelength for the waves propagating along the x direction is $\lambda = \lambda_0/\sqrt{\varepsilon_y \mu} = \lambda_0/\sqrt{f(x, y)} << \infty$, where λ_0 is the wavelength in free space. This is different from the cases of zero-index media with $\varepsilon_x \approx \varepsilon_y \to 0^+$ or $\varepsilon_y \to 0^+$, inside of which, the wavelength tends to be infinite [34–36].

If the inhomogeneity profile $f(x, y)$ is simply a constant in the whole system, then the value of ε_x does not have any influence on wave propagation unless the propagation direction deviates from the x direction. However, if $f(x, y)$ is a spatially varying function, then the near zero ε_x would play a crucial role in the scattering mechanism.

4.2.1 *Scatterings in Highly Anisotropic Media and EM Flux Redistribution*

It is well known that in inhomogeneous media composed of traditional materials (e.g., dielectrics), incident waves will be scattered into various directions when the wave encounters scatterers. When the sizes of the defects are comparable with or larger than the incident wavelength, most scattered waves will be propagating waves, as illustrated in Fig. 4.1a. As a consequence, the transmission is largely reduced and the wave front of the wave is dramatically distorted.

However, the scattering characteristics in highly anisotropic media are totally different. In Fig. 4.1b, we show the schematic picture of multiple scatterings in highly anisotropic media with $\varepsilon_x \to 0^+$. Instead of producing propagating waves, the scatterings produce waves (shown as blue wavy arrows) that are evanescent in the forward and backward directions, and only capable of transferring energy flux in the perpendicular directions.

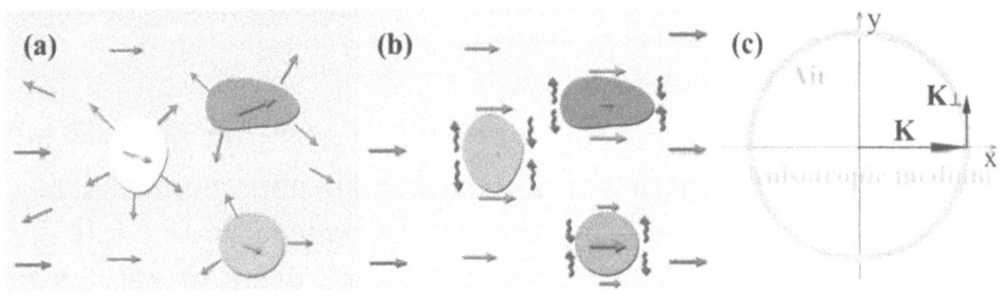

Figure 4.1 Schematic illustration of wave propagation behaviors in (a) a normal inhomogeneous medium and (b) an anisotropic inhomogeneous medium satisfying Eq. 4.1. (c) Equifrequency contours of the background media in (a) and (b). Reprinted (figure) with permission from Ref. [62]. Copyright (2014) by the American Physical Society.

The differences can be comprehended from the equifrequency contours of an isotropic medium, for example, air (green) and a highly anisotropic medium with $\varepsilon_x \rightarrow 0^+$ (blue), as illustrated in Fig. 4.1c. Suppose that the scatterings are induced by a defect with a characteristic perpendicular length $L_\perp$, then a series of waves would be excited with perpendicular wave vectors $K_\perp$. For scatterings in air, when $L_\perp \geq \lambda$, $K_\perp$ mostly fall within the maximum values of k_y of equifrequency contour (green circle). Excited waves are thus mostly propagating waves. While for scatterings in highly anisotropic medium of $\varepsilon_x \rightarrow 0^+$, even when $L_\perp \geq \lambda$, $K_\perp$ will fall outside of the maximum values of k_y of the equifrequency contour (blue flat ellipse). Evanescent waves are thus excited. Such a difference in the physical nature of scattering leads to totally different transmission behaviors.

Here, numerical simulations are performed to directly demonstrate the differences in scatterings and transmission by using the finite element software, COMSOL Multiphysics. First, we consider an inhomogeneous system composed of defects embedded in a background. For comparison purposes, the first case considered is composed of three isotropic defects embedded in air. The defects have square, triangular and circular shapes, and they have relative permeability $\mu = 1$ and permittivities are $\varepsilon = 1.5$, 2.5 and 0.5, respectively. Periodic boundary conditions are set on the upper and lower boundaries in the simulation, and a TM-polarized plane wave is incident from the left, with a wavelength about the same size of the defects. It is seen from the magnetic field H_z shown in Fig. 4.2a that strong scattered waves are excited, leading to distorted wave fronts and interference, which results in a reduced transmittance.

Next, let us change the system to be highly anisotropic by setting $\varepsilon_x = 0.001$, while keeping ε_y and μ unchanged in the whole system. The wavelength $\lambda = \lambda_0/\sqrt{\varepsilon_y \mu} = \lambda_0$ and impedance $Z_r = Z_0\sqrt{\mu/\varepsilon_y} = Z_0$ in the background is unchanged. In Fig. 4.2b–f, as shown, respectively, the magnetic and electric fields are in the y and x directions and the power flows are in the x and y directions under the same incident wave. The transmission behavior is dramatically altered. In Fig. 4.2b, the magnetic field H_z is almost uniform in the y direction. Such uniformity is a result of the large anisotropy. In Fig. 4.2b, it is also found that the transmittance is near unity, despite

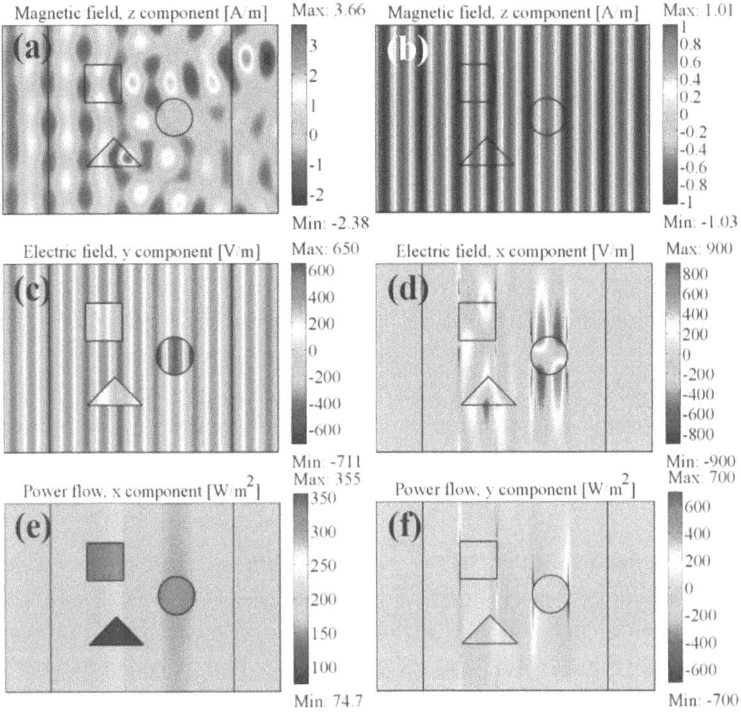

Figure 4.2 (a) The field map of the magnetic field H_z in a system of three isotropic defects with $\varepsilon = 1.5$ (square), 2.5 (triangular), and 0.5 (circular) and $\mu = 1$ in an isotropic background, under an incident TM wave from the left. (b–f) Field maps of the magnetic field H_z, the electric field E_y, the electric field E_x, energy flux S_x, and energy flux S_y for the same system in (a) except for $\varepsilon_x = 0.001$ throughout the whole system. Reprinted (figure) with permission from Ref. [62]. Copyright (2014) by the American Physical Society.

of the existence of three defects. Note that the shapes and materials of the defects have been chosen arbitrarily. This indicates that the high transmittance is not a resonance effect which would strongly depend on the details of inhomogeneity.

To understand the physical origin of such a robust high transmission, let us look at the electric field in the y direction, that is, E_y, in Fig. 4.2c. It is seen that E_y is almost inversely proportional to ε_y, indicating that the displacement field $D_y = \varepsilon_y E_y$

is almost uniform in the y direction. As a result, the time-averaged energy flux in the x direction, that is, $\bar{S}_x = \frac{1}{2}\mathrm{Re}\left(E_y H_z^*\right)$, is also inversely proportional to ε_y, as shown in Fig. 4.2e where we plot the x component of the Poynting vector. Since the total transmission is near unity, from the conservation law of EM flux, a proper amount of perpendicular flux, that is, $\bar{S}_y$, must send flux to where $\bar{S}_x$ is increased and from where $\bar{S}_x$ is decreased. This is accomplished by evanescent waves. From Fig. 4.2d which shows the x component of the E field, it is found that evanescent waves are induced when waves encounter defects. In Fig. 4.2f, one can find that strong flux $\bar{S}_y = -\frac{1}{2}\mathrm{Re}\left(E_x H_z^*\right)$ in the direction perpendicular to wave propagation emerges at the front and end positions of defects where the $\bar{S}_x$ distribution changes. The evanescent-wave-induced $\bar{S}_y$ is essential to keep the total flux $\int \bar{S}_x dy$ almost conserved throughout the whole media, leading to transmittance of almost unity.

In order to obtain a better understanding, analytical expressions are derived to describe wave behaviors in Fig. 4.1b. We assume that the perpendicular length of the background medium is a in the y direction, which satisfies $a << \sqrt{\varepsilon_y/\varepsilon_x}\lambda$. When $L_\perp \geq \lambda$, most of the scattered waves turn to be evanescent waves propagating in the y direction. And most waves propagating in the y direction undergo quite small phase change in the scale of a, implying that magnetic fields are nearly uniform in the y direction for waves propagating in the x direction. Of course, the scattered evanescent waves with large wave vectors affect the homogeneity of magnetic fields. But if the distribution function $f(x, y)$ varies relatively smoothly and is positive throughout the system, the magnetic fields are almost constant in the y direction. It is worth noting that when the anisotropy approaches infinity, that is, $\varepsilon_y/\varepsilon_x \rightarrow \infty$, any finite wavelength will satisfy the condition $\lambda >> a\Big/\sqrt{\varepsilon_y/\varepsilon_x}$.

Without loss of generality, suppose that relative permeability is also the function of space coordinates, that is,

$$\mu = g\,(x, y)\,. \tag{4.2}$$

The system can be divided into many thin layers in the x direction. When the thickness tends to zero, ε_y and μ in each layer only rely on

y. In each layer, one can use the approximation

$$H_z \approx H_0 e^{ik_x x}, \tag{4.3}$$

in which H_0 is a constant, to describe the amplitudes of magnetic fields. The time variation term $e^{-i\omega t}$ is omitted for convenience.

Using Maxwell's equation $\nabla \times \mathbf{H} = -i\omega\mathbf{D}$, one can immediately obtain $E_y = \frac{k_x}{\omega\varepsilon_0\varepsilon_y(y)} H_0 e^{ik_x x}$, implying that E_y is proportional to $1/\varepsilon_y(y)$, that is,

$$E_y \approx 1/\varepsilon_y(y). \tag{4.4}$$

Consequently, $\bar{S}_x$ is also proportional to $1/\varepsilon_y(y)$, that is,

$$\bar{S}_x \approx 1/\varepsilon_y(y). \tag{4.5}$$

This means that the energy flux $\bar{S}_x$ would concentrate to where ε_y is small. This simple relation between $\bar{S}_x$ and ε_y also indicates that energy flux can be easily manipulated by assigning a proper profile of ε_y.

Likewise, E_x can be obtained as

$$E_x = \frac{i}{\omega\varepsilon_0\varepsilon_x}\frac{\partial H_z}{\partial y}. \tag{4.6}$$

It is found that if H_z is strictly uniform in the y direction, then $E_x = 0$. However, when encountering inhomogeneity, evanescent waves are induced, and variations in H_z in the y direction will appear. Since $\omega\varepsilon_x << \frac{2\pi c}{a}\sqrt{\frac{\varepsilon_x}{\mu}} \rightarrow 0$, a large E_x can be induced. Moreover, such E_x turns out to be very sensitive to inhomogeneity. This provides a perfect mechanism to produce the necessary E_x and EM flux $\bar{S}_y$, so as to conduct the proper amount of energy flux in the vertical direction of propagation. Note that large E_x does not necessarily induce large $\bar{S}_y$, as E_x may form standing waves in the y direction. $\bar{S}_y$ is only induced where there is a change in $\bar{S}_x$, as can be seen in Fig. 4.2d. From Maxwell's equation $\nabla \times \mathbf{E} = i\omega\mathbf{B}$, one can see that fast decaying or increasing E_x in the y direction is capable of producing fast decaying or increasing E_y in the x direction. As a result, E_y can be matched on the left and right interfaces of the defects.

By inserting the approximate solutions of H_z and E_y into the integral form of the Maxwell–Faraday equation, $\int \mathbf{E}d\mathbf{l} = i\omega \int \mathbf{B}d\mathbf{S}$, one can obtain the effective refractive index n_{eff} for the waves propagating in the x direction in such a y dependent system,

$$n_{\text{eff}} = \frac{k_x}{k_0} = \sqrt{\int \mu(y)dy \Big/ \int \frac{dy}{\varepsilon_y(y)}}. \tag{4.7}$$

The effective refractive index in Eq. 4.7 can well describe the phase changes of the magnetic fields inside the thin layer having translational symmetry in the x direction. Considering the inhomogeneity in the x direction, the effective refractive index is also a function of x. Furthermore, the effective permittivity and permeability can be achieved as, respectively,

$$\varepsilon_{\text{eff}}(x) = \frac{a}{\int \frac{dy}{\varepsilon_y(x,y)}} \quad \text{and} \quad \mu_{\text{eff}}(x) = \frac{\int \mu_z(x,y)dy}{a}. \tag{4.8}$$

Equation 4.8 can be derived from the fact that the magnetic field H_z and electric displacement $D_y[= \varepsilon_y(x,y)E_y]$ are almost uniform in the y direction. This situation is quite similar to the quasi-static limit which generally requires much longer wavelength compared with the inhomogeneity size. Here the long wavelength is guaranteed by the near-zero ε_x for the waves propagating along the y direction. But for the waves propagating along the x direction, the effective wavelength can be comparable with and even smaller than the size of the inhomogeneity. Therefore, on the macroscopic scale, the system turns out to be an x-dependent system due to the extreme anisotropy. On the microscopic scale, evanescent waves are induced by anisotropy and inhomogeneity and guide the flux towards where ε_y is small.

4.2.2 *Robust High Transmission*

With Eq. 4.8, the effective refractive index and impedance are expressed as, respectively,

$$n_{\text{eff}}(x) = \sqrt{\varepsilon_{\text{eff}}(x)\mu_{\text{eff}}(x)} \quad \text{and} \quad Z_{\text{eff}}(x) = Z_0\sqrt{\mu_{\text{eff}}(x)/\varepsilon_{\text{eff}}(x)}. \tag{4.9}$$

where Z_0 is the impedance of air.

Figure 4.3a presents the case of a rectangular defect with $\varepsilon_x = 0.001$, $\varepsilon_y = 4$ and $\mu = 1$ embedded in a background of $\varepsilon_x = 0.001$, $\varepsilon_y = 1$ and $\mu = 1$, showing that the effective wavelength

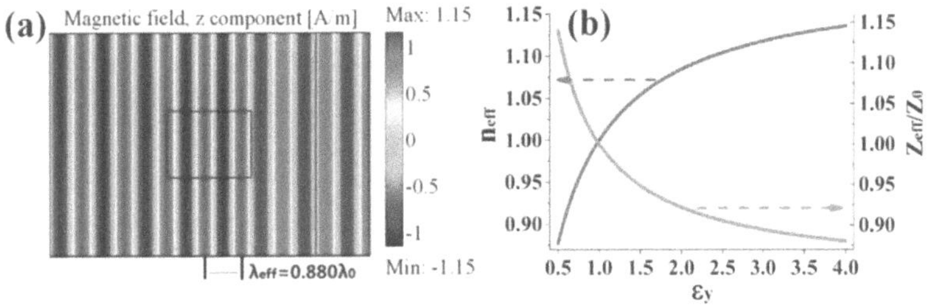

Figure 4.3 (a) Field maps of the magnetic field H_z in a system of a rectangular defect of $\varepsilon_y = 4$ embedded in a background of $\varepsilon_y = 1$. $\varepsilon_x = 0.001$ and $\mu = 1$ throughout the whole system. (b) Dependence of effective refractive index n_{eff} and ratio of effective impedance to the impedance of air Z_{eff}/Z_0 on the ε_y value of the rectangular defect. Reprinted (figure) with permission from Ref. [62]. Copyright (2014) by the American Physical Society.

$\lambda_{\text{eff}} = \lambda_0/n_{\text{eff}}$ is changed to be $0.880\lambda_0$ due to the existence of the defect. One can easily see from the plotted n_{eff} and Z_{eff}/Z_0 in Fig. 4.3b that although the parameter change of the defect is relatively large, the impedance change is rather small. This can be understood as the result of averaging effect in ε_{eff} and μ_{eff}, which can be seen from Eq. 4.8. Therefore, the robust high transmittance through the inhomogeneous anisotropic media originates from the averaging effect due to the homogeneity of fields in the perpendicular direction of propagation.

The averaging effect has interesting implications, especially in the cases of random media which usually lead to localization of waves and low transmittance. Here, we give an example of high transmittance through discrete random media. The system is composed of random blocks with random shapes and constitutive parameters $\mu = 1$, $\varepsilon_x = 0.001$, and $\varepsilon_y = f(x, y)$, with $f(x, y)$ ranging from 1 to 2.5, as shown in Fig. 4.4a. The incident waves and periodic boundary conditions are the same as those in Fig. 4.2a–f. The magnetic field H_z, the energy flux $\bar{S}_x$ and $\bar{S}_y$ are plotted in Figs. 4.4b, 4.4c, and 4.4d, respectively. It is seen that the transmittance is still quite high even in such a random system with wavelength comparable to the size of inhomogeneity. The energy flux $\bar{S}_x$ and $\bar{S}_y$ are distributed accordingly with the random profile, indicating that

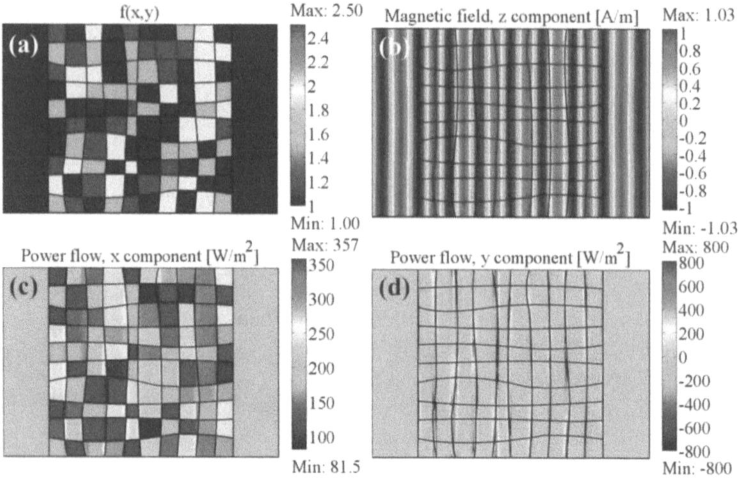

Figure 4.4 (a) The $\varepsilon_y = f(x, y)$ distribution map in a random medium composed of discrete blocks with various shapes. $\varepsilon_x = 0.001$ and $\mu = 1$ throughout the whole medium. (b–d) Field maps of the magnetic field H_z, energy flux S_x, and energy flux S_y. Reprinted (figure) with permission from Ref. [62]. Copyright (2014) by the American Physical Society.

the energy flux actually undergoes complicated routes to ensure the high transmission.

4.2.3 *Examples of EM Flux Control*

Equation 4.5 indicates an interesting conclusion that the energy flux would be concentrated to regions where ε_y is small. By utilizing this property, we can control how the EM flux flows in an almost arbitrary way by designing the ε_y profile.

Now, we provide designs for three kinds of gradient ε_y profiles. Let $f(x, y)$ be smooth functions of inhomogeneity profiles within a square domain of length a, as shown in Figs. 4.5a, 4.5c, and 4.5e. In Figs. 4.5b, 4.5d, and 4.5f, we show the energy flux and Poynting vectors for light coming in from the left passing through anisotropic gradient media of $\varepsilon_y = f(x, y)$ and a constant $\varepsilon_x = 0.001$ and $\mu = 1$, respectively. The angular frequency is chosen to be $\omega = 2\pi c/\lambda_0 = 10\pi c/a$, where c is the velocity of light in free space.

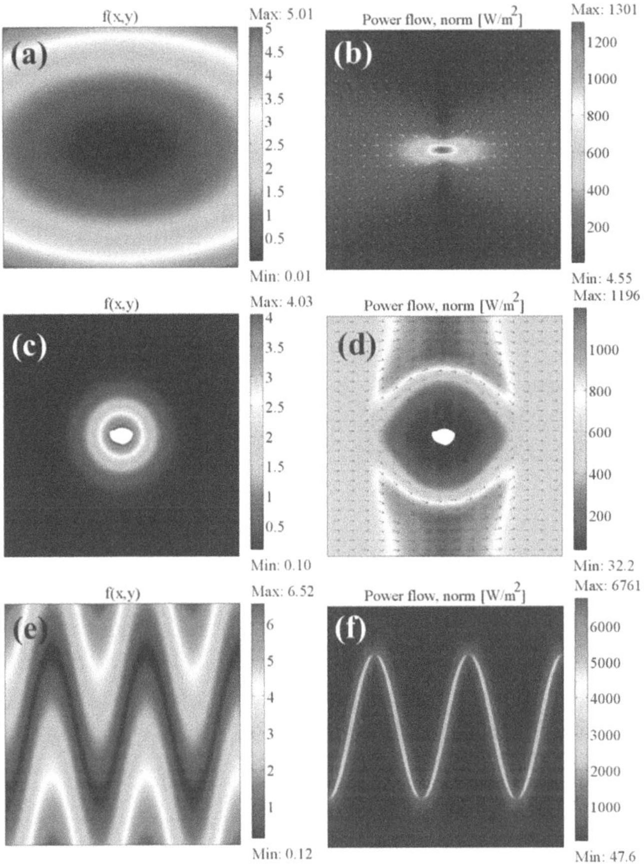

Figure 4.5 The distribution of (a, c, e) $\varepsilon_y = f(x, y)$ and (b, d, f) total energy flux (color) and Poynting vectors (arrows) in three different gradient media, which demonstrate (a, b) focusing of energy power, (c, d) flux splitting, and (e, f) free routing of energy flux in a designed sinusoidal path, respectively. $\varepsilon_x = 0.001$ and $\mu = 1$ throughout the whole medium. The distribution function for the three examples: (a) $f(x, y) = 5\left[(x/a)^2 + 3(y/a)^2\right] + 0.01$, (c) $f(x, y) = 0.1e^{250\left[(x/a)^2+(y/a)^2-0.35^2\right]^2}$ for the region $x^2 + y^2 \leq 0.35^2a^2$, and $f(x, y) = 0.1$ for the region $x^2 + y^2 > 0.35^2a^2$, (e) $f(x, y) = 8\left|y/a - 0.3\sin(5\pi x/a)\right| + 0.12$. And the computation region is restricted in $-a/2 \leq |x| \leq a/2$ and $-a/2 \leq |y| \leq a/2$. Reprinted (figure) with permission from Ref. [62]. Copyright (2014) by the American Physical Society.

Perfect electric conductor (PEC) boundary conditions on upper and lower boundaries have been applied.

In the first example, the inhomogeneity profile $f(x, y)$ exhibits a minimal value at the central area, as shown in Fig. 4.5a. As a consequence, the EM flux is focused into a subwavelength area, as shown in the flux and Poynting vector plot in Fig. 4.5b. It is interesting to point out that the perpendicular size of the focused flux spot is about 0.27 times of the free-space wavelength. If we increase the wavelength, the focused flux spot would be even sharper. Therefore, subwavelength focusing can be realized with this method.

In the second example, the inhomogeneity profile $f(x, y)$ exhibits a maximal value at the central area where an irregular PEC shell is embedded, as shown in Fig. 4.5c. As a result, the energy flux is spitted and recombined to go around the central area, rendering anything inside the PEC shell "cloaked", as shown in Fig. 4.5d.

In the third example, the inhomogeneity profile $f(x, y)$ exhibiting a sinusoidal pathway of minimal ε_y is designed, as shown in Fig. 4.5e. As a consequence, the energy flux indeed concentrates and flows along the designed sinusoidal path, as displayed in Fig. 4.5f. Note that the free-space wavelength in Fig. 4.5f is about 10 times larger than the path width, which is about $0.02a$, indicating that this is a subwavelength control of flux. In this way, the energy flux can be conveniently controlled to flow along any desired curve in a subwavelength area.

In addition to the gradient examples shown above, we present two more examples consisting of gradient media with discontinuity. And we will show that the discontinuity, however, does not affect the almost perfect control of EM flux. Still, the relevant parameters are chosen as $\varepsilon_x = 0.001$, $\varepsilon_y = f(x, y)$ and $\mu = 1$.

In the first example shown in Figs. 4.6a and 4.6b, a beam splitter is designed. Different from the example shown in Figs. 4.5c and 4.5d, the EM flux is confined in slim lines. Since the ratio $\varepsilon_y/\varepsilon_x$ is large, the magnetic field is forced to be almost constant in the y direction. The energy flux can still be tuned effectively to obtain almost perfect flux control, as presented in Fig. 4.6b.

In the second example shown in Figs. 4.6c and 4.6d, the EM flux is first confined into a slim line. Then, the flux is uniformly dispersed

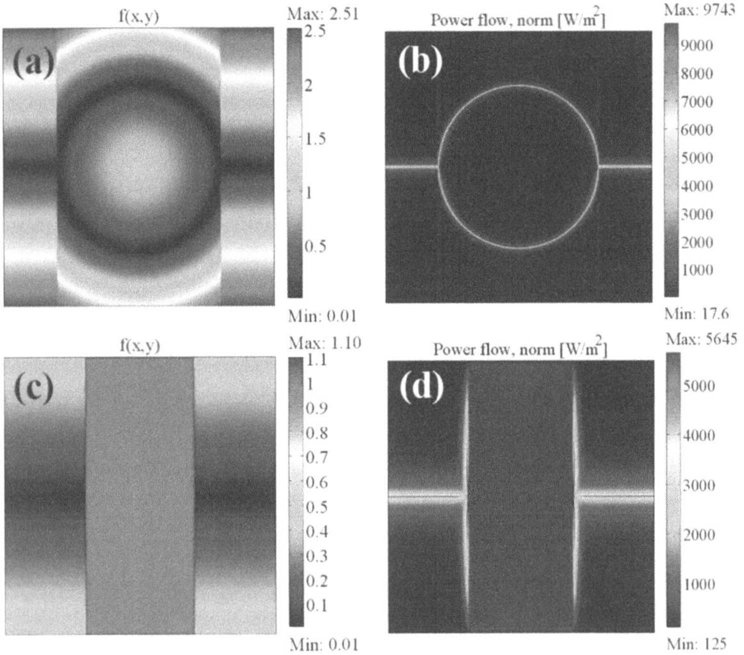

Figure 4.6 (a, c) Inhomogeneity profiles $f(x, y)$ of two inhomogeneous materials with $\varepsilon_y = f(x, y)$, $\varepsilon_x = 0.001$ and $\mu = 1$. (b, d) Time-averaged total energy flux in the two materials. The distribution function for the three examples: (a) $f(x, y) = 10\left|(x/a)^2 + (y/a)^2 - 0.3^2\right| + 0.01$ for the region $|x| \leq 0.3a$, and $f(x, y) = 5|y/a| + 0.01$ for the region $|x| > 0.3a$, (c) $f(x, y) = 1$ for the region $|x| \leq 0.2a$, and $f(x, y) = |y/a| + 0.01$ for the region $|x| > 0.2a$. And the computation region is restricted in $-a/2 \leq |x| \leq a/2$ and $-a/2 \leq |y| \leq a/2$. Reprinted (figure) with permission from Ref. [62]. Copyright (2014) by the American Physical Society.

into a homogeneous block, and finally collected into a slim line again. At the surfaces of the homogenous block, we can clearly see strong vertical EM flux, which can disperse and collect the energy flux quickly.

4.2.4 *Effect of Anisotropy*

From the above discussion, we have seen that the EM flux control behavior largely rely on the strong anisotropy ($\varepsilon_y/\varepsilon_x$). Ideally, for

$\varepsilon_x \rightarrow 0$, the magnetic fields are uniform enough to ensure that the energy flow $\bar{S}_x$ is proportional to $1/\varepsilon_y(y)$. But in practice, the ε_x values cannot be exactly zero. To quantitatively show the effect of the anisotropy, we revisit the focusing example (Figs. 4.5a and 4.5b).

In the vertical line across the center point shown in the right inset of Fig. 4.7a, the maximal EM flux $\bar{S}_{x,\max}$ and the minimal EM flux $\bar{S}_{x,\min}$ will appear at the center point and the boundary, respectively, which can be concluded from the inhomogeneity profile $f(x, y)$ presented in Fig. 4.5a. In Fig. 4.7a, the relationship between $\bar{S}_{x,\max}/\bar{S}_{x,\min}$ and $\varepsilon_{y,\min}/\varepsilon_x$ is plotted, where $\varepsilon_{y,\min}$ is the minimal value of ε_y in the vertical line. For the ideal case with $\varepsilon_x \rightarrow 0$, the maximum value of $\bar{S}_{x,\max}/\bar{S}_{x,\min}$ is about 370, which is plotted in Fig. 4.7a as the green dashed lines. And one can clearly observe that as the increase of the anisotropy $(\varepsilon_y/\varepsilon_x)$, $\bar{S}_{x,\max}/\bar{S}_{x,\min}$ tends to the ideal value, that is, $\left[\frac{5}{a^2} \times 3\left(\frac{a}{2}\right)^2 + 0.01\right] / 0.01 = 376$.

The distributions of electric fields and time-averaged EM flux are also simulated to show the effect of anisotropy. In Figs. 4.7b and 4.7c, the ratio $\varepsilon_{y,\min}/\varepsilon_x$ is set to be 50, and the magnetic fields are already rather uniform along the y direction. Meanwhile, the focusing performance is quite good, as displayed in Fig. 4.7c. However, as the ratio $\varepsilon_{y,\min}/\varepsilon_x$ is decreased to be 2, we can see that magnetic fields are less uniform along the y direction (see Fig. 4.7d), resulting in weak decrease of focusing ability (see Fig. 4.7e).

After many numerical simulations, we found that in order to observe relatively ideal flux control behavior, $\varepsilon_{y,\min}/\varepsilon_x$ should be larger than 10 for this case. But for lower frequencies, the requirement of anisotropy could be largely relieved.

4.2.5 *Effect of Loss and Failure of Effective Medium Theory*

In practical implementations, loss is inevitable. The loss effect can be magnified especially when the real part is small. So we focus on the loss in ε_x to show the influences of material loss.

For an easy comparison, we take the focusing model (Figs. 4.5a and 4.5b) as an example to examine the lossy cases in which ε_x not only has a near zero real part but also carries an imaginary part. The real part of ε_x fixed to be 0.001, while the imaginary part

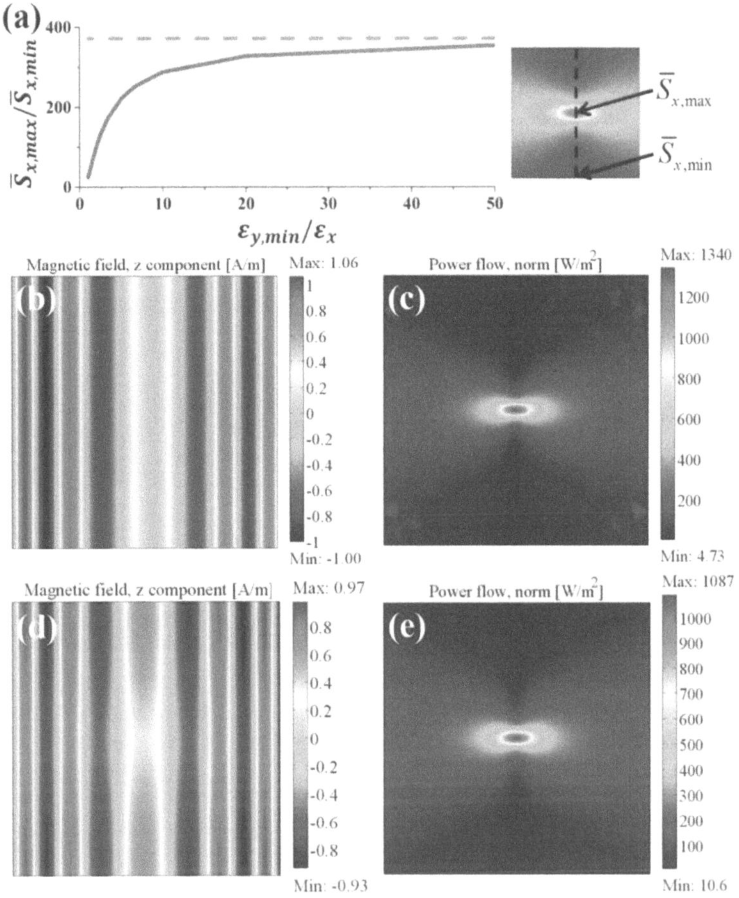

Figure 4.7 (a) Dependence of $\bar{S}_{x,\max}/\bar{S}_{x,\min}$ on $\varepsilon_{y,\min}/\varepsilon_x$. The dashed blue line corresponds to the ideal value of $\bar{S}_{x,\max}/\bar{S}_{x,\min}$ (i.e., $\varepsilon_{y,\max}/\varepsilon_{y,\min}$). The right inset illustrates the definition of $\bar{S}_{x,\max}$ and $\bar{S}_{x,\min}$. The distribution of (b, d) magnetic fields and (c, e) time-averaged EM flux in the focusing example for the cases with (b, c) $\varepsilon_{y,\max}/\varepsilon_{y,\min} = 50$ and (d, e) $\varepsilon_{y,\max}/\varepsilon_{y,\min} = 2$.

is adjusted. Figures 4.8a,b and 4.8c,d relate to the lossy systems with $\mathrm{Im}(\varepsilon_x)/\mathrm{Re}(\varepsilon_x) = 1$ and $\mathrm{Im}(\varepsilon_x)/\mathrm{Re}(\varepsilon_x) = 10$, respectively. Interestingly, we find that loss does not compromise the flux controlling effect, but the major effect is to reduce the transmission rate, which can be seen from the distributions of magnetic fields

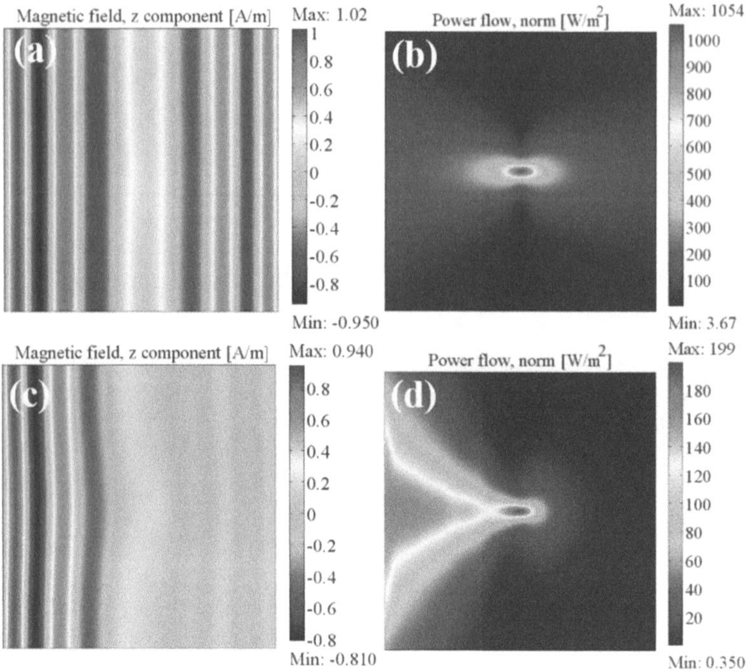

Figure 4.8 The field maps of (a, c) the magnetic field H_z and (b, d) time-averaged power flow $\bar{S}$ in lossy systems. (a, b) and (c, d) correspond to the lossy cases with $\text{Im}(\varepsilon_x)/\text{Re}(\varepsilon_x) = 1$ and $\text{Im}(\varepsilon_x)/\text{Re}(\varepsilon_x) = 10$, respectively.

(Figs. 4.8a and 4.8c) and time-averaged power flow (Figs. 4.8b and 4.8d). Even if the imaginary part of ε_x is 10 times larger than its real part, a strong ability to control the flux in the vertical direction of propagation is still preserved in Fig. 4.8d.

Another effect induced by the material loss is that the effective parameters in Eq. 4.8 would fail to describe the system. The validity of Eq. 4.8 is guaranteed by the perfect redistribution of EM flux, which is enabled by the excited evanescent waves at where there exists inhomogeneity. However, once the loss part is introduced to ε_x, the evanescent waves propagating in the vertical direction will be partially absorbed, therefore the EM flux redistribution ability will be restricted. As a result, the effective parameters predicted by Eq. 4.8 will fail to describe the wave propagation, especially the wave absorption. As we can see from Eq. 4.8, the effective parameters are still real values even when ε_x is a complex value, implying that the analysis of wave absorption cannot be accomplished based on Eq. 4.8. But in most cases that the real part of ε_x is 10 times larger than its imaginary part, we can approximately use Eq. 4.8 to study the EM flux behaviors.

Finally, we note that the issue of material loss has been extensively discussed in the field of plasmonics and various schemes have been proposed to mitigate the problem [64]. Another way to avoid the loss problem is to use zero-index materials with less loss, for example, dielectric materials [30, 33, 65, 66], or to operate in the microwave regime in which zero-index metamaterials have been experimentally verified to exhibit less loss and long transmission distances [26, 31, 36].

4.3 Applications in Waveguides

In the above section, we have discussed the basic mechanism to control EM flux in inhomogeneous AENZ media. In this section, we will show a series of almost perfect waveguide devices that benefit from the EM flux control behavior. It is found that an almost arbitrarily shaped waveguide filled homogeneous AENZ media in physical space is equivalent to a straight waveguide filled with inhomogeneous AENZ media in the virtual space according to transformation optics. And with the understanding of the inhomogeneous AENZ media in the above section, we can well explain the good performances of the waveguide devices in the physical space. In the following section, we will show three typical waveguides, that is, straight waveguide with irregular boundaries in which the AENZ media are straight, bending waveguide in which the AENZ media are bent accordingly, and bending waveguides with irregular boundaries.

4.3.1 *Waveguides with Irregular Boundaries*

Firstly, we consider a waveguide constructed with the irregular PEC boundaries characterized by the functions of $y_1 = g_1(x)$ and $y_2 = g_2(x)$, respectively, as illustrated in Fig. 4.9a. The input and output

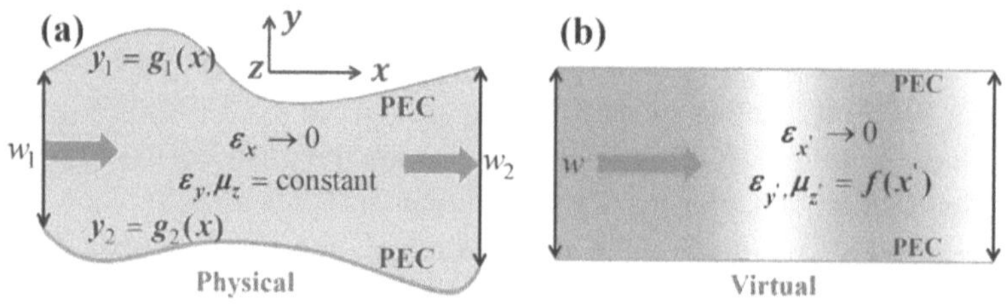

Figure 4.9 Schematic drawing of the coordinate transformation that relates (a) a physical waveguide with irregular boundaries into (b) a virtual straight waveguide. The physical waveguide is filled with homogeneous AENZ media, while the equivalent virtual waveguide is filled with inhomogeneous AENZ media. Adapted with permission from Ref. [63]. Copyright © 2014 by Nature Publishing Group.

ports are parallel and have the width of w_1 and w_2, respectively, as shown in Fig. 4.9a. The waveguide is filled with homogeneous AENZ media which have a near-zero permittivity in the propagation direction, that is, $\varepsilon_x \to 0$. The other two parameters, that is, the y component of the relative permittivity ε_y and relative permeability μ, are nonzero and positive constants. TM-polarized waves incident from the left port are considered.

To analyze the transmission properties of such a waveguide with wavy boundaries, transformation optics is applied to turn such a waveguide into a straight waveguide having a constant width of w, as illustrated in Fig. 4.9b, based on the coordinate mapping of $x' = x$, $y' = w\frac{y-g_2(x)}{g_1(x)-g_2(x)}$, $z' = z$. According to transformation optics, the material parameters of the virtual media are derived as,

$$\varepsilon' = \frac{1}{a_{22}}\begin{pmatrix} \varepsilon_x & a_{21}\varepsilon_x & 0 \\ a_{21}\varepsilon_x & a_{21}^2\varepsilon_x + a_{22}^2\varepsilon_y & 0 \\ 0 & 0 & \varepsilon_z \end{pmatrix} \text{ and } \mu_{z'} = \frac{\mu}{a_{22}} \tag{4.10}$$

with $a_{21} = \frac{\partial y'}{\partial x} = -\frac{\partial}{\partial x}\frac{g_2(x)}{g_1(x)-g_2(x)}$ and $a_{22} = \frac{\partial y'}{\partial y} = \frac{w}{g_1(x)-g_2(x)}$.

Since $\varepsilon_x \to 0$, the off-diagonal term of the permittivity tensor $a_{21}\varepsilon_x$ tends to vanish, and the diagonal terms are left with $\varepsilon_{x'} = \frac{\varepsilon_x}{a_{22}} \to 0$ and $\varepsilon_{y'} = a_{22}\varepsilon_y$. Thus, we can see that the original AENZ media are still AENZ media in the virtual space, but $\varepsilon_{y'}$ and $\mu_{z'}$ are the functions of x'. The wave impedance can be written as,

$$Z = Z_0\sqrt{\frac{\mu_{z'}}{\varepsilon_{y'}}} = \frac{Z_0}{a_{22}}\sqrt{\frac{\mu}{\varepsilon_y}} \tag{4.11}$$

Based on Eq. 4.11, we can evaluate the transmittance of the waves through the waveguide with arbitrary boundary shapes. Interestingly, one may notice that when $g_1(x) - g_2(x) = w$ is a constant, we have $a_{22} = 1$ and $Z = Z_0\sqrt{\frac{\mu}{\varepsilon_y}}$, implying that the wave impedance of the wavy waveguide equals that of a uniform medium with relative permittivity ε_y and relative permeability μ. Therefore, when the waves enter such an anisotropic zero-index media waveguide with irregular boundaries described by $g_1(x)$ and $g_2(x)$, which satisfy $g_1(x) - g_2(x) = w$, the transmission will always be unity irrespective of the form of $g_1(x)$ or $g_2(x)$. This is an amazing property.

In Fig. 4.10a, a waveguide with zigzag-shaped boundaries is simulated. The incident and transmitted media are both air, and the bending region is filled with the AENZ medium with $\varepsilon_x = 0.001$ and $\varepsilon_y = \mu = 1$. A TM-polarized wave comes from the

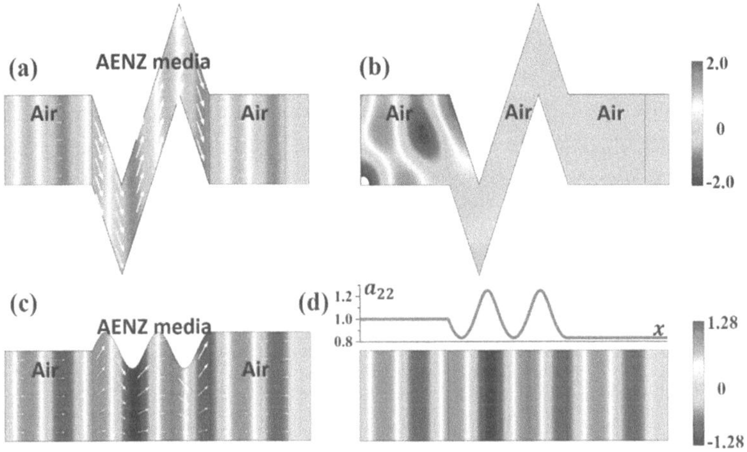

Figure 4.10 (a, b) Field maps of the magnetic field H_z in a waveguide with sharply bent walls when the bending region is filled with AENZ media and air, respectively. (c, d) Field maps of the magnetic field H_z in a waveguide expander with a trigonometric boundary and a larger output port, and in the virtual waveguide obtained by transformation optics, respectively. The arrows in (a) and (c) correspond to the Poynting vectors. The upper inset of (d) shows the distribution of a_{22}, which determines the distribution of $\varepsilon_{y'}$ and $\mu_{z'}$. Adapted with permission from Ref. [63]. Copyright © 2014 by Nature Publishing Group.

left port. Apparently, such a waveguide possesses constant width, that is, $g_1(x) - g_2(x) = w$. As a result, a perfect transmission is seen, as shown by the distributions of magnetic fields (color map) and Poynting vectors (arrows) in Fig. 4.10a. The almost uniform magnetic field in the y direction is a result of the near-zero ε_x. For comparison, the magnetic field distribution in a hollow waveguide is also simulated in Fig. 4.10b, showing that almost all the incident waves are reflected.

Generally, the width of a waveguide is not always a constant. For instance, we consider another wavy-shaped waveguide whose output port is larger than its input port, as presented in Fig. 4.10c. The upper boundary of the waveguide has a trigonometric function. And the relevant parameters of the AENZ media are $\varepsilon_x = 0.001$ and $\varepsilon_y = \mu = 1$. After transformation to a straight waveguide as shown in Fig. 4.10d, $\varepsilon_{x'}$ is still set to be 0.001, while the other parameters are changed, which can be calculated from Eq. 4.10. The related parameter a_{22} is plotted in the upper figure in Fig. 4.10d. Figure 4.10c displays the magnetic fields (color map) and Poynting vectors (arrows), showing a transmittance greater than 99% even though the impedance is not exactly matched. And the distribution of magnetic fields is inconsistent with that inside the straight waveguide shown in Fig. 4.10d. The high transmission can be explained by the small impedance variation in the straight waveguide. Since a_{22} changes smoothly around unity, the impedance of the system is also varying around the impedance of air. In such cases, high transmission can be generally obtained.

In 2013, Ma et al. [51] has experimentally proposed a zigzag-shaped waveguide to obtain a one-dimensional cloak. In their experiment, TE-polarized waves are considered instead of TM-polarized waves, therefore, AMNZ media and perfect magnetic conductor (PMC) boundaries are required. The AMNZ media are designed by utilizing split-ring resonators (SRRs), which are fabricated on the printed circuit board. When the magnetic fields are polarized along propagating direction, that is, the x direction, strong magnetic responses will be generated by SRRs. By tuning the sizes of the SRRs, the plasma frequency of the design of SRRs is tuned to be about 10 GHz, leading to $\mu_x = 0$ at 10 GHz. Moreover, the other components of parameters, that is, μ_y and ε_z, can be obtained

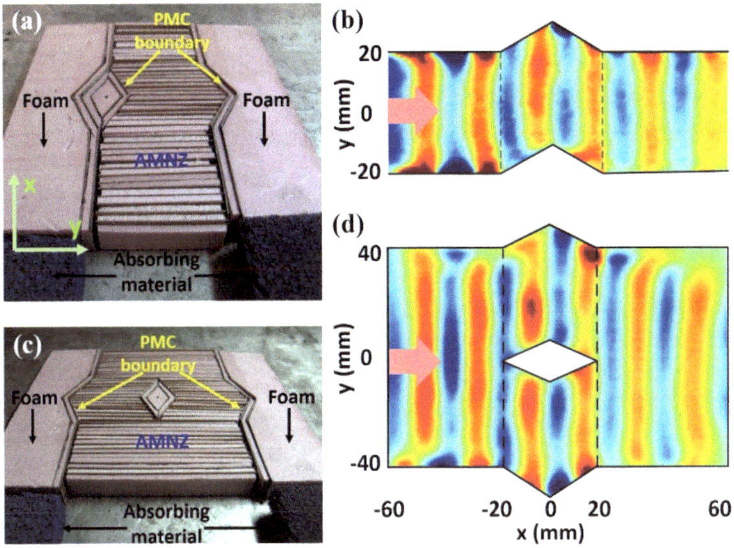

Figure 4.11 The fabricated samples of (a) a waveguide with zigzag-shaped boundaries and (b) a waveguide splitter. The measurement results of electric field distribution in (b) the zigzag-shaped waveguide and (b) the waveguide splitter. Reprinted with permission from Ref. [51]. Copyright © 2013, AIP Publishing LLC.

by applying the scattering parameter retrieval method [67]. It is found that μ_y is nearly equal to 1 from 7 to 11 GHz, and ε_z is equal to 3.2 at 10 GHz. In the experimental setup, the waveguide boundaries are constructed of artificial PMC, which is fabricated by using metal patches. And the artificial PMC is carefully designed to make it perform as a PMC at 10 GHz.

In the experiment, two kinds of waveguides are designed, as illustrated in Figs. 4.11a and 4.11c. Similar to the theoretical analysis and the simulation prediction, the measured electric field shown in Figs. 4.11b and 4.11d clearly demonstrate that the EM waves are nearly totally transmitted through the zigzag-shaped waveguides. Note that there exists a diamond-shaped region with PMC boundaries in the center of the second setup, creating a small region that the waves cannot penetrate in. As a consequence, a one-dimensional cloak is obtained.

4.3.2 *Bending Waveguides*

Second, we consider a bending waveguide illustrated in Fig. 4.12a whose input and output ports are not parallel. To study such a waveguide, it is more convenient to utilize the cylindrical coordinate instead of the Cartesian coordinate, so we have $\varepsilon_\theta \to 0$ instead of $\varepsilon_x \to 0$. Through transformation optics, the bending waveguide can be transformed into a straight waveguide with the mapping $x' = R\theta$, $y' = g(r)$, $z' = z$ [60, 63, 68]. We assume that the outer and inner walls are characterized by $r_1 = b$ and $r_2 = a$, respectively, and the width of the waveguide is fixed to be w. The parameters of the straight waveguide are derived as,

$$\varepsilon' = \begin{pmatrix} \varepsilon_\theta R/[g'(r)r] & 0 & 0 \\ 0 & \varepsilon_r g'(r)r/R & 0 \\ 0 & 0 & \varepsilon_z r/[g'(r)R] \end{pmatrix} \text{ and}$$

$$\mu_{z'} = \mu r/[g'(r)R] \tag{4.12}$$

Since $\varepsilon_\theta \to 0$, we have $\varepsilon_{x'} = \varepsilon_\theta R/[g'(r)r] \to 0$. On the other hand, $\varepsilon_{y'}$ and $\mu_{z'}$ are now functions of r, and therefore functions of y', as illustrated in Fig. 4.12b. As we have discussed in the above section, evanescent waves would be induced in such an inhomogeneous system, which can efficiently redistribute the energy flow to smoothly match the waveguide modes everywhere [60]. As a consequence, robust high transmission can be obtained, which can be also understood from the averaging effect. Combined with Eqs.

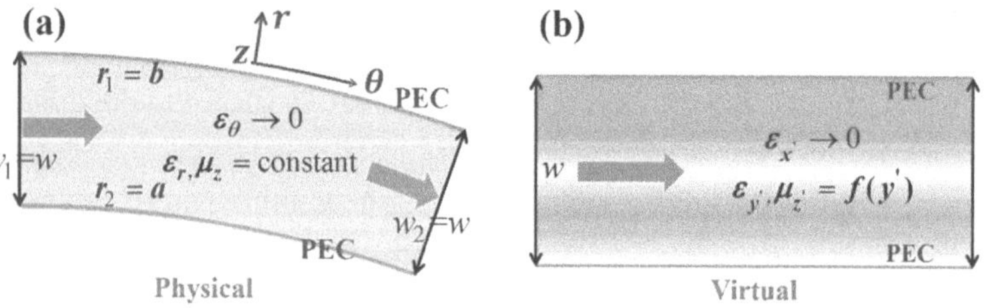

Figure 4.12 Schematic drawing of the coordinate transformation that relates (a) a physical waveguide with unparallel boundaries into (b) a virtual straight waveguide. The physical waveguide is filled with homogeneous AENZ media, while the equivalent virtual waveguide is filled with inhomogeneous AENZ media. Adapted with permission from Ref. [63]. Copyright © 2014 by Nature Publishing Group.

4.8 and 4.9, we can derive the wave impedance of the bending waveguide as,

$$Z = Z_0 \sqrt{\frac{\int_{r=a}^{r=b} \mu_{z'} dr \Big/ w}{w \Big/ \int_{r=a}^{r=b} dr/\varepsilon_{y'}}} = \frac{1}{w}\sqrt{\frac{1}{2}\left(b^2 - a^2\right) \ln \frac{b}{a}} Z_0 \sqrt{\frac{\mu}{\varepsilon_r}} \quad . \tag{4.13}$$

For small variations, the wave impedance is generally close to the impedance of air. For instance, if $b/a = 2$ or 20, we will get the ratio of impedance $\eta = \frac{Z}{Z_0}\sqrt{\frac{\varepsilon_r}{\mu}} = \frac{1}{w}\sqrt{\frac{1}{2}\left(b^2 - a^2\right) \ln \frac{b}{a}} = 1.02$ or 1.29. When a film with impedance Z placed in the background medium with impedance $Z_0\sqrt{\frac{\mu}{\varepsilon_r}}$, the minimal transmittance can be derived as $\left|\frac{2\eta}{\eta^2+1}\right|^2 = 0.999$ or 0.938. Thus, it is apparent that the bending waveguides possess robust high transmittance irrespective of the propagation distance.

To verify the good performance, numerical simulations are performed. Figure 4.13a presents the magnetic field distribution in a 180° bending waveguide whose bending region is filled with the AENZ medium with $\varepsilon_\theta = 0.001$ and $\varepsilon_r = \mu = 1$. It is evident that almost all the waves incident from air are transmitted through the bending waveguide. Meanwhile, the planar wave front shape is persevered. However, if the waveguide is hollow, most of the incident waves will be blocked and only a very small part can be transmitted, as shown in Fig. 4.13b.

The experimental verification was performed by Ma et al. [61]. Figure 4.13c illustrates such an experimental setup of a 90° bending waveguide. In the experiment, AMNZ metamaterial is used to replace the AENZ medium. So, the incident wave is chosen as TE polarizations, and the waveguide boundaries is changed as PMC accordingly. The AMNZ metamaterials and the PMC boundaries are constructed by SRRs and metal patches, respectively. And the operating frequency is tuned to be 10 GHz. From the measured electric field distribution shown in Fig. 4.13d, we can see that the incident waves can be bent and transmitted very well. We notice that the phase surfaces of waves always keep perpendicular to the PMC boundaries without any distortions.

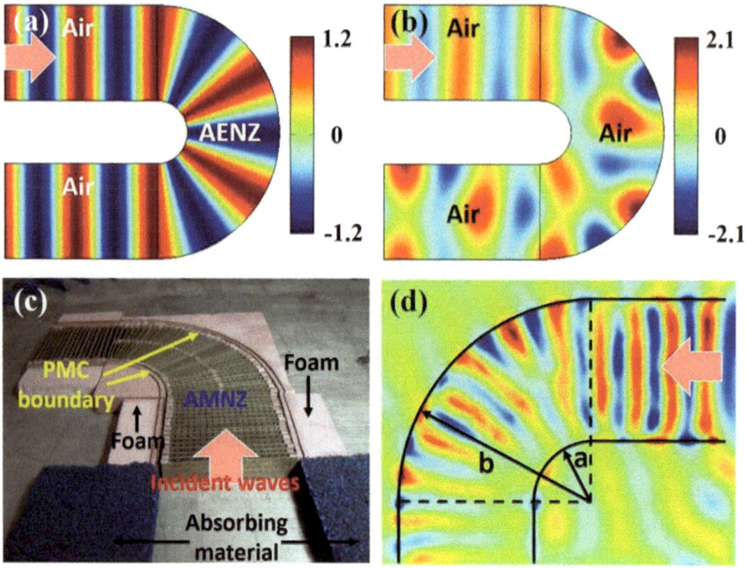

Figure 4.13 Magnetic field distribution in a waveguide with the bend part filled with (a) AENZ media ($\varepsilon_\theta = 0.001$, $\varepsilon_r = \mu = 1$) and (b) air. (c) The experimental setup of a 2D bending waveguide in which the AMNZ metamaterials are used instead of AENZ media. (d) The measured electric fields in the bending waveguide filled with AMNZ. (a, b) Reprinted with permission from Ref. [60]. Copyright © 2012, AIP Publishing LLC. (c, d) Reprinted with permission from Ref. [61]. Copyright © 2012, AIP Publishing LLC.

4.3.3 *Bending Waveguides with Irregular Boundaries*

Third, we consider a general waveguide with both irregular boundaries and unparallel ports, that is, the combination of the above two kinds of waveguides. Performing the two steps of transformation described above, the physical waveguide filled with homogeneous AENZ media (shown in Fig. 4.14a) would be transformed into a virtual straight waveguide filled with inhomogeneous AENZ media (shown in Fig. 4.14b), whose parameters are functions of both x' and y' in the virtual space. According to the EM flux control mechanism discussed in the above section, we can expect robust high transmission in such an unusual waveguide. Since the limitations of the waveguide boundaries and the orientation of the

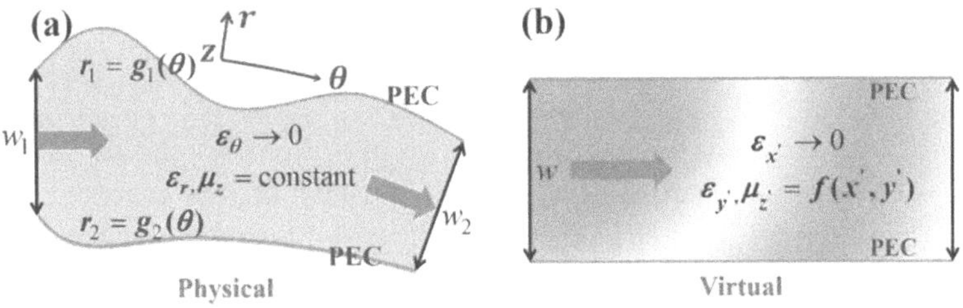

Figure 4.14 Schematic drawing of the coordinate transformation that relates (a) a physical waveguide with both irregular boundaries and unparallel ports into (b) a virtual straight waveguide. The physical waveguide is filled with homogeneous AENZ media, while the equivalent virtual waveguide is filled with inhomogeneous AENZ media. Adapted with permission from Ref. [63]. Copyright © 2014 by Nature Publishing Group.

ports are released, a series of almost perfect waveguide devices can be achieved. Next, we present two examples to show interesting applications.

The first example is a bending wave compressor in Figs. 4.15a and 4.15b. We impose a TM-polarized wave incident from the left port. The incident and transmitted regions are both air. And the relevant parameters of the AENZ media filled in the bending region are $\varepsilon_\theta = 0.001$, $\varepsilon_r = 1.25$ and $\mu = 1$. From the distributions of magnetic fields (color map) and Poynting vectors (arrows) in Fig. 4.15a, we can clearly see that almost all the incident waves are transmitted and compressed into the narrow output waveguide. Numerically, the transmittance is found out to be as high as 0.99.

We note that the ε_r is chosen to be 1.25 because such a ε_r larger than unity is easier to achieve in practice. It is known that the AENZ media can be readily realized by designing subwavelength multilayers composed of dielectric and negative permittivity materials [32, 42, 69–74]. Suppose that the relative permittivity and filling ratio of the two materials are represented by ε_d, ε_n, p_d and p_n. The effective permittivities parallel and perpendicular to the layers are, respectively [32, 42, 69–74],

$$\varepsilon_{||} = p_d\varepsilon_d + p_n\varepsilon_n \quad \text{and} \quad \varepsilon_\perp = \frac{\varepsilon_n\varepsilon_d}{p_d\varepsilon_n + p_n\varepsilon_d}. \tag{4.14}$$

If we insist $\varepsilon_{\|} = 0$ and $\varepsilon_{\perp} = 1$, the parameters will have to satisfy

$$\varepsilon_d = 2 - \frac{1}{p_d} \quad \text{and} \quad \varepsilon_n = 2 - \frac{1}{p_n}. \tag{4.15}$$

Since $0 < p_d, p_n < 1$, the relative permittivities should satisfy $\varepsilon_d, \varepsilon_n < 1$, indicating that the permittivity of one material has to be negative, and the other one has to be less than unity. Here, for simplicity, we choose $\varepsilon_r = 1.25$, which can be realized by the materials with $\varepsilon_d = 1$, $\varepsilon_n = -5 + 0.5i$, $p_d = 5/6$ and $p_n = 1/6$. The material with $\varepsilon_n = -5 + 0.5i$ can be realized by metamaterials composed of metallic mesh of thin wires in the microwave regime [26, 75]. The lattice constant a is chosen as $\lambda_0/30$ to guarantee the validity of the effective parameters. In Fig. 4.15b, the lossy multilayer structure is exploited to replace the lossless AENZ media in Fig. 4.15a. One can see from magnetic field maps that the wave

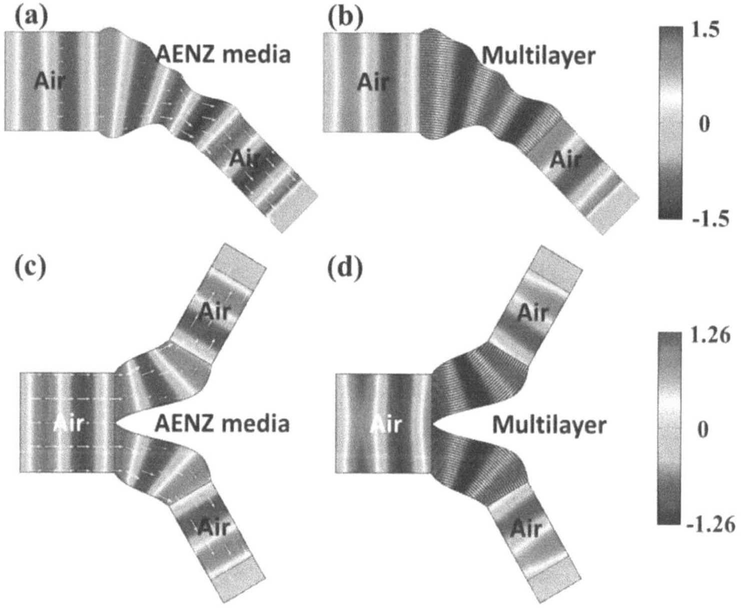

Figure 4.15 The field map of the magnetic field H_z in bending wave (a, b) compressor and (c, d) splitter with the bending region filled with (a, c) lossless AENZ media and (b, d) lossy multilayers. The multilayers are composed of dielectric and negative permittivity materials. Adapted with permission from Ref. [63]. Copyright © 2014 Nature Publishing Group.

front shape is always very good during the propagation, but the transmittance is reduced a little due to the material loss.

The second example is a bending wave splitter. The distributions of magnetic fields (color map) and Poynting vectors (arrows) in Fig. 4.15c indicate that more than 98% of the incident waves can be transmitted with good wave front shapes. Moreover, such good performance persists when we use lossy multilayers instead, which can be seen from the magnetic field distribution in Fig. 4.15d. Due to the existence of material loss, the total transmittance is reduced to about 0.75.

Finally, we would like to emphasize that unlike normal waveguides in which the transmission rate strongly depends on the shapes and sizes, waveguides composed of AENZ media can achieve high transmission with almost arbitrary shapes and sizes. Such an amazing property makes AENZ materials perfect as wave couplers, with potential applications such as wave expanders and compressors, splitters, bends, and devices with combined purposes, etc.

As we known, some works have also proposed the perfect waveguides by applying transformation optics to transform air into inhomogeneous media [76–79]. And we have to design different inhomogeneity profiles of parameters for different waveguide devices. However, here we can utilize the same homogeneous anisotropic media to realize a series of different waveguide devices. This property enormously simplifies the design.

In addition, almost perfect waveguides based on isotropic zero-index media also have been proposed [34–36]. In such waveguides, the either electric or magnetic fields are nearly uniform throughout the isotropic zero-index medium due to the infinitely long wavelength. This makes the physics of EM flux control in the AENZ media totally different from that in the isotropic zero-index media. And also we notice that the areas of waveguides are required to be vanishingly small to achieve high transmission when the waveguide is filled with single-zero media. Although the area can be large when the double-zero media are used instead, the simultaneous near-zero permittivity and permeability make the design more complicated.

4.4 Inhomogeneous Anisotropic High-Index Media

In the above sections, we have shown that EM fluxes can be almost arbitrarily controlled by using inhomogeneous anisotropic zero-index media. It is worth noting that although near-zero permittivity is applied to obtain high anisotropy (i.e., $\varepsilon_y = f(x, y) >> \varepsilon_x \to 0^+$), it is actually not the necessary condition to obtain EM flux control. To achieve the high anisotropy, we can also utilize isotropic high-index media, that is,

$$\varepsilon_y = f(x, y) >> \varepsilon_x \approx 1, \quad \mu = 1 \tag{4.16}$$

Such anisotropic high-index media can be constructed by subwavelength multilayer composed of dielectric and negative permittivity materials [80–82]. From Eq. 4.14, we can see that when the term $f_d\varepsilon_n + f_n\varepsilon_d$ is tuned to be small, the $\varepsilon_\perp$ can be very large. In addition, the anisotropic high-index media can also be realized by using metallic rod array in microwave regime [83–85]. Strong nonlocality may lead to very large effective permittivity [86]. We note that the anisotropic high-index media have been found to have the ability in routing deep-subwavelength optical beams [81, 82], but the physics is different from the EM flux control discussed here. In the previous works, the waves propagate along the large-permittivity direction, and the propagating direction is always perpendicular to the flat equifrequency contour. But, here the waves are propagating along the small-permittivity direction, and the waveguiding effect is obtained by the redistribution of EM flux. Moreover, in the previous works [81, 82], both electric and magnetic fields are manipulated, but here we only manipulate the electric fields.

Without loss of generality, we revisit the example of free routing of EM flux shown in Figs. 4.5e and 4.5f, but the relevant parameters are changed as $\varepsilon_x = \mu = 1$ and $\varepsilon_y = f(x, y)$ (see Fig. 4.16a). The operating wavelength in free space is increased to $\sqrt{1000}$ times of that in Fig. 4.5f. The magnetic field H_z, amplitude of the electric field $|E|$, and time-averaged energy flux $\bar{S}$ are presented in Figs. 4.16b, 4.16c, and 4.16d, respectively. It is seen from the magnetic field distribution that the uniformity of the magnetic field in the y direction is quite good, and the transmittance is quite high. On the other hand, strong electric fields are induced in the inhomogeneous

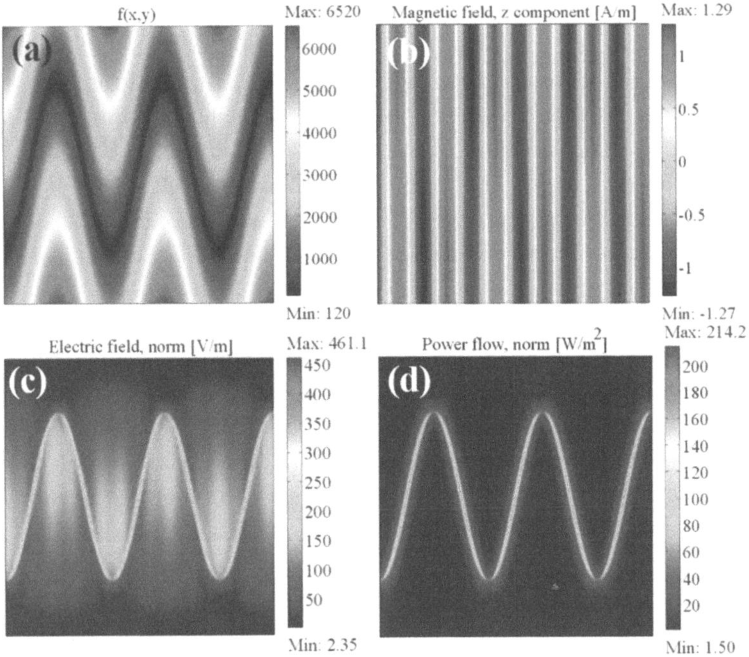

Figure 4.16 (a) The $\varepsilon_y = f(x, y)$ distribution map in the example of free routing of energy flux in a designed sinusoidal path. (b–d) Field maps of the magnetic field H_z, amplitude of the electric field $|E|$, and time-averaged energy flux $\bar{S}$. $\varepsilon_x = \mu = 1$ throughout the whole medium.

system, resulting in the redistribution of EM fluxes. And it is clearly seen from Fig. 4.16d, the EM fluxes are indeed confined and guided in the designed subwavelength sinusoidal pathway just as the one of Fig. 4.5f by using anisotropic zero-index media, confirming the feasibility of anisotropic high-index media in EM flux control. The feasibility would provide more choices for the materials used and facilitate the realization of EM flux control in practice.

4.5 Summary

We introduced a simple mechanism to almost arbitrarily control EM flux in both wavelength and subwavelength scales by using inhomogeneous anisotropic media. It is found that when the propagation direction is along the direction with much smaller index of the inhomogeneous anisotropic media, the scattered waves induced by inhomogeneity are evanescent in the propagation direction. This enables a unique way to arbitrarily control EM flux in wavelength and subwavelength scales. By designing the spatial profile of the other permittivity/permeability component in such inhomogeneous media, the distribution of EM flux can be conveniently manipulated. Moreover, robust high transmissions can be obtained attributing to averaging effect of the parameters.

Based on this mechanism, a series of almost perfect waveguide devices have been designed and analytically analyzed. By using transformation optics, we transform the physical waveguides filled with homogeneous anisotropic media into virtual waveguides filled with inhomogeneous anisotropic media for analyses. We find that the robust high transmission effect in inhomogeneous anisotropic media is the origin of high transmittance in such waveguides. One advantage of such waveguides is that the same homogeneous anisotropic media can be used to realize a series of different waveguide devices, which enormously simplifies the design.

The inhomogeneous anisotropic media can be zero-index or high-index materials, and also nonmagnetic media. The new mechanism of EM flux control may inspire ideas leading to more functional flux controlling applications and more compact devices.

Acknowledgments

This work is supported by the State Key Program for Basic Research of China (Nos. 2014CB360505, 2012CB921501), the National Natural Science Foundation of China (Nos. 11104196, 11374224, 11435009), the Program for New Century Excellent Talents in University (NCET), and a project funded by the Priority Academic Program Development of Jiangsu Higher Education Institutions (PAPD). Part of the work is supported by the Hong Kong Research Grant Council through AOE/P-02/12 and CUHK1/CRF/12G.

References

1. John, S. (1987). Strong localization of photons in certain disordered dielectric superlattices, *Phys. Rev. Lett.*, **58**, pp. 2486–2489.
2. Yablonovitch, E. (1987). Inhibited spontaneous emission in solid-state physics and electronics, *Phys. Rev. Lett.*, **58**, pp. 2059–2062.
3. Joannopoulos, J. D., Johnson, S. G., Winn, J. N., and Meade, R. D. (2008) *Photonic Crystals: Molding the Flow of Light* 2nd Ed. (Princeton University Press, Princeton, USA).
4. Barnes, W. L., Dereux, A., and Ebbesen, T. W. (2004). Surface plasmon subwavelength optics, *Nature*, **424**, pp. 824–830.
5. Ozbay, E. (2006). Plasmonics: Merging photonics and electronics at nanoscale dimensions, *Science*, **311**, pp. 189–193.
6. Maier, S. A. (2007) *Plasmonics: Fundamentals and Applications* (Springer, New York, USA).
7. Pendry, J. B., Holden, A. J., Stewart, W. J., and Youngs, I. (1996). Extremely low frequency plasmons in metallic mesostructures, *Phys. Rev. Lett.*, **76**, pp. 4773–4776.
8. Pendry, J. B., Holden, A. J., Robbins, D. J., and Stewart, W. J. (1999). Magnetism from conductors and enhanced nonlinear phenomena, *IEEE Trans. Microwave Theory Tech.*, **47**, pp. 2075–2084.
9. Smith, D. R., Padilla, W. J., Vier, D. C., Nemat-Nasser, S. C., and Schultz, S. (2000). Composite medium with simultaneously negative permeability and permittivity, *Phys. Rev. Lett.*, **84**, pp. 4184–4187.
10. Oulton, R. F., Sorger, V. J., Zentgraf, T., Ma, R.-M., Gladden, C., Dai, L., Bartal, G., and Zhang, X. (2009). Plasmon lasers at deep subwavelength scale, *Nature*, **461**, pp. 629–632.
11. Shelby, R. A., Smith, D. R., and Schultz, S. (2001). Experimental verifcation of a negative index of refraction, *Science*, **292**, pp. 77–79.
12. Smith, D. R., Pendry, J. B., and Wiltshire, M. C. K. (2004). Metamaterials and negative refractive index, *Science*, **305**, pp. 788–792.
13. Pendry, J. B. (2000). Negative refraction makes a perfect lens, *Phys. Rev. Lett.*, **85**, pp. 3966–3969.
14. Fang, N., Lee, H., Sun, C., and Zhang, X. (2005). Sub-diffraction-limited optical imaging with a silver superlens, *Science*, **308**, pp. 534–537.
15. Liu, Z., Lee, H., Xiong, Y., Sun, C., and Zhang, X. (2007). Far-field optical hyperlens magnifying sub-diffraction-limited objects, *Science*, **315**, 1686.

16. Smolyaninov, I. I., Hung, Y.-J., and Davis, C. C. (2007). Magnifying superlens in the visible frequency range, *Science,* **315**, pp. 1699–1701.
17. Lu, D. and Liu, Z. (2012). Hyperlenses and metalenses for far-field super-resolution imaging, *Nat. Commun.,* **3**, 1205.
18. Moore, D. T. (1980). Gradient-index optics: a review, *Appl. Opt.,* **19**, pp. 1035–1038.
19. Maxwell, J. C. (1965) *The Scientific Papers of James Clerk Maxwell* (Dover, New York, USA).
20. Pendry, J. B., Schurig, D., and Smith, D. R. (2006). Controlling electromagnetic fields, *Science,* **312**, pp. 1780–1782.
21. Schurig, D., Mock, J. J., Justice, B. J., Cummer, S. A., Pendry, J. B., Starr, A. F., and Smith, D. R. (2006). Metamaterial electromagnetic cloak at microwave frequencies, *Science,* **314**, pp. 977–980.
22. Leonhardt, U. (2006). Optical conformal mapping, *Science,* **312**, pp. 1777–1780.
23. Chen, H., Chan, C. T., and Sheng, P. (2010). Transformation optics and metamaterials, *Nat. Mater.,* **9**, pp. 387–396.
24. Lai, Y., Ng, J., Chen, H., Han, D., Xiao, J., Zhang, Z.-Q., and Chan, C. T. (2009). Illusion optics: The optical transformation of an object into another object, *Phys. Rev. Lett.,* **102**, 253902.
25. Lai, Y., Ng, J., Chen, H., Zhang, Z., and Chan, C. T. (2010). Illusion optics, *Front. Phys. China,* **5**, pp. 308–318.
26. Enoch, S., Tayeb, G., Sabouroux, P., Guerin, N., and Vincent, P. (2002). A metamaterial for directive emission, *Phys. Rev. Lett.,* **89**, 213902.
27. Alù, A., Silveirinha, M. G., Salandrino, A., and Engheta, N. (2007). Epsilon-near-zero metamaterials and electromagnetic sources: Tailoring the radiation phase pattern, *Phys. Rev. B,* **75**, 155410.
28. Cheng, Q., Jiang, W. X., and Cui, T. J. (2010). Radiation of planar electromagnetic waves by a line source in anisotropic metamaterials, *J. Phys. D: Appl. Phys.,* **43**, 335406.
29. Cheng, Q., Jiang, W. X., and Cui, T. J. (2011). Multi-beam generations at pre-designed directions based on anisotropic zero-index metamaterials, *Appl. Phys. Lett.,* **99**, 131913.
30. Huang, X., Lai, Y., Hang, Z. H., Zheng, H., and Chan, C. T. (2012). Dirac cones induced by accidental degeneracy in photonic crystals and zero-refractive-index materials, *Nat. Mater.,* **10**, pp. 582–586.
31. Cheng, Q., Jiang, W. X., and Cui, T. J. (2012). Spatial power combination for omnidirectional radiation via anisotropic metamaterials, *Phys. Rev. Lett.,* **108**, 213903.

32. Luo, J., Xu, P., and Gao, L. (2012). Directive emission based on one-dimensional metal heterostructures, *J. Opt. Soc. Am. B*, **29**, pp. 35–39.

33. Luo, J. and Lai, Y. (2013). Epsilon-near-zero or mu-near-zero materials composed of dielectric photonic crystals, *Sci. China Inform. Sci.*, **56**, 120414.

34. Silveirinha, M. and Engheta, N. (2006). Tunneling of electromagnetic energy through subwavelength channels and bends using ε-near-zero materials, *Phys. Rev. Lett.*, **97**, 157403.

35. Edwards, B., Alù, A., Young, M. E., Silveirinha, M., and Engheta, N. (2008). Experimental verification of epsilon-near-zero metamaterial coupling and energy squeezing using a microwave waveguide, *Phys. Rev. Lett.*, **100**, 033903.

36. Liu, R., Cheng, Q., Hand, T., Mock, J. J., Cui, T. J., Cummer, S. A., and Smith, D. R. (2008). Experimental demonstration of electromagnetic tunneling through an epsilon-near-zero metamaterial at microwave frequencies, *Phys. Rev. Lett.*, **100**, 023903.

37. Alù, A., Bilotti, F., Engheta, N., and Vegni, L. (2006). Metamaterial covers over a small aperture, *IEEE Trans. Antennas Propag.*, **54**, pp. 1632–1643.

38. Adams, D. C., Inampudi, S., Ribaudo, T., Slocum, D., Vangala, S., Kuhta, N. A., Goodhue, W. D., Podolskiy, V. A., and Wasserman, D. (2011). Funneling light through a subwavelength aperture with epsilon-near-zero materials, *Phys. Rev. Lett.*, **107**, 133901.

39. Engheta, N., Salandrino, A., and Alù, A. (2005). Circuit elements at optical frequencies: nanoinductors, nanocapacitors, and nanoresistors, *Phys. Rev. Lett.*, **95**, 095504.

40. Engheta, N. (2007). Circuits with light at nanoscales: optical nanocircuits inspired by metamaterials, *Science*, **317**, pp. 1698–1702.

41. Edwards, B. and Engheta, N. (2012). Experimental verification of displacement-current conduits in metamaterials-inspired optical circuitry, *Phys. Rev. Lett.*, **108**, 193902.

42. Salandrino, A. and Engheta, N. (2006). Far-field subdiffraction optical microscopy using metamaterial crystals: Theory and simulations, *Phys. Rev. B*, **74**, 075103.

43. Silveirinha, M. G. and Engheta, N. (2009). Transporting an image through a subwavelength hole, *Phys. Rev. Lett.*, **102**, 103902.

44. Castaldi, G., Savoia, S., Galdi, V., Alù, A., and Engheta, N. (2012). Analytical study of subwavelength imaging by uniaxial epsilon-near-zero metamaterial slabs, *Phys. Rev. B*, **86**, 115123.

45. Luo, J., Chen, H., Hou, B., Xu, P., and Lai, Y. (2013). Nonlocality-induced negative refraction and sub-wavelength imaging by parabolic dispersions in metal-dielectric multilayered structures with effective zero permittivity, *Plasmonics*, **8**, pp. 1095–1099.
46. Hao, J., Yan, W., and Qiu, M. (2010). Super-reflection and cloaking based on zero index metamaterial, *Appl. Phys. Lett.*, **96**, 101109.
47. Jin, Y. and He, S. (2010). Enhancing and suppressing radiation with some permeability-near-zero structures, *Opt. Express*, **18**, pp. 16587–16593.
48. Nguyen, V. C., Chen, L., and Halterman, K. (2010). Total transmission and total reflection by zero index metamaterials with defects, *Phys. Rev. Lett.*, **105**, 233908.
49. Xu, Y. and Chen, H. (2011). Total reflection and transmission by epsilon-near-zero metamaterials with defects, *Appl. Phys. Lett.*, **98**, 113501.
50. Luo, J., Xu, P., Gao, L., Lai, Y., and Chen, H. (2012). Manipulate the transmissions using index-near-zero or epsilon-near-zero metamaterials with coated defects, *Plasmonics*, **7**, pp. 353–358.
51. 5Ma, H. F., Shi, J. H., Cai, B. G., and Cui, T. J. (2012). Total transmission and super reflection realized by anisotropic zero-index materials, *New J. Phys.*, **14**, 123010.
52. Ma, H. F., Shi, J. H., Cheng, Q., and Cui, T. J. (2013). Experimental verification of supercoupling and cloaking using mu-near-zero materials based on a waveguide, *Appl. Phys. Lett.*, **103**, 021908.
53. Wang, T., Luo, J., Gao, L., Xu, P., and Lai, Y. (2013). Hiding objects and obtaining Fano resonances in index-near-zero and epsilon-near-zero metamaterials with Bragg-fiber-like defects, *J. Opt. Soc. Am. B*, **30**, pp. 1878–1883.
54. Wang, T., Luo, J., Gao, L., Xu, P., and Lai, Y. (2014). Equivalent perfect magnetic conductor based on epsilon-near-zero media, *Appl. Phys. Lett.*, **104**, 211904.
55. Jin, Y., Xiao, S., Mortensen, N. A., and He, S. (2011). Arbitrarily thin metamaterial structure for perfect absorption and giant magnification, *Opt. Express*, **19**, pp. 11114–11119.
56. Feng, S. and Halterman, K. (2012). Coherent perfect absorption in epsilon-near-zero metamaterials, *Phys. Rev. B*, **86**, 165103.
57. Zhong, S. and He, S. (2013). Ultrathin and lightweight microwave absorbers made of mu-near-zero metamaterials, *Sci. Rep.*, **3**, 2083.
58. Halterman, K. and Elson, J. M. (2014). Near-perfect absorption in epsilon-near-zero structures with hyperbolic dispersion, *Opt. Express*, **22**, pp. 7337–7348.

59. Luo, J., Li, S., Hou, B., and Lai, Y. (2014). A unified theory for perfect absorption in ultra-thin absorptive films with constant tangential electric or magnetic fields, *Phys. Rev. B,* **90**, 165128.

60. Luo, J., Xu, P., Chen, H., Hou, B., Gao, L., and Lai, Y. (2012). Realizing almost perfect bending waveguides with anisotropic epsilon-near-zero metamaterials, *Appl. Phys. Lett.,* **100**, 221903.

61. Ma, H. F., Shi, J. H., Jiang, W. X., and Cui, T. J. (2012). Experimental realization of bending waveguide using anisotropic zero-index materials, *Appl. Phys. Lett.,* **101**, 253513.

62. Luo, J., Lu, W., Hang, Z., Chen, H., Hou, B., Lai, Y., and Chan, C. T. (2014). Arbitrary control of electromagnetic flux in inhomogeneous anisotropic media with near zero index, *Phys. Rev. Lett.,* **112**, 073903.

63. Luo, J. and Lai, Y. (2014). Anisotropic zero-index waveguide with arbitrary shapes, *Sci. Rep.,* **4**, 5878.

64. Xiao, S., Drachev, V. P., Kildishev, A. V., Ni, X., Chettiar, U. K., Yuan, H.-K., and Shalaev, V. M. (2010). Loss-free and active optical negative-index metamaterials, *Nature,* **466**, pp. 735–738.

65. Chan, C. T., Huang, X., Liu, F., and Hang, Z. H. (2012). Dirac dispersion and zero-index in two dimensional and three dimensional photonic and phononic systems, *Prog Electromagn. Res. B,* **44**, 163.

66. Moitra, P., Yang, Y., Anderson, Z., Kravchenko, I. I., Briggs, D. P., and Valentine, J. (2013). Realization of an all-dielectric zero-index optical metamaterial, *Nat. Photon.,* **7**, pp. 791–795.

67. Smith, D. R., Schultz, S., Markoš, P., and Soukoulis, C. M. (2002). Determination of effective permittivity and permeability of metamaterials from reflection and transmission coefficients, *Phys. Rev. B,* **65**, 195104.

68. Liang, Z. and Li, J. (2011). Scaling two-dimensional photonic crystals for transformation optics, *Opt. Express,* **19**, pp. 16821–16829.

69. Shen, L., Yang, T.-J., and Chau, Y.-F. (2007). 50/50 beam splitter using a one-dimensional metal photonic crystal with parabolalike dispersion, *Appl. Phys. Lett.,* **90**, 251909.

70. Luo, J., Xu, P., Sun, T., and Gao, L. (2011). Tunable beam splitting and negative refraction in heterostructure with metamaterial, *Appl. Phys. A,* **104**, pp. 1137–1142.

71. Luo, J., Chen, H., Hou, B., Xu, P., and Lai, Y. (2013). Nonlocality-induced negative refraction and subwavelength imaging by parabolic dispersions in metal-dielectric multilayered structures with effective zero permittivity, *Plasmonics,* **8**, pp. 1095–1099.

72. Subramania, G., Fischer, A. J., and Luk, T. S. (2012). Optical properties of metal-dielectric based epsilon near zero metamaterials, *Appl. Phys. Lett.,* **101**, 241107.
73. Gao, J., Sun, L., Deng, H., Mathai, C. J., Gangopadhyay, S., and Yang, X. (2013). Experimental realization of epsilon-near-zero metamaterial slabs with metal-dielectric multilayers, *Appl. Phys. Lett.,* **103**, 051111.
74. Maas, R., Parsons, J., Engheta, N., and Polman, A. (2013). Experimental realization of an epsilon-near-zero metamaterial at visible wavelengths, *Nat. Photon.,* **7**, pp. 907–912.
75. Song, Z., He, Q., Xiao, S., and Zhou, L. (2012). Making a continuous metal film transparent via scattering cancellations, *Appl. Phys. Lett.,* **101**, 181110.
76. Roberts, D. A., Rahm, M., Pendry, J. B., and Smith, D. R. (2008). Transformation-optical design of sharp waveguide bends and corners, *Appl. Phys. Lett.,* **93**, 251111.
77. Rahm, M., Roberts, D. A., Pendry, J. B., and Smith, D. R. (2008). Transformation-optical design of adaptive beam bends and beam expanders, *Opt. Express,* **16**, pp. 11555–11567.
78. Jiang, W. X., Cui, T. J., Zhou, X. Y., Yang, X. M., and Cheng, Q. (2008). Arbitrary bending of electromagnetic waves using realizable inhomogeneous and anisotropic materials, *Phys. Rev. E,* **78**, 066607.
79. Vasić, B., Isić, G., Gajić, R., and Hingerl, K. (2009). Coordinate transformation based design of confined metamaterial structures, *Phys. Rev. B,* **79**, 085103.
80. Belov, P. A. and Hao, Y. (2006). Subwavelength imaging at optical frequencies using a transmission device formed by a periodic layered metal-dielectric structure operating in the canalization regime, *Phys. Rev. B,* **73**, 113110.
81. Han, S., Xiong, Y., Genov, D., Liu, Z., Bartal, G., and Zhang, X. (2008). Ray optics at a deep-subwavelength scale: A transformation optics approach, *Nano. Lett.,* **8**, pp. 4243–4247.
82. Catrysse, P. B. and Fan, S. (2013). Routing of deep-subwavelength optical beams and images without refl ection and diffraction using infi nitely anisotropic metamaterials, *Adv. Mater.,* **25**, pp. 194–198.
83. Belov, P. A., Hao, Y., and Sudhakaran, S. (2006). Subwavelength microwave imaging using an array of parallel conducting wires as a lens, *Phys. Rev. B,* **73**, 033108.

84. Silveirinha, M. G., Fernandes, C. A., and Costa, J. R. (2008). Superlens made of a metamaterial with extreme effective parameters, *Phys. Rev. B*, **78**, 195121.

85. Silveirinha, M. G., Medeiros, C. R., Fernandes, C. A., and Costa, J. R. (2010). Experimental verification of broadband superlensing using a metamaterial with an extreme index, *Phys. Rev. B*, **81**, 033101.

86. Simovski, C. R., Belov, P. A., Atrashchenko, A. V., and Kivshar, Y. S. (2012). Wire metamaterials: physics and applications, *Adv. Mater.*, **24**, pp. 4229–4248.

Chapter 5

Transmission-Line Metamaterials for Surface- to-Leaky-Wave Transformation

Chung-Tse Michael Wu,[a] Pai-Yen Chen,[a] and Tatsuo Itoh[b]
[a]*Department of Electrical and Computer Engineering, Wayne State University, 5050 Anthony Wayne Drive, Detroit, MI 48202, USA*
[b]*Department of Electrical Engineering, University of California, Los Angeles 420 Westwood Plaza, Los Angeles, CA 90095, USA*
ctmwu@wayne.edu

In this chapter, we will review recent research progress on radio-frequency (RF) and microwave beam-forming techniques using planar metamaterials and metamaterial surfaces (metasurfaces). This chapter starts with discussions of principles and physics of guided- and leaky-wave characteristics of periodic planar structures, such as transmission-line-based metamaterials, followed by their active versions that enable novel smart antennas with beam-steering functions. We include detailed descriptions of their practical realizations and the integration with active elements and circuits, which enables adaptively tailoring electromagnetic waves. We clarify a wide variety of planar metamaterial-/metasurface-based beam-forming techniques, giving a critical assessment and comparative analysis in the RF and microwave bands. These techniques, which are ideally low cost, low profile, and fully integrated with monolithic circuits, will offer promising applications

Transformation Wave Physics: Electromagnetics, Elastodynamics, and Thermodynamics
Edited by Mohamed Farhat, Pai-Yen Chen, Sebastien Guenneau, and Stefan Enoch

ISBN 978-981-4669-95-5 (Hardcover), 978-981-4669-96-2 (eBook)
www.panstanford.com

in modern wireless communication, radar, remote sensing, and medical and security imaging.

5.1 Introduction

Metamaterials (MTMs), in a general term, refer to artificial structures engineered to have exotic and anomalous electromagnetic properties that may not be found in nature, such as the negative refractive index [1–3] and the magnetism above terahertz (THz) frequencies [4]. The word *meta*, in MTMs, is a Greek word that means beyond/after. Typical MTMs consist of arrays of scatterers with electric and magnetic dipolar resonances, such as magnetodielctric spheres, and a periodicity that is much smaller than the impinging and dominant Bloch wavelengths. The first bulk MTM was realized in the microwave regime using the periodically arranged continuous metal wires and split-ring resonators (SRRs) [2–4]. Such structure has been characterized with, for the first time, the double negative material property, namely a simultaneous negative permittivity and permeability. A double-negative MTM is also named as a "left handed" (LH) medium, in which the electromagnetic wave propagates with an exotic negative phase velocity or a group velocity of opposite sign, making the pulse appears to propagate backward. In an LH medium, Snell's law is inverted and evanescent waves can be amplified to realize a "perfect lens" with subdiffraction imaging capability [2, 3].

Since 2002, an alternative approach based on the printed circuit transmission line (TL) has been proposed to effectively realize MTMs in the radio-frequency (RF) and microwave regimes [5–7]. The TL-based MTMs, also termed as composite right-/left-handed transmission lines (CRLH-TLs), have become a popular way to realize MTMs in RF and microwave regimes due to their simplicity and compatibility with the printed circuit technology. In addition, since the TL structures are circuit oriented and nonresonant, they may provide less loss and broader bandwidth than those bulk MTMs that are composed of resonant inclusions. The TL MTMs also allow for the straightforward integration with active and/or passive lumped components, such as varactors or transistors, to make adaptive and reconfigurable electromagnetic responses in the RF and microwave systems. This general TL approach provides useful physical insights into the phenomena of single/double negative material properties and also an efficient design tool for MTM applications. The exotic guided-wave properties in the CRLH-TL medium, that is, the negative phase velocity, have been used to implement a wide variety of compact planar microwave components, including the multiband and enhanced bandwidth power combiners/splitters, resonators, couplers [8], phase shifters [9], bandpass filters [10], and subwavelength focusing devices [11, 12], which are readily integrated with the planar monolithic circuits. Furthermore, the periodic CRLH-TL structures can exhibit leaky-wave radiations, due to the unique fast-wave and dual RH-LH characteristics, which have been used to realize 1D and 2D leaky-wave antennas (LWAs) [13–16]. Except from being ultracompact, such MTM-based LWA can achieve a superior beam-scanning performance with radiation angles varying from backfire to endfire, including also a broadside scanning (which is, however, not available for conventional uniform or periodic LWAs). Unlike conventional LWAs that require complex and narrow-band feeding networks [17], an LWA based on the CRLH-TL can operate at the dominant mode that can be excited by a simple microstrip line, thus significantly alleviating the design complexity. Further, by loading tunable reactive elements, such as PIN diodes (which are diodes with a wide, undoped intrinsic semiconductor region between a p-type semiconductor and an n-type semiconductor region) or varactor diodes [18], this MTM LWA is not necessarily to be frequency dependent, like most LWAs. In this scenario, the beam-scanning capabilities are enabled by tuning the variable reactive loads via the biasing circuitry. As a result, the beam scanning can function properly at a fixed frequency, being of great interest for the effective channelizing in the modern communication systems. In recent years, there have been significant efforts toward the realization of frequency-independent LWAs. Many electronic scanning approaches have been proposed, including the use of PIN diodes as switches to control the period of the leaky-wave structure [18] (which provides only few discrete radiation angles), the magnetically scannable antenna built on a ferrite substrate [19], and the CRLH-TL comprising varactor diodes whose voltage-dependent capacitance can allow continuously varying the beam angle [15]. It has been reported that by tuning the reverse-bias voltages of loaded varactors, the beam can be scanned over a wide range of angles ($\sim 99^\circ$), with high antenna gain ~ 18 dBi; this MTM-based leaky-wave antenna consists of 30 cells with a subwavelength unit cell size of 1.2 cm $\times$ 1.2 cm in the ultrahigh-frequency (UHF) band [15].

In the past few years, the RF and microwave beam-forming techniques have attracted intensive research interest in wireless communication (i.e., spatial multiplexing multiple access [SDMA] and space shift keying [SSK]), microwave imaging, and radar for the directional and/or reconfigurable signal transmission or reception. The beam-forming technique can be used at both the transmitting and receiving ends to achieve spatial selectivity. Typically, adaptive beam forming is achieved with phased arrays consisting of multiple active antennas, which launch electromagnetic waves at particular angles by properly phasing the wave front of individual antennas that experience constructive or destructive interference. Nonetheless, typical phased arrays are composed of lots of radiating elements and at least one stage of amplifier and phase shifter, and are expensive to purchase due to their technical complexity. An MTM-based LWA with a much simplified design and lower cost may emerge as a promising solution to the RF/microwave beam-forming and beam-shaping applications. In this chapter, we present an overview of TL-based MTMs and their applications in reconfigurable LWAs. We will review analysis techniques based on the equivalent TL circuit models, and discuss a variety of new and novel planar MTM and metasurface antennas with beam-forming applications that have been developed in recent years. We begin by discussing the background of planar TL MTMs that lead to the development of passive and active printed LWAs, as well as the state-of-the-art surface antenna technologies, such as the metasurface-based holographic antennas.

5.2 Principle of Transmission-Line Metamaterials

To begin with the illustration of MTMs using the TL approach, we start with a standard lossless RH-TL model, as shown in

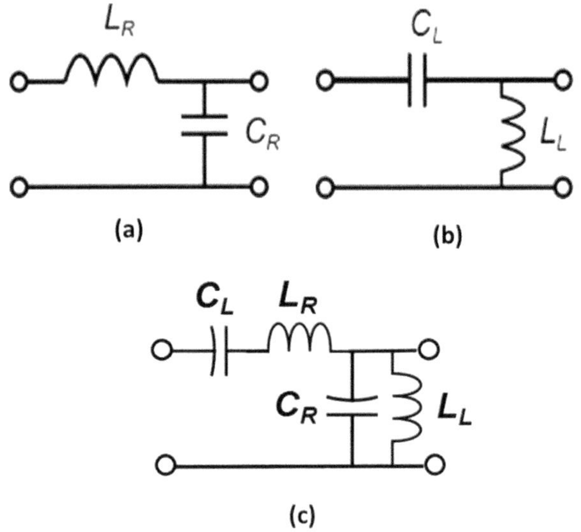

Figure 5.1 Unit cells of various transmission-line models: (a) right-handed (RH)-TL, (b) left-handed (LH)-TL, and (c) composite right-/left-handed (CRLH)-TL.

Fig. 5.1a. The equivalent circuit consists of series inductance and shunt capacitance, which is assumed to represent a small section of the TL, much less than $\lambda_g/4$ (λ_g is the guided wavelength). The equivalent constitutive parameters can be found by mapping Maxwell's equation onto the telegrapher's equation, leading to the similar dispersion diagram [20]. The propagation constant (γ) can therefore be derived in terms of the series impedance (Z') and shunt admittance (Y'), in which $Z' = j\omega L_R$ and $Y' = j\omega C_R$ [5]:

$$\gamma = j\beta = \sqrt{Z'Y'} = j\omega\sqrt{L_R C_R} \qquad (5.1)$$

Here, we adapt the $e^{j\omega t}$ notation throughout this chapter. From Eq. 5.1, we obtain a dispersion relationship that is linearly proportional to the frequency, as described in Eq. 5.2 and plotted in Fig. 5.2:

$$\beta_{RH} = j\omega\sqrt{L_R C_R} \qquad (5.2)$$

An LH-TL can be obtained by flipping the inductor and capacitor locations. Now consider the unit cell as depicted in Fig. 5.1b, in which the series impedance becomes a capacitance, and the shunt admittance becomes an inductance. Following the similar derivation

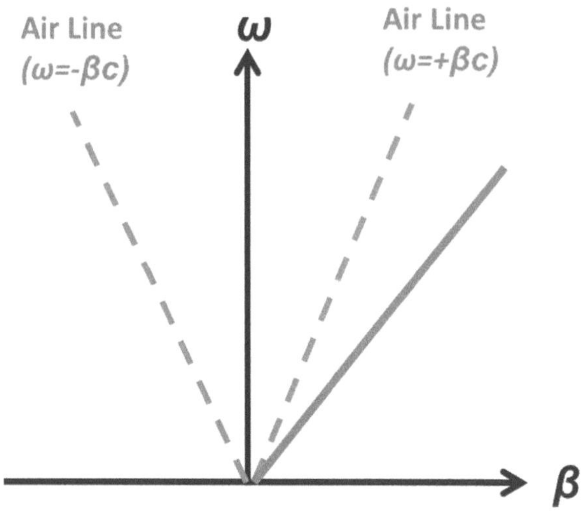

Figure 5.2 Dispersion diagram of the RH-TL model; the dashed lines represent the air lines.

as in Eq. 5.1, however, in this case substituting Z' with $1/j\omega C_L$ and Y' with $1/j\omega L_L$, the propagation constant can be written as:

$$\beta_{LH} = -\frac{1}{\omega\sqrt{L_L C_L}} \tag{5.3}$$

Since the propagation constant is always negative, which implies the phase velocity in this case is opposite to the direction of that for an RH-TL and is governed by the "left hand" rule. This LH-TL can be considered as a compact planar MTM, as it has novel negative-phase-velocity property that can only be artificially engineered. The corresponding dispersion diagram of such LH-TL is shown in Fig. 5.3.

Unlike the conventional case, the dispersion of an LH-TL is nonlinear and is always negative; the dashed lines represent the air lines.

In reality, due to the inevitable parasitics, that is, the parasitic effect of host medium, it is rather difficult to realize the perfect LH-TL MTMs. Indeed, the ordinary RH wave propagation is supported at higher frequencies. Therefore, a more generalized circuit model as shown in Fig. 5.1c must be adopted to accommodate both RH and LH propagations. This configuration is the so-called CRLH-TL. The dispersion relation can be derived using the same procedure

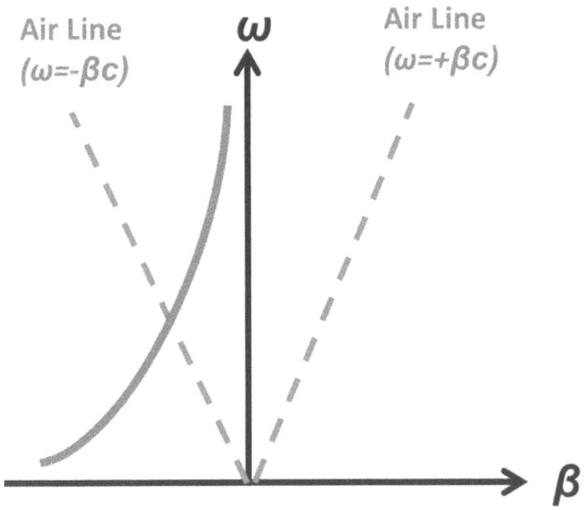

Figure 5.3 Dispersion diagram of the LH-TL model.

by letting $Z' = j\omega L_R + 1/j\omega C_L$ and $Y' = j\omega C_R + 1/j\omega L_L$. After some mathematical manipulations, the propagation constant can be written as:

$$\beta_{\text{CRLH}} = s(\omega)\sqrt{\omega^2 L_R C_R + \frac{1}{\omega^2 L_L C_L} - \left[\frac{L_R}{L_L} + \frac{C_R}{C_L}\right]} \tag{5.4}$$

where

$$s(\omega) = \begin{cases} -1, \ \omega<\omega_2 = \min\left\{\frac{1}{\sqrt{L_R C_L}}, \frac{1}{\sqrt{L_L C_R}}\right\} \\ 1, \ \omega>\omega_1 = \max\left\{\frac{1}{\sqrt{L_R C_L}}, \frac{1}{\sqrt{L_L C_R}}\right\} \end{cases} \tag{5.5}$$

Figure 5.4 plots the corresponding dispersion curve. From Fig. 5.4, we can observe that the CRLH-TL can support both of the LH mode ($\beta < 0$) in the low frequency region and the RH mode ($\beta > 0$) in the high frequency region. Depending on the operating frequency, CRLH-TLs can have different LH or RH characteristics. In addition, if we define two airlines that have a slope of the speed of light, c, we can divide the dispersion diagram into two regions. The region sandwiched by the two air lines is called the *fast-wave* region, as the phase velocity ($v_p = \omega/\beta$) is greater than c. On the other hand, the rest is defined as the *slow-wave* region as v_p is

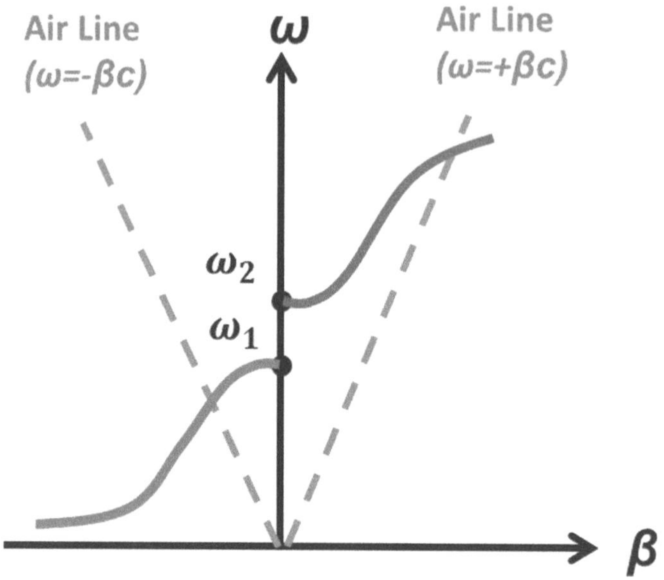

Figure 5.4 Dispersion diagram of a CRLH-TL model.

slower than c. CRLH-TLs behave differently in these two regions. As will be shown in the next section, CRLH-TLs will radiate in the fast-wave region, corresponding to a "leaky" mode, which is necessary for an LWA. On the other hand, when operated in the slow-wave region, corresponding to a "guided" mode, CRLH-TLs behave like waveguiding structures with ignorable radiation loss. The dual RH/LH and guided-to-leaky-wave characteristics make CRLH-TLs fascinating structures offering significantly more flexibility in the design of guided-wave and radiated-wave microwave components.

5.3 Guided and Radiated Modes of CRLH-TLS

Figure 5.5 shows the comparison of dispersion characteristics between a CRLH-TL and a conventional RH-TL. As discussed in the previous section, the dispersion region can be divided into radiation (leaky-wave) and guided-wave regions. Moreover, depending on the polarity of the propagation constant β the dispersion diagram can

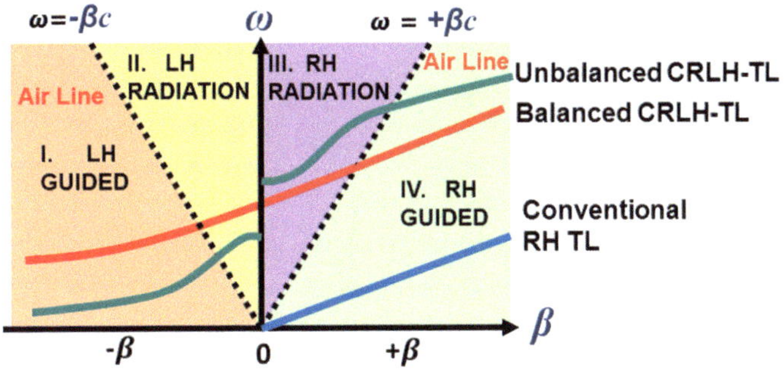

Figure 5.5 Dispersion diagram of a CRLH-TL model.

be categorized to RH ($\beta > 0$) and LH ($\beta < 0$) regions. A conventional RH-TL will lie in the RH-guided region, as $\beta > \omega/c$ or $v_p < c$ and will not radiate. On the other hand, the dispersion curve of CRLH-TL continues across all the four regions in the dispersion diagram, which are LH-guided-, LH-radiated-, RH-guided-, and RH-radiated-wave regions. As a result, CRLH-TL structures may provide a great deal of flexibility to tailor the electromagnetics responses at will. The band gap at $\beta = 0$ ($v_p \to \infty$) will disappears, if the following condition is satisfied:

$$\sqrt{\frac{L_R}{C_R}} = \sqrt{\frac{L_L}{C_L}} \text{ or } L_R C_L = L_L C_R \tag{5.6}$$

We call this a "balanced" condition for a CRLH-TL. A balanced CRLH-TL does not have a band gap and therefore has a smooth transition at the center frequency where the propagation constant is zero. This property is especially useful when it operates as a leaky-wave antenna, as it allows a continuous frequency-dependent beam-forming capability.

If the CRLH-TL operates in the radiated or fast-wave region with $\beta < \omega/c$ or $v_p > c$, it will leak out power to the free space as the wave travels along the TL. Therefore, a leaky-wave antenna is sometimes referred to as a fast-wave traveling-wave antenna.

To further explain the leaky-wave characteristics, Fig. 5.6 illustrates the relationship between the vertical wavenumber k_z, the free-space wavenumber k_0, and the modal propagation (phase) constant inside the waveguiding structure β_x. Here k_z can be

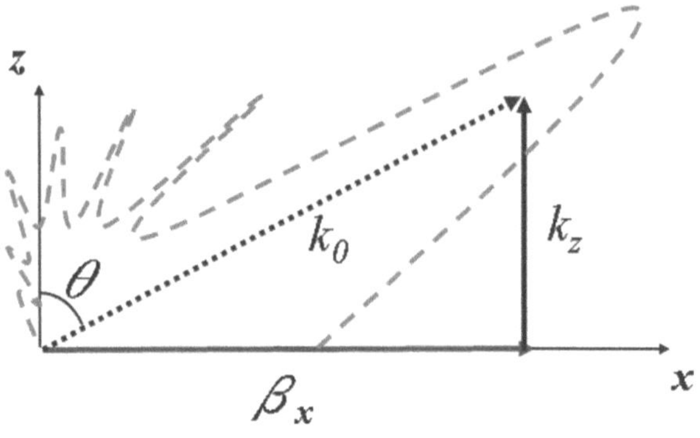

Figure 5.6 Illustration of the radiation phenomenon.

expressed as:

$$k_z = \sqrt{k_0^2 - \beta_x^2} \tag{5.7}$$

When the wave propagates through the leaky-wave structure with a complex wavenumber $\beta_x - j\alpha$, the fields outside the structure have the general form [17]:

$$\varphi(x, z) = \varphi_0 \left(e^{-\alpha x} e^{-j\beta_x x}\right) e^{-jk_z z} \tag{5.8}$$

The attenuation constant $\alpha = \alpha_{\text{loss}} + \alpha_{\text{rad}}$, where α_{loss} is due to the material (dielectric and conduction) loss inside the guiding structure and α_{rad} is due to the radiative energy leakage (leakage rate). Therefore, one can observe that if k_z is a real number, that is, $k_0 > \beta_x$, the wave can propagate along the z direction and hence the leakage radiation occurs. This is consistent with the aforementioned statement that when the CRLH-TL operates in the fast-wave region ($v_{\text{p}} = \omega/\beta_x > c = \omega/k_0$), the propagating wave will be coupled into the free-space radiation, with a real value of k_z. On the other hand, if k_z is imaginary, the wave will decay exponentially along the z direction, which results in an evanescent mode of the field.

From the vectorial relation shown in Fig. 5.6, the direction of the main beam of radiation can be determined as

$$\theta = \sin^{-1}\left(\frac{\beta_x}{k_0}\right) \tag{5.9}$$

Equation 5.9 is approximately valid for many practical antennas, provided that $\alpha \ll \beta_x$. It can be known from Fig. 5.6 that since

β_x is a function of frequency, the angle θ also varies with respect to the frequency, thereby enabling the frequency-dependent beam forming. Such property is useful in many wireless systems that require the beam-scanning function, as it does not need complex feeding networks and phase shifters used in conventional phased arrays. Equation 5.9 also indicates that an ideal backfire-to-endfire (-90° to $+90^\circ$) radiation may be possible, provided that the propagation constant is properly tailored.

5.4 Free-Space Scanning and Adaptive CRLH-LWAS

As discussed above, CRLH-LWAs can provide frequency-dependent beam-scanning capabilities for continuously varying the beam from backfire to endfire directions by frequency tuning. Using this unique property, novel tunable and adaptive LWAs can be realized. This section will discuss the practice of MTM-based LWAs and the experimental demonstrations for free-space scanning capability with enhanced functionality.

5.4.1 *1D and 2D Beam Scanning*

Figure 5.7 shows a typical microstrip realization for a CRLH-TL using the printed circuit board (PCB) technology. The unit cell of the CRLH-TL contains an interdigital capacitor and a stub inductor to form the desired LH elements. The size of the unit cell must be much smaller than the guided wavelength such that the structure behaves

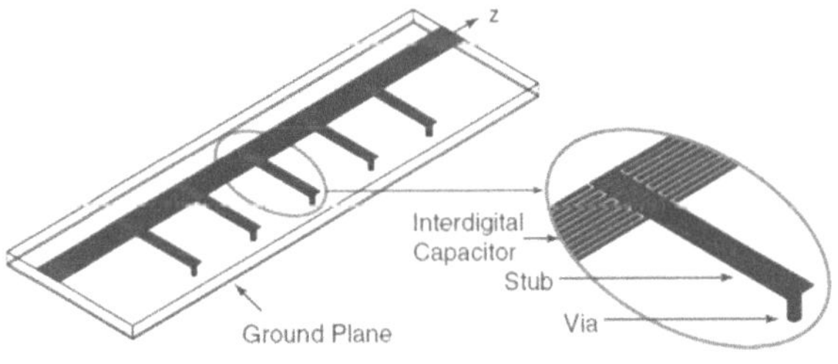

Figure 5.7 Microstrip implementation for a 1D CRLH-TL metamaterial structure comprising periodically repeated subwavelenght unit cells.

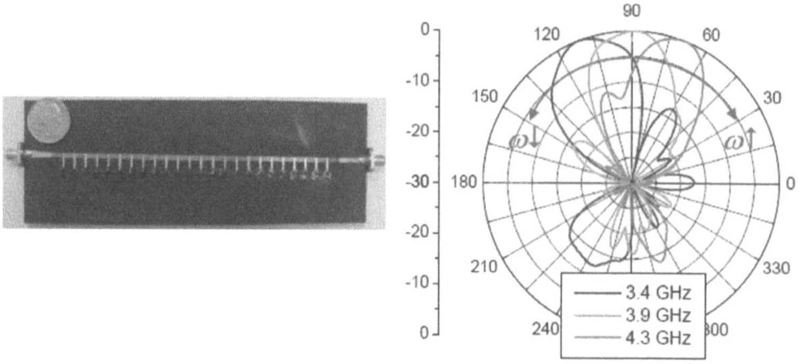

Figure 5.8 Prototype of Fig. 5.7 and its measured radiation patterns for different operating frequencies.

as effectively homogenized in the frequency range of interest. Assuming a lossless scenario, the unit cell model of this planar MTM can be described by the equivalent circuit in Fig. 5.1c. Due to the nature of inevitable parasitics from the substrate, the RH series inductance and shunt capacitance also exist, forming an entire CRLH unit cell, which exhibit a modal dispersion as shown in Fig. 5.5: LH components dominate at lower frequency, while RH components dominate at higher frequencies. A more rigorous unit cell model should also take into account the radiation loss (leakage rate), which are associated with resistance R and conductance G in the TL circuit model [15]. It should be noted that the characteristic impedance is nearly frequency independent for both RH and LH operations if properly designed, which allows the broadband matching to any quasi-TEM port with a constant impedance (i.e., 50 Ω in most RF and microwave systems).

Figure 5.8 shows a 24-cell CRLH-LWA prototype and the measured radiation pattern. It is evident that, as the frequency increases, the main beam direction scans from backward to forward, achieving a 1D frequency scanning [5]. Ideally, a CRLH-LWA can scan a broad range of spatial region from -90° to $+90^\circ$, as frequency is varied from $\omega = -\beta c$ to $\omega = +\beta c$, including the broadside direction ($\theta = 0^0$), which is usually cutoff for conventional uniform and periodic LWA [17]. By forming an LWA array with additional delay lines or phase shifters, the antenna array can sweep in both

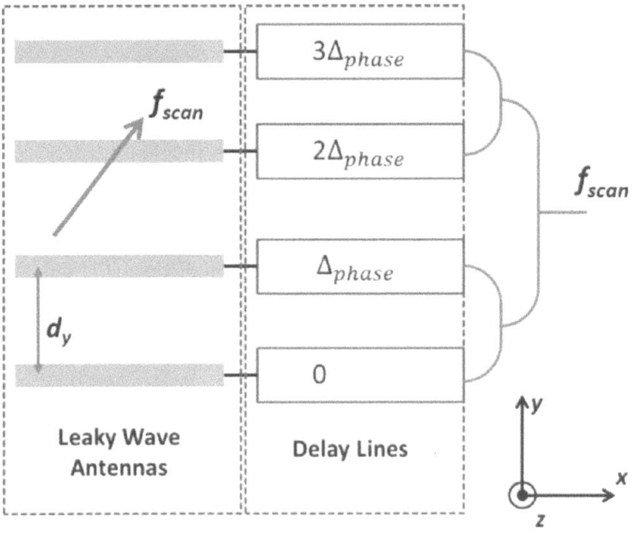

Figure 5.9 2D scanning LWA arrays using delay lines that provide the necessary phase shift.

xz and yz planes, as indicated in Fig. 5.9, thereby forming a 2D full-space frequency scanning [14].

5.4.2 *Tunable LWA*

If the unit cell of LWA is loaded with varactor diodes (see Fig. 5.10a), the series or/and shunt capacitance (C_L, C_R) can be made tunable. Originally, the propagation constant of a fully passive CRLH-LWA is a function of frequency, that is, $\beta(\omega)$. Therefore, the beam scanning can only be achieved by detuning the frequency. Now, if we use the unit cell consisting of varactors at a fixed frequency, the propagation constant $\beta(V)$ can be controlled by simply changing the bias voltage V across the varactor, thanks to the voltage-dependent capacitance of varactors, as shown in Fig. 5.10a. Figure 5.10b presents the concept of shifting the dispersion curve by varying the bias voltage of varactors ($V_1 = 0$ V, $V_2 = 5$ V, $V_3 = 15$ V), showing an electrically tunable $\omega - \beta$ diagram. Unlike the frequency-controlled beam steering in Section 5.4.1, a varactor-loaded CRLH-LWA allows dynamically steering the beam at a fixed frequency by an external

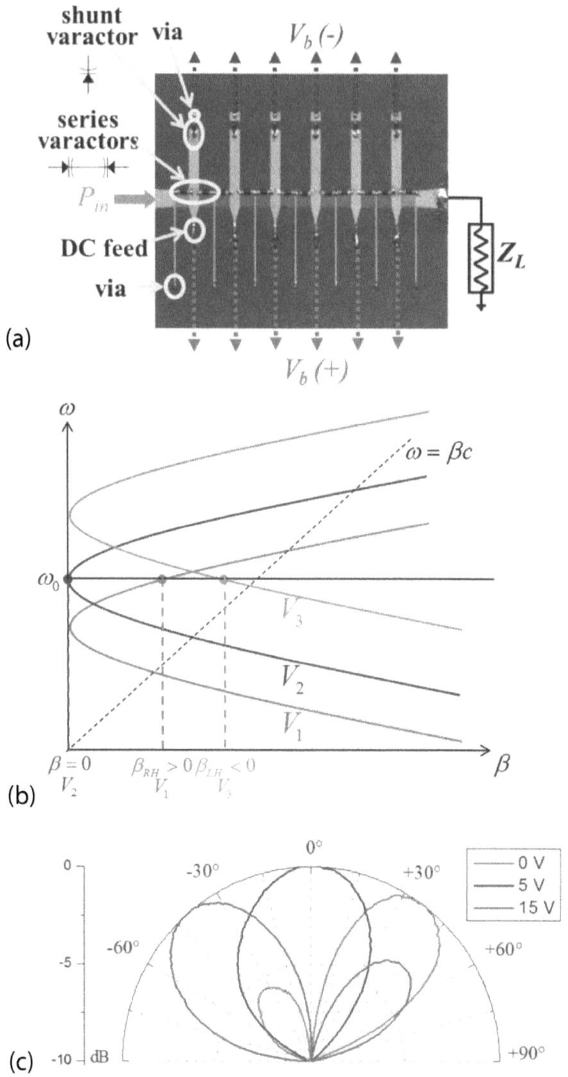

Figure 5.10 Electrically scanned CRHL-TL LWA: (a) prototype, (b) dispersion diagram, and (c) radiation patterns.

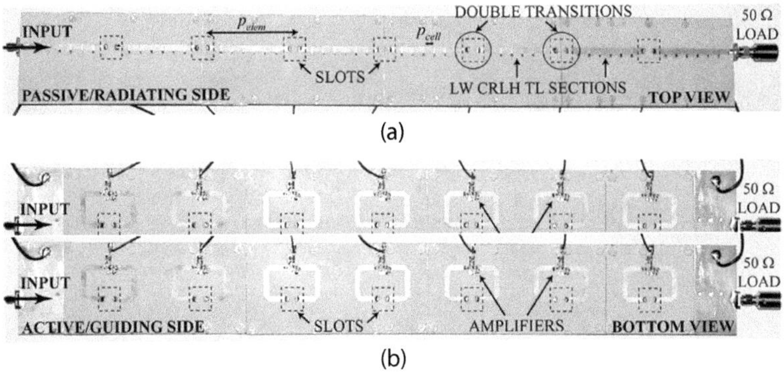

Figure 5.11 Active LWA with cascaded amplifiers.

circuitry. Such property is particularly useful when only certain carrier frequencies (channels) are permitted in wireless protocols. The radiation pattern versus voltages is plotted in Fig. 5.10c [15].

5.4.3 *Active CRLH-LWA*

5.4.3.1 Cascaded amplifiers

With the aid of active circuitry, the gain of LWAs is expected to be further enhanced [16]. Figure 5.11 shows the prototype of the active CRLH-TL LWA integrated with the amplifiers loaded in a cascaded fashion. The leakage at each section of LWA will then be compensated by the gain provided by amplifiers. In addition, the optimum current distribution can be tapered to achieve the maximum directivity of the antenna. In this section, we will report current progress on active-loaded CRLH-TL antennas with superior performance and radiation properties compared to those passive ones.

5.4.3.2 Distributed amplifier-based LWAs

Although the power gain of the amplifiers can compensate the loss (mainly the radiation leakage) and enhances the antenna gain, the operating band is only limited to a narrow frequency band. Therefore, it is necessary to have a scheme that can provide broadband amplification for LWAs. Fortunately, in RF and

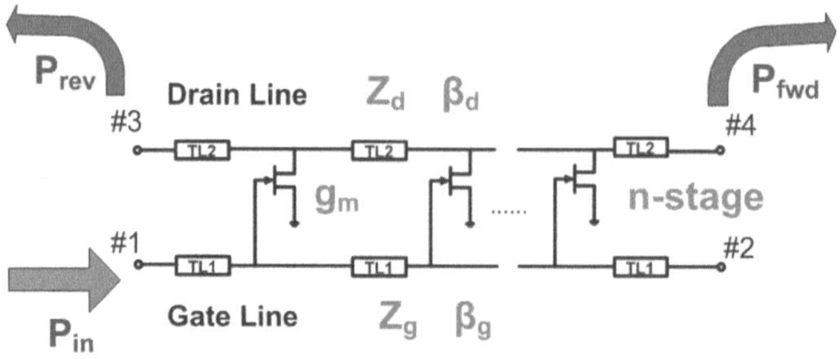

Figure 5.12 Topology of a generic distributed amplifier.

microwave regimes, one type of amplifier schemes called distributed amplifier (DA) can suitably do the job. Figure 5.12 depicts the diagram of a generic DA [20, 21]. The DA is well known for its broadband amplification features and contains two sets of TLs: (1) the input TL is connected to the gate of field-effect transistors (FETs), which is therefore also termed gate line, and (2) the output TL is connected to the drain of FETs and is termed drain line. When the input signal travels along the gate line, each FET will tap off some of the input power. On the output side, the amplified signals from the FETs will form a traveling wave on the drain line. The propagation constants of the gate and drain lines must be properly designed to achieve the constructive phasing of the output signal.

In addition, the parasitic capacitance at gate (C_{gs}) and drain (C_{ds}), and the input resistance of the FETs will effectively become parts of the gate and drain TLs. This type of circuitry is also known as a traveling wave amplifier.

As shown in Fig. 5.12, we can then define the forward gain (G_{f}) and reverse gain (G_{r}) as follows:

$$G_{\mathrm{f}} = \frac{g_{\mathrm{m}}^2 Z_{\mathrm{g}} Z_{\mathrm{d}}}{4} \left[\frac{\sin \frac{n}{2} \left(\beta_{\mathrm{g}} - \beta_{\mathrm{d}} \right)}{\sin \frac{n}{2} \left(\beta_{\mathrm{g}} - \beta_{\mathrm{d}} \right)} \right]^2 \tag{5.10}$$

$$G_{\mathrm{r}} = \frac{g_{\mathrm{m}}^2 Z_{\mathrm{g}} Z_{\mathrm{d}}}{4} \left[\frac{\sin \frac{n}{2} \left(\beta_{\mathrm{g}} + \beta_{\mathrm{d}} \right)}{\sin \frac{n}{2} \left(\beta_{\mathrm{g}} + \beta_{\mathrm{d}} \right)} \right]^2 \tag{5.11}$$

where n denotes the number of stages, g_{m} is the transconductance, Z_{g} and Z_{d} are the characteristic impedance of gate and drain lines,

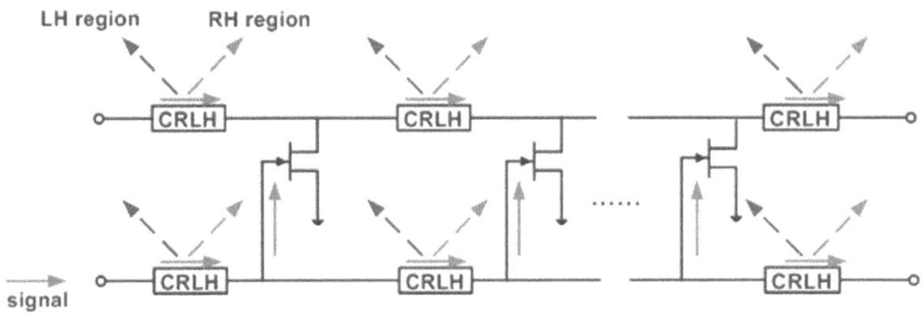

Figure 5.13 Distributed amplifier integrated with CRLH-LWAs.

respectively, and β_g and β_d are the propagation constants of gate and drain lines, respectively.

From Eqs. 5.10 and 5.11, we can observe that if the dispersion is the same for both gate and drain lines, that is, $\beta_g = \beta_d$ the forward gain (G_f) will become independent of the frequency. Therefore, wide band amplification can be achieved. Furthermore, if we replace conventional TLs with CRLH-TLs, DA-based LWAs can be realized and exhibit the frequency dependent beam-scanning capability with additional amplification as shown in Fig. 5.13 [22, 23].

5.4.3.3 Power-recycling schemes for DA-based CRLH-LWAs

In typical DAs, both lines are terminated with the properly matched load resistors to absorb waves traveling in the reverse direction. Although DAs provide CRLH-TL LWAs with broadband amplification, much power is dissipated in the termination, especially on the drain side of the DA if there is no additional feedback structure in the system. To address this issue, several design schemes have been proposed to recycle the power of active antennas [24–28]. For instance, one recycling scheme utilizing the ring resonator concept has been developed [24]. The topology is shown in Fig. 5.14. In this scheme, a section of microstrip line is inserted at the drain line to form a feedback unit; the unradiated power on the drain-side antenna can be redirected and injected back into the drain line, and finally radiate to the free space again. This power-recycling scheme can be regarded as a ring resonator that is actively coupled by the DA as shown in Fig. 5.15.

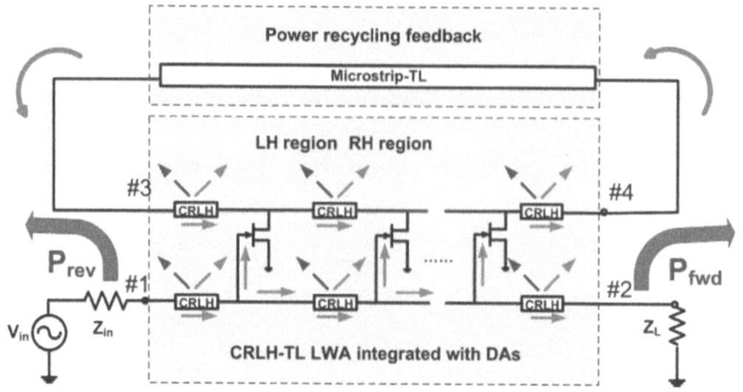

Figure 5.14 Ring-resonator-inspired power-recycling scheme for a distributed amplifier integrated with CRLH-LWAs.

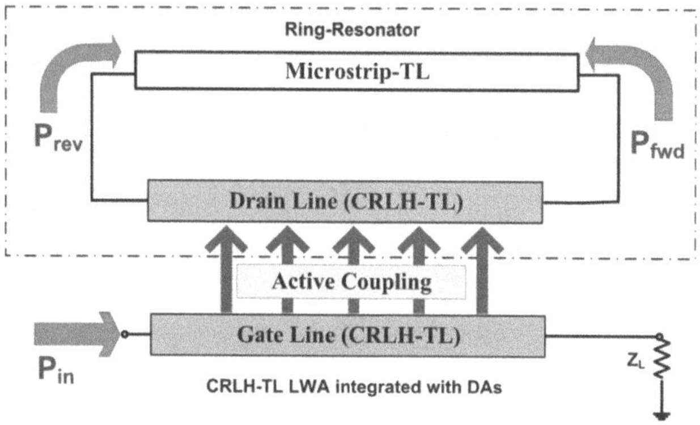

Figure 5.15 Distributed amplifier–based CRLH-LWA as an active coupling ring resonator.

One can then infer when the closed loop formed by the microstrip line and drain line has a phase equal to a multiple of 2π, the fields will build up in the closed loop. In other word, the resonance will occur when:

$$\varphi_{\text{CRLH\_TL\_drain}} + \varphi_{\text{MSTL}} = 2m\pi \qquad m = 0, 1, 2 \ldots \in N \qquad (5.12)$$

where $\varphi_{\text{CRLH\_TL\_drain}}$ and φ_{MSTL} are the phase of the CRLH-TL at the drain side and that of the microstrip line, respectively. Figure 5.16 plots the power flow inside the closed loop and its phase. One can observe that the power flow reaches its local maximum when Eq. 5.12 is satisfied.

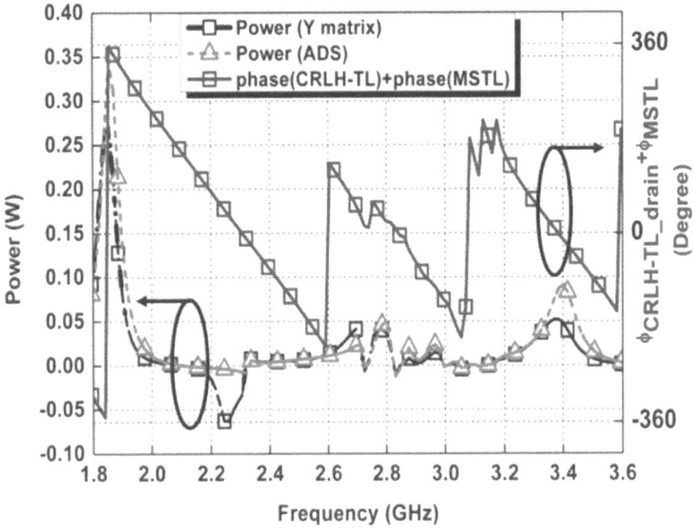

Figure 5.16 Distributed amplifier–based CRLH-LWA as an active coupling ring resonator.

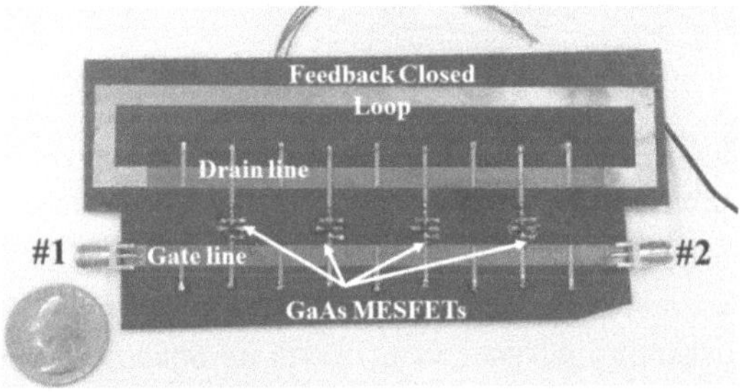

Figure 5.17 Prototype of distributed amplifier–based CRLH-LWA with the ring resonator power-recycling scheme.

Figure 5.17 shows the prototype of a DA-based CRLH-LWA with ring-resonator-inspired feedback used to verify the proposed power-recycling scheme. The prototype contains four discrete transistors (GaAs metal–semiconductor field-effect transistors [MESFETs]) and is fabricated using PCB technology.

The simulated and measured radiation patterns are plotted in Fig. 5.18 in comparison to the active antenna without any feedback

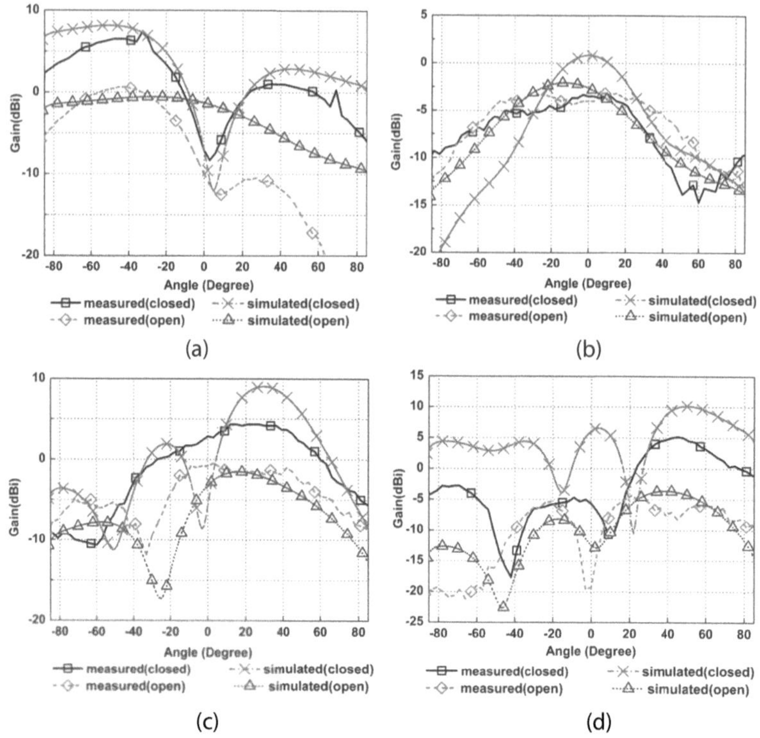

Figure 5.18 Simulated and measured antenna gain comparison with (closed) and without (open) the power-recycling feedback. (a) 1.9 GHz, (b) 2.35 GHz, (c) 2.8 GHz, and (d) 3.4 GHz.

unit. A substantial increase in gain can be observed (8 dB increase at most in the measurement), thereby confirming the mechanism of power recycling. In addition, the proposed scheme maintains the frequency dependent beam-scanning capability as can be found in a passive CRLH-LWA.

The configuration presented in Fig. 5.15 is based on a single-fed DA, of which the power is fed from one of the two ports of the DA. In fact, a DA can also be fed from both of the two input ports. It has been reported that a dual-fed DA has several advantages over a conventional single-fed DA, such as the increased gain and the reduced noise figure. Furthermore, for the dual-fed configuration, it is possible to make each device have the identical load impedance

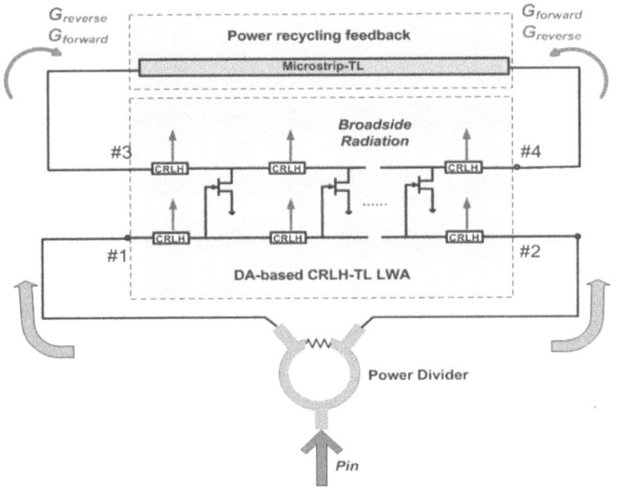

Figure 5.19 Dual-fed DA-based CRLH-LWA with power-recycling scheme.

and achieve power equalization. Therefore, a dual-fed DA-LWA has more superior power equalization among the active devices than a single-fed DA-LWA, especially when the loss is present. It can utilize the active devices to their full extent, due to the equally distributed power along the antenna, and thus is suitable to handle high power [24–28].

Figure 5.19 illustrates the schematic of a dual-fed DA-based CRLH-LWA with the ring-resonator-inspired power-recycling scheme. The active antenna is fed by a power divider and is designed at the transition frequency of the CRLH-TL, where the electrical spacing (θ) between active devices is equal to 0°. In this case, each active device inside the DA contributes nearly the same amount of power. Moreover, since the antenna is operated at its transition frequency, the radiation pattern is in the broadside direction. Thanks to the dual-fed configuration, the radiating power of the active antenna can be increased around twice as much as the single-fed structure.

Figure 5.20 shows the prototype of the dual-fed DA with the ring-resonator-based power-recycling scheme using a Wilkinson power divider to provide two feeds to the gate line. Figure 5.21 compares the simulated and measured return loss of the fabricated dual-fed

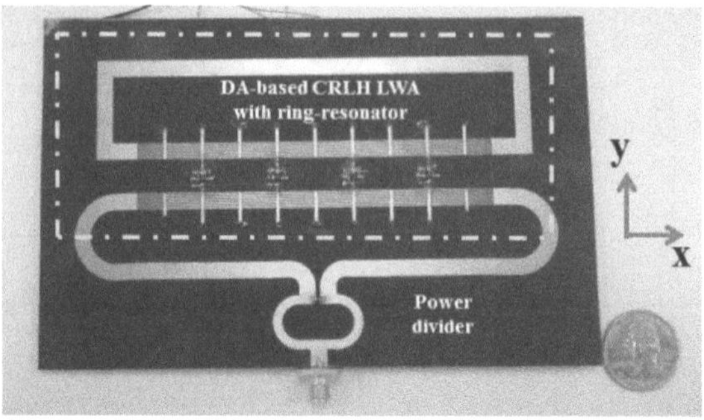

Figure 5.20 Dual-fed DA-based CRLH-LWA with power-recycling scheme.

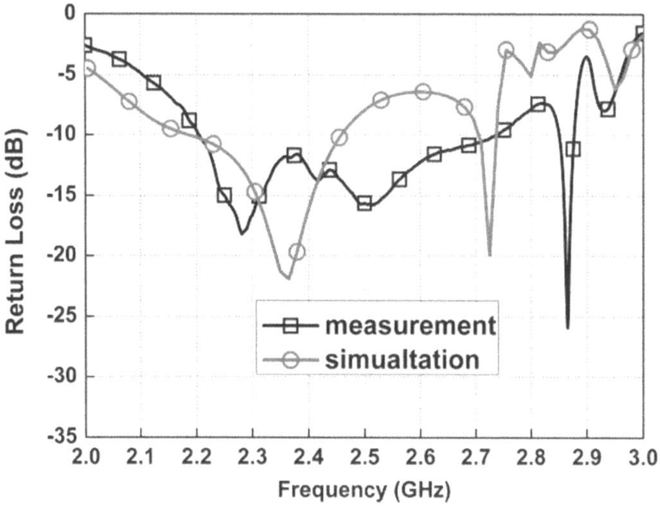

Figure 5.21 Dual-fed DA-based CRLH-LWA with power-recycling scheme.

DA. The operating frequency is 2.28 GHz, which has a below −10 dB return loss for both simulated and measured results. The radiation patterns for single- and dual-fed configurations are illustrated in Fig. 5.22. By adopting the dual-fed configuration, the gain is increased compared with the single-fed active antenna by 3.4 dB and 2.5 dB in the simulation and the measurement, respectively. The magnitude of the gain of the proposed structure can achieve up to 14.4 dBi in the measurement [27].

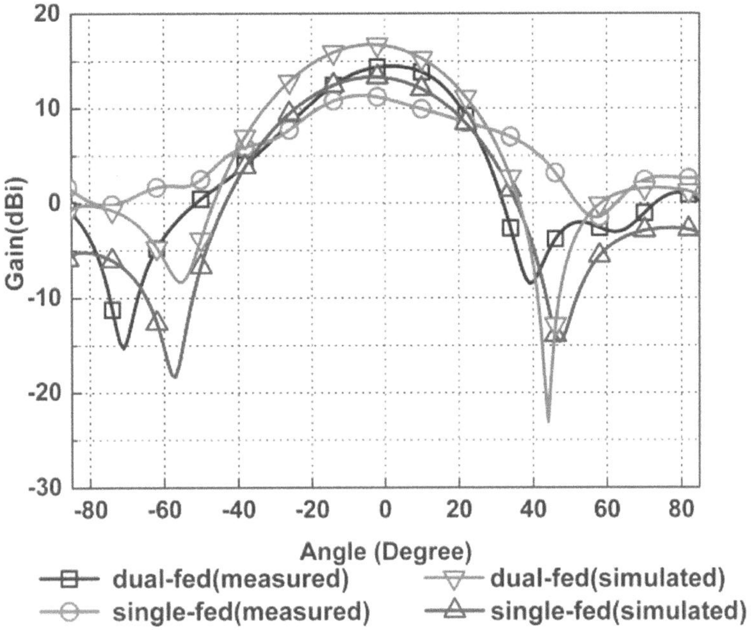

Figure 5.22 Dual-fed DA-based CRLH-LWA with power-recycling scheme.

The dual-fed DA-based LWA can be further extended into an array fashion as shown in Fig. 5.23 [29]. In this configuration, a coupling-enhanced coupler also based on CRLH-TLs is used to couple the energy from the loop of the first element into the both end of the gate line of the second element [5]. The two antenna elements are placed by half free-space wavelength apart in the y direction at the operating frequency, where the two radiating elements are in phase. It is noted that in order to have constructive addition at the broadside, the phase of the two elements is synchronized by adjusting the phase response of the CRLH coupler. The simulated 3D radiation patterns of the single element and two-element antenna array are plotted in Fig. 5.24. It is obvious that the radiation beam becomes more directive for the two-element array; the two-element array has more than 3 dB increase in the directivity compared to the single element that has a fan beam radiation pattern.

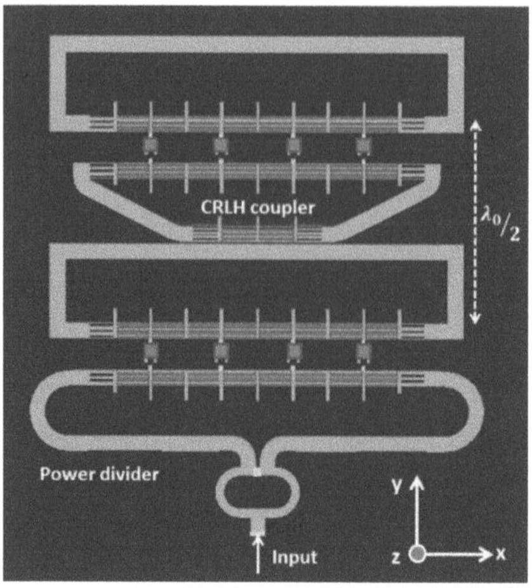

Figure 5.23 Array form of the dual-fed DA-based CRLH-LWAs with the ring resonator power feedback scheme.

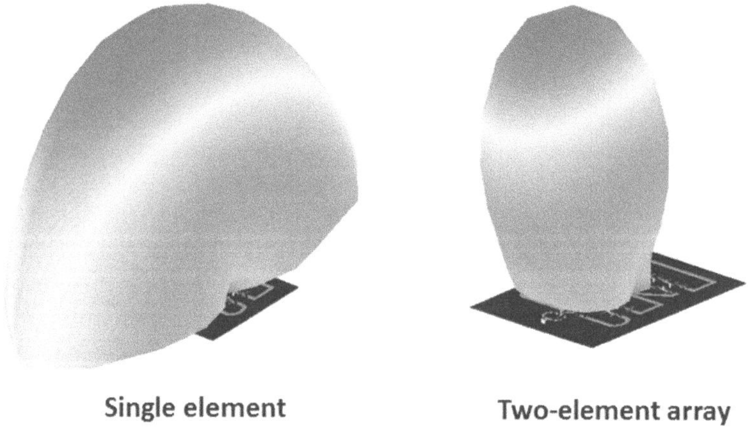

Figure 5.24 Simulated 3D radiation patterns.

5.5 Holographic Antennas Based on Metasurfaces

This section describes a new holographic antenna concept using the MTM surface (also termed as metasurface or metafilm) hologram, which is fed by a surface wave. Similar to the transformation optics, the radiation mechanism is based on the transformation of guided waves in the thin structure into the leaky waves, due to the interaction with the anisotropic reactance tensor of the metasurface that can have an analytical formula. The inhomogeneous and anisotropic surface reactance can be similarly achieved with dense, textured subwavelength metallic structures printed on a grounded dielectric slab. In this design, the beam angle with respect to the propagation direction of the surface wave depends on the modulated, anisotropic reactance tensor of the metasurface. Like the TL-MTM antenna, the metasurface antennas can be built with the PCB process for the standard printed circuit and be excited by a simple in-plane feeder. Therefore, the metasurface-based holographic antennas show significant potential for making simple, cost-effective, compact, and low-profile antennas, which are particularly useful for high-gain, beam-forming and beam-shaping antennas in satellite and space applications. Unlike the transformation optics approaches [30–33], where the control on propagation often requires highly anisotropic values for $\overline{\overline{\varepsilon}}$ and $\overline{\overline{\mu}}$ tesnors, the applicability of transformation techniques to 2D surface structures can avoid the complex transformation structures.

5.5.1 *Introduction to Metasurface Technology*

A metasurface can be seen as a 2D version of MTM. A metasurface is formed by arranging subwavelength scatterers or apertures into a 2D surface or interface [34–36]. Through the design of size, period and shape of conductive textured surface, the quasi-homogeneous surface impedance can be varied [35–39]. For many applications, metasurfaces can be used in place of MTMs, which have the advantage of occupying less physical space than 3D MTM structures. Hence, metasurfaces may offer the possibility of less lossy and moderate-bandwidth structures. Metasurfaces have been demonstrated for a wide range of potential applications, including miniaturized cavity resonators, novel waveguiding structures, compact filters, polarizers and absorbers, and terahertz switches [34], to name a few. In this section, we will discuss the development in recent years of metasurfaces, bringing the exciting applications into the RF and microwave beam-forming technology. Using reconfigurable metasurfaces in conjunction with the holography principles may enable new antennas that can electronically steer the beam, with high gain and moderate bandwidth.

5.5.2 *Principle and Practice of Holographic Antennas*

A microwave holographic antenna was first demonstrated by Checcacci, Russo, and Scheggi in 1970 [40]. Holographic antennas are a class of antennas whose radiation apertures are specifically formed by the conductive metallic patterns (diffraction apertures) generated by the interference of the desired beam with that of the reference beam, somehow similar to its counterpart in optics (i.e., optical holography). Since the idea is originated from optical holographic imaging, it is named as holographic antenna. RF and microwave holographic antennas have been investigated over 40 years [41]. Holographic antennas typically comprise a main surface-wave-carrying impedance surface, which is separated from the ground plane by a dielectric slab. The electrically thin, artificial-impedance surfaces have been known for their capabilities of controlling the phase and amplitude fronts of surface waves [41–49]. Therefore, an impedance surface can be designed to serve as a guiding structure or to support leaky waves that create highly directive radiation beams by controlling the modulation period and the depth of an impedance surface [40–46, 49].

The distribution pattern of surface perturbations can be designed by a holography method [42]. Holography involves producing an interference pattern using two waves, and then using the interference pattern to scatter one wave to produce the other. The interference pattern formed by these two waves is recorded as a *hologram*. When the reference wave illuminates the hologram, it is scattered by the recorded interference pattern to produce a copy of the original object wave. For a reference wave Ψ_{ref} and an object wave Ψ_{obj}, the interference pattern contains a term proportional to $\Psi_{\text{obj}}\Psi^*_{\text{ref}}$. When the interference pattern is illuminated by the reference wave, it renders $(\Psi_{\text{obj}}\Psi^*_{\text{ref}})\,(\Psi_{\text{ref}} = \Psi_{\text{obj}}|(\Psi_{\text{ref}}|^2$, forming a copy of the original object wave. For the case of an LWA excited by the bound transverse magnetic (TM) surface modes, we may define the geometrical surface perturbation as the interference pattern between these two waves [42]:

$$Z(x, y) = j\left|X + M\,\mathrm{Re}\left[\Psi_{\mathrm{rad}}\Psi^{*}_{\mathrm{ref}}\right]\right|, \tag{5.13}$$

where X and M are the arbitrary real average value and the real modulation depth at the position (x, y) on the impedance surface; we assume the surface lies in the xy plane. The radiated (object) wave results from the scattering of the surface wave from the modulated surface, given by $\Psi_{\mathrm{obj}}|(\Psi_{\mathrm{ref}}|^2$.

For the previous 1D case in 5.7, the hologram is formed by the interference pattern of a 1D surface wave and a plane wave propagating at θ_0 degrees from the normal direction, resulting in the sinusoidally modulated surface reactance [44, 49]. The relationship between θ_0 and the periodicity of M follows exactly the relationship of a 1D LWA, with the modulated surface reactance described in Ref. [44]. Assuming that we would like to have a pencil beam in a specific direction θ_0 the interference between the fields associated with the surface wave (i.e., $\Psi_{\mathrm{surf}} = \Psi_{\mathrm{ref}} = e^{-j\beta r}$) and the plane-wave radiation pattern (i.e., $\Psi_{\mathrm{rad}} = e^{-jkx\sin(\theta_{\mathrm{L}})+j\varphi}$) on the metasurface will define a hologram pattern, where β is the propagation constant of the surface wave (which can be obtained from the rigorous eigenmodal analysis [34–36, 42]), r is the radial distance from the antenna, and θ_{L} is the beam angle in the xz plane, φ is an arbitrary phase offset that sets the impedance values at the origin. A scalar surface impedance can therefore be defined as the interference pattern between these two waves, as shown in Eq. 5.13. In this design, X in Eq. 5.13 is set to the average impedance value for the patch geometry, and M in Eq. 5.13 is set to span the entire available impedance range of geometry of choice. An antenna gain of 20 dBi is measured with a beam angle of 60 degrees [41], as predicted by the holography theory. Sampling the hologram impedance function at each unit cell center gives the local impedance value and hence the geometry and size of local metal patch.

The generation of arbitrary polarization requires an *anisotropic* (tensor) and inhomogeneous metasurface. Assume that a uniform and anisotropic metasurface lies in the xy plane, the averaged tangential electric field at the surface $\bar{\mathbf{E}}|_{z=0^+}$ is related to the averaged induced surface current $\bar{\mathbf{J}}_{\mathbf{s}}$ as: $\bar{\mathbf{E}}|_{z=0^+} = \bar{\bar{\mathbf{Z}}} \cdot \left(\hat{\mathbf{z}} \times \bar{\mathbf{H}}\right)|_{z=0^+} = \bar{\bar{\mathbf{Z}}} \cdot \bar{\mathbf{J}}_s$, where the surface impedance dyad $\bar{\bar{\mathbf{Z}}}$ is written in the matrix form as [42, 43]:

$$\begin{pmatrix} E_x \\ E_y \end{pmatrix} = \begin{pmatrix} Z_{xx} & Z_{xy} \\ Z_{yx} & Z_{yy} \end{pmatrix} \begin{pmatrix} J_x \\ J_y \end{pmatrix} \tag{5.14}$$

In the absence of loss, the impedance matrix $\overline{\overline{\mathbf{Z}}}$ is anti-Hermitian (i.e., $\overline{\overline{\mathbf{Z}}} = -\overline{\overline{\mathbf{Z}}}^{*T}$), and the reciprocity requires that $\overline{\overline{\mathbf{Z}}}$ be purely imaginary, which implies that $Z_{xy} = Z_{yx}$ The tensor surface impedance is constructed from the outer product of the expected surface current vectors and the desired outgoing electric field vectors. The outer product generalizes the simple multiplicative pattern described earlier. In the tensor case, the desired radiated vector wave term $\bar{\mathbf{E}}_{\text{rad}}|\bar{\mathbf{J}}_{\text{surf}}|^2$ is obtained from the scattering of a vector surface wave from a modulated tensor impedance surface, where $|\bar{\mathbf{J}}_{\text{surf}}|^2 = \bar{\mathbf{J}}^{\dagger}_{\text{surf}} \cdot \bar{\mathbf{J}}_{\text{surf}}$. When illuminated by the reference surface wave, a modulated tensor impedance proportional to $\bar{\mathbf{E}}_{\text{rad}} \otimes \bar{\mathbf{J}}^{\dagger}_{\text{surf}}$ creates this radiated vector wave term: $\overline{\overline{\mathbf{Z}}} \cdot \bar{\mathbf{J}}_{\text{surf}} \propto \bar{\mathbf{E}}_{\text{rad}} \otimes \bar{\mathbf{J}}^{\dagger}_{\text{surf}} \cdot \bar{\mathbf{J}}_{\text{surf}} = \bar{\mathbf{E}}_{\text{rad}} \cdot |\bar{\mathbf{J}}_{\text{surf}}|^2$ (the dagger represents the Hermitian conjugate). For an energy conserving and reciprocal metasurface, the impedance tensor must additionally be made anti-Hermitian and pure imaginary as:

$$\overline{\overline{\mathbf{Z}}} = j \begin{pmatrix} X & 0 \\ 0 & X \end{pmatrix} + j \frac{M}{2} \text{Im} \left[\bar{\mathbf{E}}_{\text{rad}} \otimes \bar{\mathbf{J}}^{\dagger}_{\text{surf}} - \bar{\mathbf{J}}_{\text{surf}} \otimes \bar{\mathbf{E}}^{\dagger}_{\text{rad}} \right] \tag{5.15}$$

where the diagonal X in the impedance matrix binds the TM surface wave, while the modulated impedance components scatter the vector surface wave into the desired vector outgoing wave. The circularly polarized flat antenna using the holography concept and the anisotropic metasurface have been demonstrated in the recent works [45], of which the linearly polarized sources were used to generate the circularly polarized radiation through the transformation via the metasurface hologram. The designs were based on analytical results, which agree excellently with experimental results. The losses have been seen to be very low, due to the subwavelength dimension of the unit cell that implies weak currents on the antenna surface. Most importantly, the metasurface-based holographic antenna can be made extremely flat, light weight ($<$1 kg), low cost, and built with the PCB processes. With the advent in the metasurface technology, the robust design for the surface impedance of unit cells can be developed. Several approaches have been proposed to design and characterize the surface impedance dyad of metasurfaces with great accuracy [48].

Finally, we notice that using reconfigurable metasurface in conjunction with the principles of holography would allow dynamically steering the beam for high-gain and broadband antennas, while not having any mechanically moving parts [50–54]. In 2013, Kymeta Corporation has demonstrated the prototype of bidirectional high-speed Internet connectivity using an active metasurface antenna [50]. Here the reconfigurability can be achieved by using active elements that can be individually tuned to generate the dynamic hologram.

5.6 Conclusions

Planar MTM technologies manipulating the electromagnetic waves, somehow similar to the transformation optics, may suggest new types of efficient beam-forming and beam-shaping techniques for the modern wireless systems. This chapter has presented some of the most recent and representative innovations in the planar TL-MTM or metasurface antenna based on the microstrip implementation. Their operation principles basically rely on tailoring the propagation (phase) constant and impedance by using the homogenized, subwavelength reactive elements in an effective medium. Further, the tunability in phase velocity, propagation constant, and impedance can be achieved with the combinational responses associated with the periodic nature of MTM/metasurface structure and with the loaded lumped components (i.e., diodes, varactors, switches, and transistors) in the unit cell. This enables the beam-forming and beam-steering antennas to be made thinner, lighter weight, less expensive, and requiring less power and having enhanced bandwidth than the conventional alternatives; for instance, phased arrays typically include multiple active antennas with costly phase shifters, amplifiers, and the complex feeding structure. The novel MTM and surface antenna technologies, being capable of dynamically transforming the guided waves to the desired radiation patterns, show great potential to serve as beam formers and beam steerers for the next-generation wireless technologies.

References

1. Sihvola, A. (2007). Metamaterials in electromagnetics, *Metamaterials*, **1**(1), 2–11.
2. Smith, D. R., Padilla, W. J., et al. (2000). Composite medium with simultaneously negative permeability and permittivity, *Phys. Rev. Lett.*, **84**(18), 4184–4187.
3. Shelby, R. A., Smith, D. R. et al. (2001). Experimental verification of a negative index of refraction, *Science*, **292**(5514), 77–79.
4. Yen, T. J., Padilla, W. J., et al. (2004). Terahertz magnetic response from artificial materials, *Science*, **303**(5663), 1494.
5. Caloz, C., and Itoh, T. (2005). *Electromagnetic Metamaterials: Transmission Line Theory and Microwave Applications* (John Wiley & Sons).
6. Eleftheriades, G. V., Iyer, A. K., and Kremer, P. C. (2002). Planar negative refractive index media using periodically LC loaded transmission lines, *IEEE Trans. Microwave Theory Techniques*, **50**(12), 2702–2712.
7. Engheta, N., and Ziolkowski, R. W. (Eds.) (2006). *Metamaterials: Physics and Engineering Explorations* (John Wiley & Sons).
8. Lai, A., Itoh, T., and Caloz, C. (2004). Composite right/left-handed transmission line metamaterials, *IEEE Microwave Mag*,, **5**(3), 34–50.
9. Kuylenstierna, D. Vorobiev, A., Linner, P., and Gevorgian, S. (2006). Composite right/left handed transmission line phase shifter using ferroelectric varactors, *IEEE Microwave Wireless Component Lett.*, **16**(4), 167–169.
10. Tseng, C. H., and Itoh, T. (2006, June). In *IEEE MTT-S International Microwave Symposium Digest, 2006* (IEEE), pp. 931–934.
11. Eleftheriades, G. V., Grbic, A., and Antoniades, M. (2004, June). Negative-refractive-index transmission-line metamaterials and enabling electromagnetic applications. In *Antennas and Propagation Society International Symposium 2004, IEEE APS-S International* (IEEE), pp. 1399–1402.
12. Eleftheriades, G. V. (2009). EM transmission line metamaterials, *Mater. Today*, **12**(3), 30–41.
13. Liu, L., Caloz, C., and Itoh, T. (2002). Dominant mode leaky-wave antenna with backfire-to-endfire scanning capability, *Electron. Lett.*, **38**(23), 1414–1416.
14. Allen, C. A., Leong, K. M., and Itoh, T. (2006, June). 2-D frequency-controlled beam-steering by a leaky/guided-wave transmission line

array. In *IEEE MTT-S International Microwave Symposium Digest, 2006* (IEEE), pp. 457–460.

15. Lim, S., Caloz, C., and Itoh, T. (2004). Metamaterial-based electronically controlled transmission-line structure as a novel leaky-wave antenna with tunable radiation angle and beamwidth, *IEEE Trans. Microwave Theory Techniques,* **52**(12), 2678–2690.
16. Casares-Miranda, F. P., Camacho-Peñalosa, C., and Caloz, C. (2006). High-gain active composite right/left-handed leaky-wave antenna, *IEEE Trans. Antennas Propag.*, **54**(8), 2292–2300.
17. Oliner, A. A. (1993). Leaky-wave antennas. In *Antenna Engineering Handbook,* 3rd ed. John, R. C. (Ed.) (McGraw-Hill).
18. Huang, L., Chiao, J., and Lisio, P. (2000). An electronically switchable leaky wave antenna, *IEEE Trans. Antennas Propag.,* **48**(11), 1769–1772.
19. Maheri, H., Tsutsumi, M., and Kumagi, N. (1988). Experimental studies of magnetically scannable leaky-wave antennas having a corrugated ferrite slab/dielectric layer structure, *IEEE Trans. Antennas Propag.* **36**(7), 911–917.
20. Pozar, D. M. (2009). *Microwave Engineering* (John Wiley & Sons).
21. Wong, T. T. (1993). *Fundamentals of Distributed Amplification* (Artech House).
22. Mori, K., and Itoh, T. (2008, October). Distributed amplifier with CRLH-transmission line leaky wave antenna. In *38th European Microwave Conference, 2008* (IEEE), pp. 686–689.
23. Wu, C.-T. M., and Itoh, T. (2009, December). A re-radiating CRLH-transmission line leaky wave antenna using distributed amplifiers. In *Asia Pacific Microwave Conference, 2009* (IEEE), pp. 1998–2001.
24. Wu, C.-T. M., Dong, Y., Sun, J. S., and Itoh, T. (2012). Ring-resonator-inspired power-recycling scheme for gain-enhanced distributed amplifier-based CRLH-transmission line leaky wave antennas, *IEEE Trans. Microwave Theory Techniques,* **60**(4), 1027–1037.
25. Wu, C.T. M., and Itoh, T. (2011, October). Combined gain-enhanced power recycling feedbacks for distributed amplifier-based crlh-leaky wave antennas. In *41st European Microwave Conference (EuMC), 2011* (IEEE), pp. 499–502.
26. Wu, C.-T. M., and Itoh, T. (2011, June). Gain-enhanced distributed amplifier-based CRLH-leaky wave antenna for quasi-resonant power recycling scheme. In *IEEE MTT-S International Microwave Symposium Digest (MTT), 2011* (IEEE), pp. 1–4.

27. Wu, C.T. M., and Itoh, T. (2012, May). Dual-fed distributed amplifier-based CRLH-leaky wave antenna for gain-enhanced power combining. In *IEEE MTT-S International Microwave Workshop Series on Innovative Wireless Power Transmission: Technologies, Systems, and Applications (IMWS), 2012* (IEEE), pp. 87–90.
28. Wu, C.-T. M., and Itoh, T. (2011, April). CRLH-transmission line leaky wave antennas integrated with distributed amplifiers with power recycling feedback scheme. In *Proceedings of the 5th European Conference on Antennas and Propagation (EUCAP)* (IEEE), pp. 3901–3904.
29. Wu, C.-T. M., and Itoh, T. (2012, December). A High Gain Active Antenna Array using dual-fed distributed amplifier-based CRLH-leaky wave antennas. In *Asia-Pacific Microwave Conference Proceedings (APMC), 2010* (IEEE).
30. Pendry, J. B., Schurig, and D., Smith, D. R. (2006). Controlling electromagnetic fields, *Science*, **312**(5781), 1780–1782.
31. Leonhartdt, U. (2006). Optical conformal mapping, *Science*, **312**(5781), 1777–1780.
32. Chen, H., Chan, C. T., and Sheng, P. (2010). Transformation optics and metamaterial, *Nat. Mater.*, **9**, 387–396.
33. Ma, H. F., and Cui, T. J. (2010). Three-dimensional broadband and broad-angle transformation-optics lens, *Nat. Commun.*, **1**, 124.
34. Holloway, C. L., Kuester, E. F., et al. (2012). An overview of the theory and applications of metasurfaces: the two-dimensional equivalents of metamaterials, *IEEE Antennas Propag. Mag.*, **54**(2), 10–35.
35. Holloway, C. L., Mohamed, M. A., et al. (2005). Reflection and transmission properties of a metafilm: with an application to a controllable surface composed of resonant particles, *IEEE Trans. Antennas Propag.*, **47**(4), 853–865.
36. Holloway, C. L. Kabos, P. et al., (2010). Realization of a controllable metafilm/metasurface composed of resonant magnetodielectric particles: measurements and theory, *IET Microwave, Antennas, Propag.*, **4**(8), 1111–1122.
37. Chen, P. Y., and Alù A. (2011). Mantle cloaking using thin patterned metasurfaces, *Phys. Rev. B*, **84**(20), 205110.
38. Padooru, Y. R., Yakovlev, A. B., Chen, P. Y., and Alù A. (2012). Analytical modeling of conformal mantel cloaks for cylindrical objects using sub-wavelength printed and slotted arrays, *J. Appl. Phys.*, **112**(3), 034907.

39. Padooru, Y. R. Yakovlev, Chen, P. Y. and Alù A. (2012). Line-source excitation of realistic conformal metasurface cloaks, *J. Appl. Phys.*, **112**(10), 104902.
40. Checcacci, P., Russo, V., and Scheggi, A. (1970). Holographic antennas, *IEEE Trans. Antennas Propag.*, **18**(6), 811–813.
41. ElSherbiny, M. Fathy, A. E., et al. (2004). Holographic antenna concept, analysis and parameters, *IEEE Trans. Antennas Propag.*, **52**(3), 830–839.
42. Fong, B. H., Colburn, J. S., et al. (2010). Scalar and tensor holographic artificial impedance surfaces, *IEEE Trans. Antennas Propag.*, **58**(10), 3212–3221.
43. Sievenpiper, D., Colburn, J., et al. (2005, July). Holographic artificial impedance surfaces for conformal antennas, *Antennas and Propagation Society International Symposium, 2005* (IEEE) pp. 256–259.
44. Patel, A. M., and Grbic, A. (2011). A printed leaky-wave antenna based on a sinusoidally-modulated reactance surface, *IEEE Trans. Antennas Propag.*, **59**(6), 2087–2096.
45. Minatti, G., Maci, S., et al. (2012). A circularly-polarized isoflux antenna based on anisotropic metasurface, *IEEE Trans. Antennas Propag.*, **60**(11), 4998–5009.
46. Minatti, G., Caminita, F., et al. (2011). Spiral leaky-wave antennas based on modulated surface impedance, *IEEE Trans. Antennas Propag.*, **59**(12), 4436–4444.
47. Eleftheriades, G. V., and Wong, A. M. H. (2008). Holography-inspired screens for sub-wavelength focusing in the near field, *IEEE Microwave Wireless Component Lett.*, **18**(4), 236–238.
48. Patel, A. M., and Grbic, A. (2013, June). Transformation electromagnetics devices using tensor impedance surface. In *IEEE MTT-S International Microwave Symposium Digest (IMS), 2013* (IEEE), pp. 1–4.
49. Chen, P. Y., Farhat, M., et al. (2014). Infrared beam-steering using acoustically modulated surface plasmons over a graphene monolayer, *J. Opt.*, **16**(9), 094008.
50. Bleicher A. (2013, May). Kymeta demos first ever satellite link with metamaterial antenna, *IEEE Spectrum*.
51. Johnson, M., Bowen, P., et al. (2014). Discrete-dipole approximation model for control and optimization of a holographic metamaterial antenna, *Appl. Opt.*, **53**(25), 5791–5799.
52. Johnson, M., Brunton, S. L., et al. (2014, August). Sidelobe canceling on a reconfigurable holographic metamaterial antenna, *International*

Conference on Electromagnetics in Advanced Applications (ICEAA), 2014, pp. 806–809.

53. Lipworth, G., Mrozack, A., et al. (2013). Metamaterial apertures for coherent computational imaging on the physical layer, *J. Opt. Soc. Am. A*, **30**(8), 1603–1612.
54. Hunt, J., Driscoll, T., et al. (2013). Metamaterial apertures for computational imaging, *Science*, **339**(6117), 310–313.

Chapter 6

Metasurfaces for Extreme Light Manipulation and Wave Control

Nasim Mohammadi Estakhri and Andrea Alù
Department of Electrical and Computer Engineering, University of Texas at Austin, 1 University Station, Austin, TX 78712, USA
estakhri@utexas.edu, alu@mail.utexas.edu

This chapter aims at providing a comprehensive insight into recent developments and applications of gradient metasurfaces to control and engineer the propagation of electromagnetic waves. We explore various aspects of this technology, starting from miniaturized metasurface building blocks at the lowest level, moving to primary optical elements for radiation patterning, and ultimately incorporating graded metasurfaces into more complex optical devices.

6.1 Introduction

Light, the narrow visible portion of the electromagnetic spectrum to which the human eye is sensitive, has fascinated mankind since ancient times, and the interest in controlling its propagation draws back to well before the development of modern electromagnetic theory. Early scientists were able to build simple optical elements

Transformation Wave Physics: Electromagnetics, Elastodynamics, and Thermodynamics
Edited by Mohamed Farhat, Pai-Yen Chen, Sebastien Guenneau, and Stefan Enoch

ISBN 978-981-4669-95-5 (Hardcover), 978-981-4669-96-2 (eBook)
www.panstanford.com

such as lenses, prisms, and mirrors to efficiently collimate, split, or redirect rays of light. The geometrical shape of these basic elements was found to be of critical importance to manipulate the light path, for example, a piece of glass would converge optical rays if carved in a convex shape, while the same material diverges light when forming a concave lens. Propagation inside different media has also been known to create distinct effects. While a glass globe filled with water shows magnifying effects, the same shape is transparent when filled with air. These, among many other empirical rules, have been developed to manipulate reflection, refraction, and scattering of light beams at a large scale and evolved through years of experimental and theoretical scientific efforts. Centuries later, similar concepts are still applied to the design of conventional optical elements, and have been scaled up to microwave and radio-frequency (RF) devices such as reflector antennas and lenses.

The major progress in electromagnetic wave theory started following the establishment of Maxwell's equations in the 19th century. Governing all forms of electromagnetic waves, Maxwell's equations enable studying wave–matter interactions at spatial ranges even smaller than the wavelength scale and explain the behavior of electric and magnetic fields in relation to the properties of the background medium. This field gradually evolved toward engineering the structures' shape, orientation, and composition to enable larger degrees of molding of the electromagnetic response.

The first notable example of radiation patterning using material engineering was proposed in 1905 by the Nobel laureate Karl Ferdinand Braun. Braun successfully transmitted a directional signal from three precisely arranged antennas, assembling the earliest phased array antenna. As he explained in his Nobel lecture, through designing the geometrical position and the relative phase between the three antennas, the emerging wave from the system exhibits a *preferred* direction for propagation in free space [1]. In another important example, also published in 1905, Robert W. Wood showed that a planar configuration may act as a converging or diverging lens if it holds a radial refractive index distribution [2]. Under this condition, the outgoing wave front from the structure is determined by varying the material distribution in the propagation medium rather than carving the element itself.

Following these early proposals, numerous methods have been introduced and examined to mold the near- or far-field electromagnetic radiation in a variety of distinct fashions, aimed at different applications. Wave front shaping through propagation inside engineered structures is amongst the most popular techniques in this context. In this technique, phase and polarization variations are gradually accumulated as the wave propagates through the component. The outgoing signal, thus, exhibits the desired pattern (in the near field or far field) by properly engineering the propagation path inside a finite volume. Conventional optical elements, such as simple gradient lenses, wave plates, and also transformation optics (TO)-based components fall into this category [3, 4]. At least theoretically, there is a broad spectrum of applications over which these methods can be applied to optimally shape the radiation pattern. However, in practice, their applicability in modern devices is challenged either because of the large size of the elements compared to the wavelength, which is undesirable for integration purposes, or due to the complicated requirements on the material distribution. In addition, natural materials exhibit only a limited range of electromagnetic properties at specific frequencies of interest. As a result, to grant full control on wave propagation in a compact, integrated platform, it is necessary to incorporate *metamaterials* in TO-based devices. Metamaterials are artificial structures in which the desired effective electromagnetic properties can be achieved through engineering the materials' constituents (typically referred to as *meta-atoms*). In general, meta-atoms are operating near resonance and have subwavelength dimensions, which are inevitably associated with substantial loss and challenging 3D fabrication at shorter wavelengths, impacting the practicality of such designs.

Another common mechanism to generate an arbitrary wave distribution is to exploit the interference from multiple, discrete or continuous, radiation sources. There is a broad range of devices proposed in the literature that work based on interference mechanisms, including phased array antennas, impedance surfaces, holograms, near-field plates, diffractive optical elements (beam shapers, beam splitters, gratings, etc.), and gradient metasurfaces [5–11]. Quite distinct from TO-based wave control, here, the appropriate spatial distribution of secondary sources are responsible for creating the desired outgoing wave front in the near or far field, depending on the spatial resolution of the component. Conventional devices implemented based on this approach are typically large compared to the operating wavelength or alternatively, the efficiency of conversion to the desired beam is limited. Gradient metasurfaces, however, add an additional near-field effect to the typical gratings and impedance surface designs, which can, in principle, eliminate such trade-off. Metasurfaces are the planarized version of metamaterials, providing strong, surface-confined interactions with the electromagnetic waves. With an appropriate design, deeply subwavelength spatial resolution of the secondary sources (surface scatterers) can be implemented in a gradient metasurface to circumvent the large-size and low-efficiency issues, at the same time. Several ideas and proposals have been suggested and developed in this line of research, unveiling the unique properties and potential applications of gradient metasurfaces in wave management, which we will discuss later in this chapter.

In addition to these major categories, other solutions have been also discussed in the literature to provide radiation pattering under particular conditions. One interesting suggestion is to employ epsilon-near-zero (ENZ) metamaterials to tailor the radiation pattern of an arbitrary source [12]. ENZ structures, effectively isolate the excitation and radiation regions and the geometrical outline of the exit face characterizes the outgoing pattern in an efficient way. ENZ metamaterials, however, are largely mismatched to free space and it is challenging to get the electromagnetic energy into an ENZ component, except by lowering its refractive index. A solution to the impedance matching problem has been discussed in recent theoretical papers [13]. Moreover, the narrow operation bandwidth and high intrinsic losses in ENZ structures restrict their practicality in realistic wave front–shaping setups.

Our group has recently proposed and presented multiple ideas for efficient and versatile infrared (IR) and optical wave management in the framework of gradient metasurfaces. In principle, gradient metasurfaces can locally alter the electromagnetic boundary condition for the wave navigating through or reflecting from their surface. They can be considered the next generation of conventional impedance surfaces, in which an arbitrary *average surface impedance* is realized on an electromagnetically thin surface. Impedance surfaces are typically realized exploiting periodic surface patterns and induce a specific phase discontinuity on the outgoing wave, related to their effective surface properties. In a gradient metasurface, however, the impedance is locally controlled on the surface and the field discontinuity induced by the structure is varying along the lateral dimensions. Accordingly, the structure exhibits a *spatially dispersive surface impedance*, offering the possibility of locally manipulating the wave flow with unprecedented efficiencies.

Next, we look at this problem in a systematic way and, after deriving the essential requirements for a gradient metasurface to grant our desired level of wave manipulation, we provide a general scheme for efficient realization of arbitrary surface functions employing hybrid plasmonic-dielectric nanoresonators. Proper realization of gradient metasurfaces is in fact a major challenge along the way of utilizing their wave-shaping potential. To create the sense of variation of impedance along the surface, subwavelength scatterers with distinct, controllable scattering properties (namely scattering phase centers) are needed. This is particularly tricky at shorter wavelengths at which the interactions between subwavelength dielectric meta-atoms and impinging electromagnetic waves are naturally imperceptible as they are largely off-resonant, allowing only a narrow range of scattering tunability. The plasmonic behavior of a number of noble metals at IR and optical regimes offer a potential solution for this problem [14]. The localized oscillations of free electrons in plasmonic materials create shorter effective wavelengths compared to the incident light, shrinking the size of near-resonant meta-atom configurations. We exploit this property to create quasi-static nanoresonators which, in principle, can be as small as the limits of our fabrication technique rather than the effective plasmonic wavelength. Our idea is based on the juxtaposition of a plasmonic and a dielectric particle [15]. This solution relies on the nanocircuit paradigm [16], which states that a plasmonic subwavelength particle plays the role of an effective inductor while the dielectric particle acts as an effective capacitor, in analogy to a low-frequency lumped resonator. Exploitation of a quasi-static resonator, not only breaks down the spatial limits forced by the finite natural plasmonic wavelength in nanoantenna-based metasurfaces, but at the same time enables an intriguing freedom in managing the polarization state of the impinging and scattered waves. This property distinguishes the class of designs we explore here from those employing some sort of geometrical phase to attain the required scattering signature (i.e., Pancharatnam–Berry phase [17, 18]). Geometrical phase is acquired through local variations in the polarization state of the wave, associated with an analogous change in the scattering phase. For the maximum conversion efficiency, a rotated half-wave plate is typically employed as the phase element and the incident and scattered waves have opposite handedness. This effect may limit the scope of applications of the method if we aim for large levels of output power, however it may be used to our advantage in specific designs limited to circularly polarized excitations. It is worth mentioning that another typical challenge in employing geometrical phase approach is the additional limits on the spatial resolution and thickness of the device, induced by the choice of the artificial uniaxial medium (i.e., the wave plate). In practice, natural birefringent crystals are inapplicable in a space-variant structure in view of the implementation restrictions and their relative size. Consequently, artificial uniaxial structures such as thick dielectric gratings or rotated nanoantennas are typically employed for this purpose [19, 20]. The first choice, however, limits the minimum resolution of the structure as we want to create an *effective uniaxial response* inside each cell, and the latter exhibits low conversion efficiencies. Interestingly, these constraints may be circumvented by employing plasmonic-dielectric meta-atoms to realize efficient, deeply subwavelength wave plates as well.

With this tool in hand, in our recent works we suggested many potential applications for wave management over an ultrathin surface in integrated, low-profile optical devices [11, 15, 21–23]. Several advanced functionalities such as polarization splitting, photon management in photovoltaic (PV) systems, beam steering, conformal cloaking, and optical computing have shown to be feasible employing ultrathin graded metasurfaces. Our goal is to move toward a functionality-based approach, showing how specific operations may be achieved with proper local surface impedances, regardless of the frequency of operation or the choice of the surface.

Still, the specific ways of realizing the gradient surface influence the performance of the ultimate device regarding bandwidth, sensitivity, and efficiency. For instance, plasmonic-dielectric meta-atoms (which we utilize here to realize the aforementioned functionalities at IR and optical regimes) have remarkably low dispersion with negligible insertion loss. At microwave, employing a strongly resonant particle such as split-ring resonators (SRRs) can result in narrow-band operation but considerable sensitivity, while a multilayer patch configuration, for instance, can provide same functionalities over much broader bandwidths and with a higher level of stability. To improve the device performance, it is crucial to distinguish between the design limitations and those corresponding to the particularly chosen implementation technique.

In the following section, we review the general concept of wave front manipulation and radiation patterning based on point-by-point scattering management in graded metasurfaces. In Section 6.2 we discuss our recently proposed idea of hybrid metal-dielectric meta-atoms to realize various types of ultrathin gradient metasurfaces. The goal of this section is to provide a handy tool at optical and IR regimes to make the use and applicability of metasurfaces simpler and more efficient. We will quantitatively characterize main aspects of these new *building blocks* regarding polarization response, possibility of tuning, and frequency dispersion. In Section 6.3, we will show how graded metasurfaces enable primary radiation pattering functionalities as in ultrathin blazed gratings and lenses, and with very large efficiencies, implying that the large majority of the input energy can funnel into the desired outgoing wave front. Relying on the controllable polarization response of metal-dielectric blocks, we will furthermore discuss the possibility of constructing ultrathin, broadband polarizing beam splitters exploiting properly designed blazed gratings. Finally, with all these design tools at our disposal, in Section 6.4 we present a number of more advanced devices implemented with graded metasurfaces. We show how to systematically design optical systems and achieve complex functionalities such as cloaking and analog signal processing over thicknesses typically much smaller than the operation wavelength. The results reviewed here are applicable in various frequency ranges and may broaden the scope of applications of gradient metasurfaces to realize novel, compact IR and optical elements and efficient integrated photonic devices.

6.2 Metasurface Design and Synthesis

Extending the concepts of impedance surfaces and phased arrays from microwave and RFs to the IR and visible range, as discussed in the introduction, is possible with gradient metasurfaces. Metasurfaces are constituted by arrays of scatterers, typically with subwavelength dimensions, which can provide a spatially varying response to impart the desired distribution on the outgoing wave. As illustrated in Fig. 6.1, multiple scattering elements (which may differ in terms of geometry or constitutive materials) are arranged in form of a metasurface, constructing a nonuniform secondary source of the electromagnetic wave. Similar to the way a programmed circuit controls the properties of radiating antennas in a phased array configuration, a gradient metasurface is driven by the incident wave and the varying electromagnetic response of each element creates the desired induced surface currents controlling the local phase, polarization, and intensity. The model shown in Fig. 6.1 exemplifies this process by considering the case when an arbitrary incident wave illuminates a reflecting metasurface from the top and creates a set of secondary sources that re-radiate a convergent wave front.

Two main features make gradient metasurfaces particularly appealing compared to other alternative approaches. First, the entire wave manipulation process occurs over an ultrathin region. While interacting with the electromagnetic wave, a metasurface (and, in general, any impedance surface) imposes an abrupt discontinuity on the properties of the impinging wave navigating through or reflecting from its surface. Namely, the metasurface may create a jump in the phase, amplitude or polarization of the scattered field enforcing a unique boundary condition to the electric and/or magnetic fields, essentially different from that of an ordinary interface between two bulk materials. In a gradient metasurface, in addition, the space gradient of this phase discontinuity may produce an effective momentum along the surface, giving rise to anomalous reflection and refraction of the impinging wave simply employing an ultrathin textured surface [9]. Second, metasurfaces based on properly designed scattering elements can provide deeply subwavelength spatial resolution of the wave front shaping. The principles of operation of gradient metasurfaces is related to conventional frequency-selective surface (FSS)-based structures such as reflectarrays, transmitarrays, and flat lens antennas, which were originally proposed at microwave and RF regimes [24–27]. Aiming at far-field radiation pattering, the size of surface inclusions in a typical FSS-based configuration is around half of the operational wavelength, merely to achieve maximum phase variation arising around a conventional resonant structure. Such low transverse resolution in implementing the surface function affects the available degree of control on the field distribution in the proximity of the FSS. While similar far-field patterns can be produced exploiting different FSS structures, the near-field distribution is a strong function of the inclusions' geometry and constitutive materials and is mostly uncontrollable. Analogous effects are happening in diffractive optical elements and gratings, generally utilized in free-space optical systems. In the modern nanophotonic devices, however, we are looking for additional control on the wave propagation with length scales comparable to or smaller than the operation wavelength. In other words, the goal is to be able to focus or bend (in general to arbitrarily mold) an incoming wave over only a few nanometers after it emerged from the surface. Gradient metasurfaces make this objective feasible, given their high transverse resolution. Arrays of deeply subwavelength scatterers suitably arranged to form a gradient metasurface provide a point-by-point control on the properties of the reflected or transmitted waves, right on top of the surface. The high spatial resolution in implementing the surface function is the essence of full control on the wave propagation, which is attainable by employing proper surface elements. It is worth mentioning that high-resolution gradient impedance surfaces have also been recently introduced at microwave and RF to improve the performance of conventional FSS-based structures, such as planar microwave lenses and reflectarrays [28, 29].

Let us consider again the structure shown in Fig. 6.1, where a gradient metasurface is illuminated by an arbitrary incident field $(\mathbf{E}_i(\vec{r}), \mathbf{H}_i(\vec{r}))$ with an $e^{j\omega t}$ time variation. $\vec{r}$ denotes the space

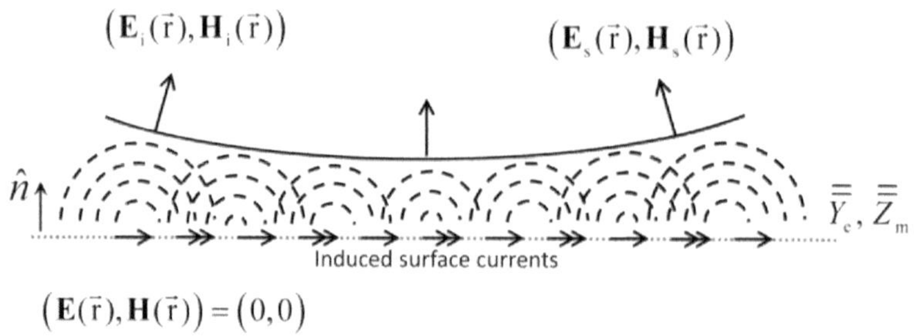

Figure 6.1 Principle of operation of graded metasurfaces. The metasurface consists of closely packed subwavelength inclusions forming a spatially varying electric/magnetic impedance surface. Dissimilar surface currents are induced along the boundary, whose emissions interfere to create the desired wave pattern in a reflecting setup.

position vector and ω is the angular frequency of operation. The goal is to design a reflecting surface which efficiently transforms the arbitrary distribution of incident field into the desired scattered wave profile $(\mathbf{E}_s(\vec{r}), \mathbf{H}_s(\vec{r}))$ in the upper half region of the space, while the total field is kept zero in the region below the metasurface. To sustain these arbitrary field distributions in both regions, the electromagnetic properties of the metasurface must be selected properly so that the fundamental boundary conditions of Maxwell's equations [30] are satisfied over the surface, that is, at $\vec{r} = \vec{r}_s$

$$\hat{n} \times (\mathbf{H}_i(\vec{r}_s) + \mathbf{H}_s(\vec{r}_s)) = \frac{1}{2}\bar{\bar{Y}}_e \cdot (\mathbf{E}_i(\vec{r}_s) + \mathbf{E}_s(\vec{r}_s))|_t$$

$$\hat{n} \times (\mathbf{E}_i(\vec{r}_s) + \mathbf{E}_s(\vec{r}_s)) = -\frac{1}{2}\bar{\bar{Z}}_m \cdot (\mathbf{H}_i(\vec{r}_s) + \mathbf{H}_s(\vec{r}_s))|_t \qquad (6.1)$$

Here $\hat{n}$ indicates the local normal unit vector of the surface as shown in Fig. 6.1 and we directly expressed the surface currents in terms of the metasurface electric admittance tensor $\bar{\bar{Y}}_e$ and the magnetic impedance tensor $\bar{\bar{Z}}_m$, through an appropriate average transition boundary condition [31]. Subscript t refers to the tangential component of the fields on the surface. Equation 6.1 represents the essential requirements at the metasurface boundary, relating the local electromagnetic properties of the surface to the desired distribution of the scattered and incident waves. Once the required admittance and impedance of the surface are obtained from Eq. 6.1, the metasurface elements must be designed to locally create these quantities, which in turn guarantees the optimal conversion of the incident pattern to the scattered one. In principle, for arbitrary incident and scattered field distributions, the required surface properties are tensorial and complex and the surface may contain polarization-dependent lossy and gain components along with lossless reactive elements. Here and in the following sections, for simplicity of notation we will discuss the general class of metasurfaces with inherently no polarization coupling. Specifically, the polarization content of the incident and scattered fields are assumed to be identical and impedance and admittance matrices are diagonal. The concept we present, however, can be easily generalized to include polarization coupling following a similar tensorial formalism as we present here with essentially the same underlying physics [32]. We note that by including intrinsic polarization coupling in the surface (e.g., by locally rotating surface component) or magneto-electric coupling (i.e., employing chiral elements), we can provide additional degrees of freedom to the design requirements, possibly to avoid extreme values and minimize local loss/gain.

Without loss of generality, let us assume that the local normal unit vector on the surface is along the z direction, that is, $\hat{n} = \hat{z}$. Following Eq. 6.1 the required surface tensors equal [11]

$$\bar{\bar{Y}}_{\mathrm{e}} = \begin{pmatrix} -2\frac{\mathbf{H}_y}{\mathbf{E}_x} & 0 \\ 0 & 2\frac{\mathbf{H}_x}{\mathbf{E}_y} \end{pmatrix}, \quad \bar{\bar{Z}}_{\mathrm{m}} = \begin{pmatrix} 2\frac{\mathbf{E}_y}{\mathbf{H}_x} & 0 \\ 0 & -2\frac{\mathbf{E}_x}{\mathbf{H}_y} \end{pmatrix} \tag{6.2}$$

Here $(\mathbf{E}, \mathbf{H})_{x,y}$ refer to the x and y components of the total fields on the surface. We note that the zero field transmission in the $z < 0$ region requires $Z_{\mathrm{m}xx} = 4/Y_{\mathrm{e}yy}$ and $Z_{\mathrm{m}yy} = 4/Y_{\mathrm{e}xx}$, imposing a fundamental relation on the electric and magnetic properties of the metasurface to operate in reflection mode.

Ideally, to ensure 100% conversion efficiency to the desired scattered wave, the exact magnetic surface impedance and electric surface admittance profile must be realized at each point on the surface (Eq. 6.1 is position dependent). The presence of both tangential electric and magnetic currents enables independent control of the incident and scattered fields on the upper half region of space (shown in Fig. 6.1), while total fields are always zero in the lower region. At the low-frequency regime the acquired surface properties in Eq. 6.2 can be directly utilized to design electromagnetic metasurfaces and generate the desired scattering pattern. A combination of electrically and magnetically polarizable particles, for example, surface capacitors and inductors [33] or electric and magnetic wire dipoles [34], can realize the optimal surface properties. In general, proper loss and gain elements may be required in the surface to account for the real parts of $\bar{\bar{Y}}_{\mathrm{e}}$ and $\bar{\bar{Z}}_{\mathrm{m}}$. For specific incident and scattered wave profiles, however, the surface can be approximated by its reactive elements (see, for example, Refs. [33, 35]), while still maintaining high conversion efficiencies to the desired scattering pattern. This is possible as long as the incident and scattered waves overlap spatially on the surface, such as it is not necessary to absorb energy in one portion of the surface and re-radiate it in another portion.

At shorter wavelengths, and especially at IR and optical frequencies, as a result of the saturation of magnetic response in optical materials [36], this technique may become unsuitable. A thin metasurface, excited from the normal direction with a plane wave, shows negligible magnetic effects [37] and Eq. 6.2 cannot be directly implemented over these frequency ranges. Indeed, achieving strong magnetic effects is also challenging at microwaves and terahertz and it is typically associated with narrow operation bandwidths and large absorption [38–41]. To circumvent these limitations and to implement the derived equivalence conditions in Eq. 6.2 solely based on the electric response of the metasurface, a grounded slab configuration, as shown in Fig. 6.2, was recently proposed in Ref. [15]. The metasurface is assumed to show an electric surface admittance tensor of $\bar{\bar{Y}}_{\mathrm{eg}}$ with negligible magnetic response. From here on, the subscript g refers to the properties of the grounded setup. A perfect mirror is placed at distance d_{s} from the substrate to ensure zero transmission and purposely break the configurations symmetry. To effectively implement the response calculated in Eq. 6.2, we first relate the local impedance and admittance elements to the scattering parameters (reflection and transmission coefficients) of a periodic surface, illuminated with a normally incident plane wave [42]

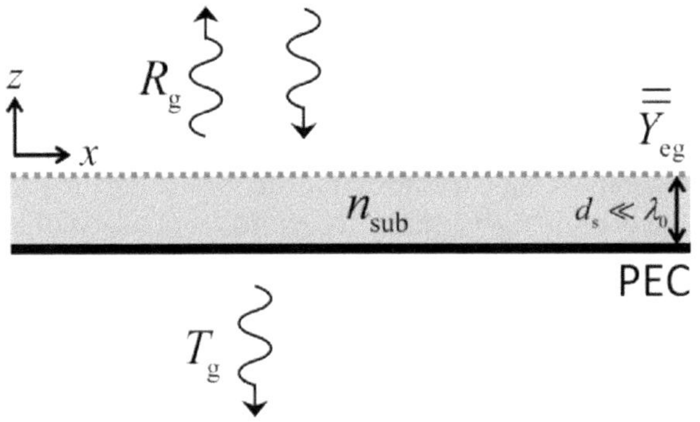

Figure 6.2 Symmetry breaking in an electric metasurface. The metasurface consists of electrically polarizable particles coupled to a perfect mirror with subwavelength separation. Adapted from Ref. [11].

$$
\begin{aligned}
R_{\mathrm{TE}} &= -\frac{2\left(\eta_0^2 Y_{\mathrm{e}yy} - Z_{\mathrm{m}xx}\right)}{\left(2+\eta_0 Y_{\mathrm{e}yy}\right)\left(2\eta_0 + Z_{\mathrm{m}xx}\right)},\\
T_{\mathrm{TE}} &= -\frac{\left(-2+\eta_0 Y_{\mathrm{e}yy}\right)}{\left(2+\eta_0 Y_{\mathrm{e}yy}\right)} + \frac{2\left(\eta_0^2 Y_{\mathrm{e}yy} - Z_{\mathrm{m}xx}\right)}{\left(2+\eta_0 Y_{\mathrm{e}yy}\right)\left(2\eta_0 + Z_{\mathrm{m}xx}\right)};\\
R_{\mathrm{TM}} &= -\frac{2\left(\eta_0^2 Y_{\mathrm{e}xx} - Z_{\mathrm{m}yy}\right)}{\left(2+\eta_0 Y_{\mathrm{e}xx}\right)\left(2\eta_0 + Z_{\mathrm{m}yy}\right)},\\
T_{\mathrm{TM}} &= -\frac{\left(-2+\eta_0 Y_{\mathrm{e}xx}\right)}{\left(2+\eta_0 Y_{\mathrm{e}xx}\right)} + \frac{2\left(\eta_0^2 Y_{\mathrm{e}xx} - Z_{\mathrm{m}yy}\right)}{\left(2+\eta_0 Y_{\mathrm{e}xx}\right)\left(2\eta_0 + Z_{\mathrm{m}yy}\right)}
\end{aligned}
\tag{6.3}
$$

Subscripts TE (transverse electric) and TM (transverse magnetic) refer to plane waves with the electric field polarized along (normal to) the y direction, and η_0 is the free-space characteristic impedance. In analogy, the scattering parameters of the grounded metasurface in Fig. 6.2 are found as

$$
\begin{aligned}
R_{\mathrm{g,TE}} &= \frac{1-\eta_0 Y_{\mathrm{eg},yy} + j n_{\mathrm{sub}} \cot\left(n_{\mathrm{sub}} k_0 d_{\mathrm{s}}\right)}{1+\eta_0 Y_{\mathrm{eg},yy} - j n_{\mathrm{sub}} \cot\left(n_{\mathrm{sub}} k_0 d_{\mathrm{s}}\right)}, \quad T_{\mathrm{g,TE}} = 0;\\
R_{\mathrm{g,TM}} &= \frac{1-\eta_0 Y_{\mathrm{eg},xx} + j n_{\mathrm{sub}} \cot\left(n_{\mathrm{sub}} k_0 d_{\mathrm{s}}\right)}{1+\eta_0 Y_{\mathrm{eg},xx} - j n_{\mathrm{sub}} \cot\left(n_{\mathrm{sub}} k_0 d_{\mathrm{s}}\right)}, \quad T_{\mathrm{g,TM}} = 0;
\end{aligned}
\tag{6.4}
$$

Here n_{sub} refers to the substrate refractive index, k_0 is the free-space wavenumber, and $Y_{\mathrm{eg},xx}$ and $Y_{\mathrm{eg},yy}$ are the diagonal elements of the electric admittance tensor $\bar{\bar{Y}}_{\mathrm{eg}}$ for the metasurface shown in Fig. 6.2.

Investigating Eq. 6.3, it is evident that as the surface characteristic impedance and admittance satisfy the previously derived conditions of $Z_{mxx} = 4/Y_{eyy}$, $Z_{myy} = 4/Y_{exx}$, the transmission coefficients are always exactly zero under both polarizations. It is easy to check that for purely imaginary admittances, the reflection coefficients are unitary and their phase span the entire 360° range by varying the magnitude of impedance and admittance. The grounded metasurface, analogously, offers full coverage of the reflection phase through varying the magnitude of the surface electric admittance and ensures zero transmission. When the overall thickness of the grounded configuration is subwavelength, the whole system can be modeled as a single metasurface [43] and it can be easily shown that Eqs. 6.3 and 6.4 are equal under the general condition [11]

$$Y_{\text{eg},yy} = \frac{Y_{\text{e}yy}}{2} + j\frac{n_{\text{sub}}\cot\left(n_{\text{sub}}k_0 d_{\text{s}}\right)}{\eta_0},$$
$$Y_{\text{eg},xx} = \frac{Y_{\text{e}xx}}{2} + j\frac{n_{\text{sub}}\cot\left(n_{\text{sub}}k_0 d_{\text{s}}\right)}{\eta_0} \qquad (6.5)$$

Note that in this scenario we do not require magnetic impedances on the interface to achieve the requirements on the local scattering from the metasurface. Similar symmetry breaking effects can be achieved in a transmitting configuration by cascading multiple electric metasurfaces [21, 32].

This technique is general and it can be applied to achieve several wave-shaping functionalities as we discuss in a series of examples in the following sections. Essentially, if the arbitrary surface reactances in Eq. 6.5 can be synthesized by the metasurface (i.e., surface reactance/resistance should be controllable between $-\infty$ and ∞), we would be able to control the distribution of the scattered field with 100% conversion efficiency. In Section 6.2.1, we will discuss a general approach to synthesize these metasurfaces employing composite plasmonic meta-atoms [15]. During our analysis, the properties of the surface are assumed to be controllable in a continuous fashion, that is, on each point on the surface. Obviously, in practice, the ideal surface impedance attained from calculations must be discretized into a finite number of steps, which in turn may slightly reduce the manipulation efficiency. We will quantitatively examine and discuss the effects of surface granularity in Section 6.3.1.

We notice that the required surface admittance $\bar{\bar{Y}}_{eg}$ and the reflection phase R_g may also be estimated employing the approximate ray optics technique (i.e., geometrical optics), typically used to design large-area reflect- and transmitarrays [9, 25]. Under this approximation, it is assumed that each surface element serves as a secondary source (induced current) whose amplitude and phase of radiation are determined solely based on the excitation field distribution, and thus the reflection phase pattern from the metasurface will be directly imprinted on the scattering phase profile, that is, $\psi_s(\vec{r}_s) = \psi_i(\vec{r}_s) + \angle R(\vec{r}_s)$ ($\psi_s(\vec{r}_s)$ and $\psi_i(\vec{r}_s)$ are the scattering and incident phase profiles along the surface). For instance, to bend an incident plane wave into a particular direction, geometrical optics indicates that surface must provide a linear scattering phase profile, or to collimate a normally incident plane wave, the surface must provide a hyperboloidal phase of reflection along its surface. To explore these methods in more detail, Fig. 6.3 compares the exact approach and the approximate technique for a metasurface designed to deflect a normally incident TE polarized plane wave toward θ° relative to x axis and in the xz plane. The field distributions of the incident and scattered waves read

$$(\mathbf{E}_i(\vec{r}), \mathbf{H}_i(\vec{r})) = E_0 e^{jk_0 z}\left(\hat{y}, \frac{\hat{x}}{\eta_0}\right);$$

$$(\mathbf{E}_s(\vec{r}), \mathbf{H}_s(\vec{r})) = \frac{E_0}{\sqrt{\cos\theta}} e^{-jk_0 x \sin\theta} e^{-jk_0 z \cos\theta}\left(\hat{y}, \frac{-\hat{x}\cos\theta + \hat{z}\sin\theta}{\eta_0}\right) \tag{6.6}$$

With the exact field distributions in hand, ideal local surface reflection coefficient R_{TE} is directly calculated based on Eqs. 6.2 and 6.3, plotted in Fig. 6.3 (solid lines) for three deflection angles of $\theta = 20^\circ$, 45°, and 70°. In general, the ideal surface may have lossy and/or gain elements, which is the case for ideal plane wave deflection. We notice that the local reflection amplitude must attain nonunitary values along the surface and a balanced loss/gain structure provides the optimal wave conversion (100% efficiency). In addition, the ideal local phase spans the entire 360° variation along the metasurface. With the growing interest in the wide range of applications of metasurfaces, many research groups have recently focused on metasurface synthesis and design, and a wide

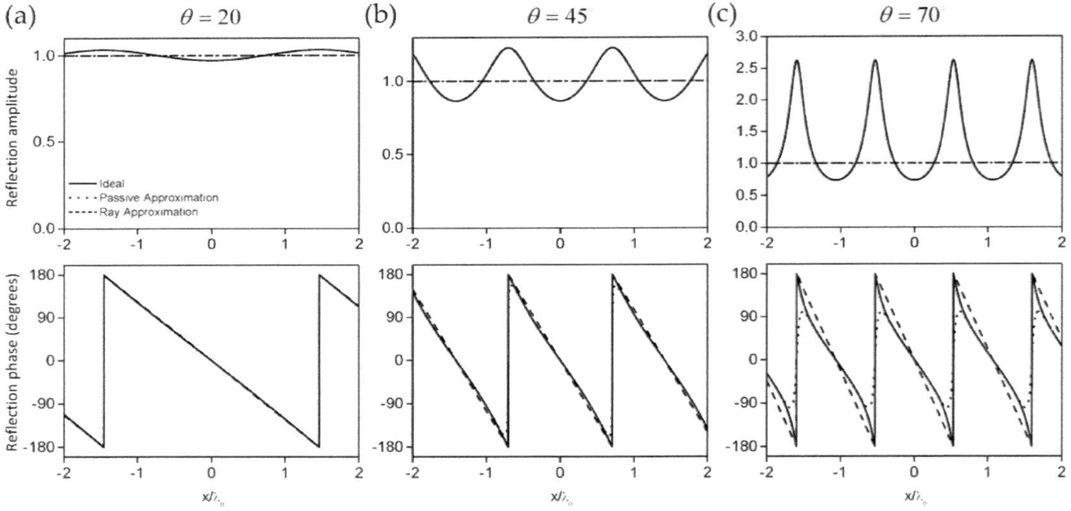

Figure 6.3 Comparison between exact and approximated local reflection coefficient in wave-bending metasurfaces. The metasurface is designed to reflect a normally incident TE plane wave toward (a) 20°, (b) 45°, and (c) 70° relative to the *x* axis. The solid lines show the exact solution, dotted lines show passive approximation by eliminating the real portion of surface admittance, and dashed lines show the ray approximation. The reflection coefficients are plotted along the lateral points on the surface, that is, the *x* axis.

range of surface elements have been introduced in the literature. To grant the full phase control, near-resonant configurations have been suggested from microwave to optical frequencies (and even for acoustic waves [44]). Some of the proposed designs include: miniaturized FSS elements [26, 29, 45, 46], rotated V-shaped nanoantennas [9], plasmonic and high index nanorod antennas [47–49], multilayer metallic printed surfaces [32], and hybrid surface resonators [15], which we look into in more detail in the following sections. Furthermore, amplitude modulation can be achieved by locally doping the constitutive materials to generate confined absorption or gain (e.g., using quantum dots), applicable to IR and optical designs. Low-frequency amplifiers and attenuators, such as transistors, diodes and resistive circuits, are also suitable candidates for microwave and RF amplitude modulation. We note, however, that unless active components are incorporated in the design, amplitude modulation is typically unfavorable as we lose a valuable portion of the input power without compensation. Accordingly, to simplify the metasurface synthesis, the goal is to maintain the lowest possible absorption. One approach to approximate the active surface is to preserve the exact local phase (solid line) while the reflection amplitude is enforced to one [11]. Alternatively, the ideal metasurface properties can be simplified to those of a lossless surface by neglecting the real portion of surface admittance, that is, $\mathrm{Re}[\bar{\bar{Y}}_\mathrm{e}] = 0$ [35]. The required surface reflections based on such passive approximation are shown with dotted lines in Fig. 6.3. Considering that only the reactive portion of the ideal surface components are used, the reflection amplitude is now always unitary, while the reflection phase is almost unaffected compared to the ideal solution. The dashed lines also show the predicted local phase based on the approximate ray approach. Following the constant phase profile of the incident wave and linear phase profile of the scattered wave along the surface, the local reflection phase is estimated as $\angle R(x) = -k_0 \sin\theta\ x$.

Inspecting the three methods shown in Fig. 6.3 we notice that if surface properties are not rapidly varying along the metasurface (i.e., in the wavelength scale), local reflection phase can be well approximated both by the ray approach and the passive approximation of the surface elements. Previous studies report around 95%

conversion efficiency numerically calculated for 41° and 45° wave bending [15, 49] employing the approximate technique. However, for extreme variations on the scattering wave front, for example, extreme angle bending or very near-field focusing, ray technique is not valid and to ensure high efficiencies, surface profile must be designed based on the exact profile derived from equivalence principle (solid and dotted lines) [11]. The passive approximations of the exact surface may be further optimized to maximize the conversion efficiency and reduce the coupling to unwanted waves. We note that the transmission coefficient is exactly zero for all three cases shown in Fig. 6.3. Following our discussions in this section, in the next part we will propose a general approach to implement arbitrary ultrathin metasurfaces, providing controllable local surface reactance and reflection phase with subwavelength lateral resolution even at IR and optical frequencies.

6.2.1 *Nanoresonators as Optical Phase Elements*

In Section 6.2, we illustrated that a complex beam-forming task can be simplified to the design of a spatially modulated reactive surface, and local capacitive and inductive elements are essential to ensure effective wave manipulation in such configuration. To realize the metasurface (as shown in Fig. 6.2), we get inspiration from RF concepts. One of the main reasons behind the rapid and widespread development of the microelectronic technology at radio frequencies is the modularization of lumped circuit elements. Basic building blocks such as chip resistors, capacitors, and inductors are readily available in this range and combined to synthesize complex functionalities, which subsequently build large operational circuits at radio frequencies. Optical and IR designs, on the other hand, are conventionally performed at the physical level where the complex light–matter interactions are directly exploited via different methods to achieve the desired effect.

Quite recently, a comprehensive nanocircuit paradigm has been proposed to bridge this gap and extend the conventional circuit theory to shorter wavelengths [16]. This paradigm allows to quantitatively describe subwavelength nanoparticles in terms of equivalent lumped circuit elements, proposing an ideal solution for what we are seeking for the gradient metasurfaces. As the wavelength shrinks, the contribution of the conduction current versus the displacement current in the material drops down rapidly and eventually becomes negligible. Based on this phenomenon, nanocircuit theory defines an intrinsic equivalent optical impedance for subwavelength particles as the ratio of the effective optical voltage $V = |\mathbf{E}|\, h$ across the particle and the flux of the electric displacement current $I = |J_\mathrm{d}|\, S$ flowing through it. In analogy to circuit theory, the voltage is related to the local electric field $\mathbf{E}$ across the particle and height of the nanoparticle along this field (h), while retardation effects are neglected in view of the small size of the structure (quasi-static approximation). Similarly, the displacement current density $J_\mathrm{d} = j\omega\varepsilon E$ integrated over the transverse cross section of the particle, S, sustains the total flux of the electric displacement current, which is in quadrature phase with the effective voltage when the material is lossless (i.e., real permittivities). Based on this concept, a dielectric nanoparticle, $\varepsilon > 0$, has a capacitive effective impedance while a plasmonic one, $\varepsilon < 0$, is characterized by an inductive effective impedance. In the presence of loss (i.e., complex permittivities) a resistive component is added to the impedance, taking absorption into account.

Alongside modeling the scattering response of quasi-static nanoparticles, the nanocircuit paradigm allows translating various low-frequency concepts to the IR and visible range. For instance, the input impedance is one of the characterizing parameters of RF antennas, typically utilized to match the antenna to arbitrary feeding networks and tuning its resonance frequency. In a similar fashion, nanocircuit theory allows defining an effective optical voltage, current, and impedance for nanoantennas (e.g., a plasmonic dipole) and adjust their scattering dispersion employing *nanoloads* [50].

The contribution of the nanocircuit modeling in realizing of gradient wave-shaping surfaces is evident now. Particles with different equivalent impedances need to be arranged properly on the surface to realize the electric surface reactance profiles derived in Eq. 6.5. These particles are inherently subwavelength in this quasi-static regime, guaranteeing a satisfactory transverse resolution on the surface. For the capacitive portion (i.e., $\mathrm{Im}[\bar{\bar{Y}}_\mathrm{eg}] > 0$), dielectric particles may be utilized, and for the inductive

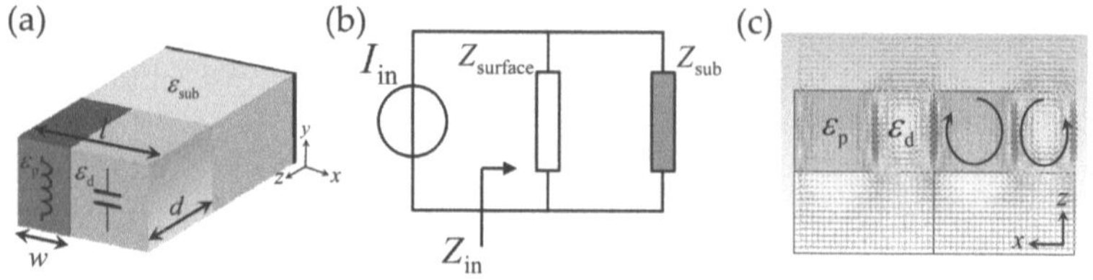

Figure 6.4 Spatial phase modulator. (a) Basic element of the metasurface made of alternated plasmonic (blue) and dielectric (orange) materials deposited on a grounded substrate layer (gray). The element lies in the *xy* plane and is invariant along the *y* axis. (b) Equivalent circuit model for the structure shown in (a). The metasurface layer is modeled by the appropriate shunt electric surface impedance $Z_{surface}$, and Z_{sub} includes the effect of the substrate and ground plane. (c) Magnetic field distribution in a sample periodic metasurface composed of the elements shown in (a) and excited with a plane wave (E_y, H_x) along the $-z$ direction. The opposite direction of rotation of the magnetic field indicates a reverse sign of local reactance in each portion.

elements ($\mathrm{Im}[\bar{\bar{Y}}_{eg}] < 0$), plasmonic nanoparticles are required, both widely available at the IR and visible range. To tailor the value of the impedance, the shape of nanoparticle (i.e., h and S) may be modified or different materials can be employed. Besides, nanoparticles may be combined to create more complex impedance responses. We explore here the idea of implementing lumped *nanoresonators* on the surface, realized with pairs of local inductors and capacitors connected in series/parallel. In the simplest format, Fig. 6.4a illustrates a 2D cubical nanoresonator ($l \ll \lambda_0$) consisting of two materials with permittivities $\varepsilon_p < 0$ and $\varepsilon_d > 0$, deposited on a grounded substrate. We will later apply these elements to implement a variety of gradient surfaces. Employing such hybrid particles to realize the required surface impedance, rather than a homogenous dielectric or plasmonic nanoparticle, brings out several advantages. First, the equivalent impedance of the particle cluster now depends on the ratio between the size of the inductive and capacitive elements (i.e., $w/(l - w)$) rather than their absolute geometrical size l [51]. This enables us to shrink the footprint of the resonator arbitrarily, basically as far as fabrication techniques allow us. Second, two materials would be essentially sufficient to sweep the entire impedance spectrum. Here, the surface impedance can be easily controlled through changing the filling ratio along the surface, rather than employing a range of different materials at different points, which is relatively impractical.

The electromagnetic response of the fundamental element shown in Fig. 6.4a can be accurately modeled employing simple transmission-line concepts together with the nanocircuit paradigm [15]. When arranged in a periodic fashion, these elements create an effective, homogenous surface impedance Z_{surface} associated to the corresponding impedance of each nanoparticle cluster. In this regard, if the flux of the electric displacement current is continuous between the two particles, that is, $\mathbf{E} = |\mathbf{E}|\,\hat{x}$, the particles form a series connection and $Z_{\text{surface}} = Z_{\text{p}}^{\text{s}} + Z_{\text{d}}^{\text{s}}$, in which $Z_{\text{p,d}}^{\text{s}}$ are equivalent series impedances of the plasmonic and dielectric portions. Alternatively, if analogous voltages are induced across the particles, that is, $\mathbf{E} = |\mathbf{E}|\,\hat{y}$, they form a parallel connection and $Z_{\text{surface}} = Z_{\text{p}}^{\text{p}}||Z_{\text{d}}^{\text{p}}$, where now $Z_{\text{p,d}}^{\text{p}}$ indicate the parallel equivalent impedance of the two particles.

For the 2D configuration considered here, these unit length impedances can be easily found as [15]

$$Z_{\text{p}}^{\text{s}} = w\left(j\omega d(\varepsilon_{\text{p}} - \varepsilon_0)\right)^{-1},\ Z_{\text{d}}^{\text{s}} = (l - w)\left(j\omega d(\varepsilon_{\text{d}} - \varepsilon_0)\right)^{-1};$$
$$Z_{\text{p}}^{\text{p}} = \left(j\omega w d(\varepsilon_{\text{p}} - \varepsilon_0)\right)^{-1},\ Z_{\text{d}}^{\text{p}} = \left(j\omega\,(l - w)\,d(\varepsilon_{\text{d}} - \varepsilon_0)\right)^{-1} \quad (6.7)$$

which are purely imaginary in the lossless case. Equation 6.4 already showed that if $Z_{\text{surface}} = (Y_{\text{surface}})^{-1}$ can be designed arbitrarily at each point on the surface, the local reflection phase is also arbitrarily controllable. With an embedded resonance in Z_{surface}, we expect to be able to efficiently control its magnitude by means of common optical dielectrics and metals. In fact, it is clear from Eq. 6.7 that if the total length of the element, l, is fixed, by simply changing the filling ratio of the plasmonic portion we move from a capacitive surface to an inductive surface, and a broad range of impedances is accessible. For a specific filling ratio, a resonance is achieved (i.e., $Z_{\text{surface}} = 0$ in series connection, and $Z_{\text{surface}} = \infty$ in the parallel case) around which a rapid variation of impedances and reflection phases are expected.

As an example of a practical implementation of such surface in the visible range ($\lambda_0 = 500$ nm), the equivalent surface impedance

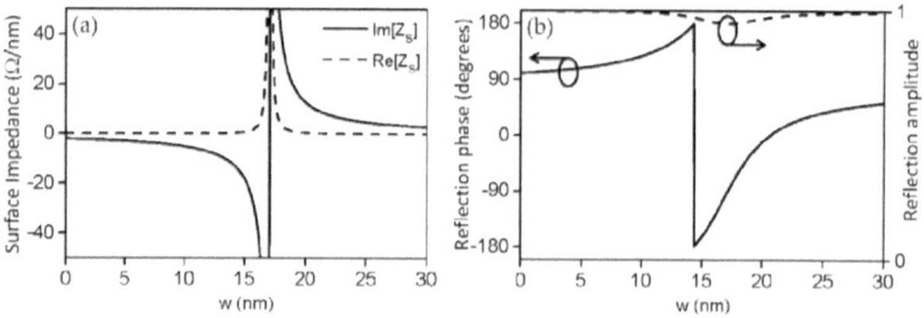

Figure 6.5 (a) Surface reactance and resistance per unit length as a function of the width of the plasmonic portion for the structure shown in Fig. 6.4a. Material and geometry specifications are included in the main text. (b) Corresponding reflection coefficient. Parallel combination of the nanoelements is considered (reprinted (figure) with permission from Ref. [15]. Copyright (2014) by the American Physical Society).

of this hybrid particle is plotted in Fig. 6.5a as a function of the width of the plasmonic portion. To construct the capacitive and inductive portions we select an optical semiconductor and silver as the constitutive materials, and particle dimensions are set at $l = \lambda_0/16 = 31.25$ nm and $d = 40$ nm. At optical frequencies silver follows a Drude-type dispersion, $\varepsilon_\mathrm{p} = \varepsilon_\infty - f_\mathrm{p}^2/f\,(f - j\gamma)$, with $\varepsilon_\infty = 5$, plasma frequency of $f_\mathrm{p} = 2175$ THz and collision frequency $\gamma = 4.35$ THz [52]. The real part of the permittivity of silver takes negative values in optical regime and at the same time it shows relatively low intrinsic loss, which makes it a suitable candidate for the inductive element. We note that due to the realistic losses included in this model, the equivalent inductor calculated from Eq. 6.7 is now complex with a resistive component. For the dielectric portion an optical semiconductor with relative permittivity of $\varepsilon_\mathrm{d} = 12$ [53], is employed. Semiconductors are particularly appealing at optical frequencies due to negligible loss, high refractive index, and the possibility of tuning the permittivity through proper doping.

Figure 6.5a, which is plotted for a parallel combination of nanoresonator elements (i.e., incident electric field is along the y axis and $Z_\mathrm{surface} = \left(Y_{\mathrm{eg},yy}\right)^{-1}$), demonstrates the capability of the proposed composite particle to achieve any arbitrary local impedance on the surface. Around the resonance, which occurs approximately at $w = 17$ nm in this example, both ranges of inductive and capacitive elements can be easily obtained by varying the filling ratio. It is particularly important that such wide range of variation of impedance is obtained over only 31.25 nm lateral dimension of the unit cell, 40 nm depth of the surface, and simply employing two materials. This implies that if embedded in a nonperiodic and gradient metasurface, these elements are capable of practically controlling the local impedance at each point, with an ultrathin depth along the direction of the propagation of wave (z axis). As the result of wave matter interaction over a surface, effect of loss is also negligible. Evident from Fig. 6.5a, except around the resonance point where an inevitable surface resistance is created, the hybrid particle is effectively reactive. We will specifically look at the effect of material absorption on the performance of the metasurface in Section 6.2.2.

The corresponding scattering coefficient, that is, $R_{\mathrm{g,TE}}$, when such impedance surface is loaded over a ground plane with $d_{\mathrm{s}} = 100$ nm spacing and $n_{\mathrm{sub}} = 1$, is plotted in Fig. 6.5b (setup shown in Fig. 6.2). Unsurprisingly, we successfully cover almost the entire 360° phase variation due to the artificial in-plane resonance created with the composite nanoparticle. This effect is clearly independent of the gap size between the metasurface and the back mirror because we don't rely on the propagation effects in this region to provide the required phase variation. Changing the gap distance will simply shift the entire phase curve [15]. The performance of the proposed metasurface regarding the covered spectrum of impedance, ultrathin thickness, spatial resolution, and efficiency, is far superior to other proposals at IR and optical frequencies [9, 45–47]. In the following section, we will refer to these surface elements as *spatial phase modulators*.

6.2.2 *Tunability, Frequency Dispersion, and Effect of Loss*

When a gradient metasurface is implemented based on the general class of conjoined nanoparticles introduced in the previous section, we theoretically predict very large conversion efficiency to the desired scattered wave. In practice, however, several parameters may affect the ideal performance of the metasurface, including loss, deviation from center design frequency, surface granularity, and fabrication defects. Here and in the following sections we will qualitatively look into these imperfections and nonidealities to estimate their effect on the performance of the phase modulator and gradient metasurfaces. In addition, we investigate the possibility of including tunability in the design, envisioning a broader set of applications for switchable and reconfigurable wave-manipulating optical elements.

Examining Eq. 6.7, there is an interesting possibility to add tunability to the introduced spatial phase modulator through controlling the material properties of the particle. This control may be attained in various manners such as varying temperature, external bias, intensity (when nonlinear materials are included), or doping. Figure 6.6a illustrates the effect of the dielectric permittivity on the phase response of the element. Dimensions and the inductive portion are kept unchanged compared to the example in Fig. 6.5, however, we assume that through one of the aforesaid methods we are capable of slightly varying the capacitive element, that is, the permittivity of the dielectric portion. Altering the nanocapacitor, the surface resonance frequency is expected to slightly shift. Indeed, as shown in Fig. 6.6a, reflection phase which is a direct function of the surface impedance (via Eq. 6.4) is moving along the frequency axis

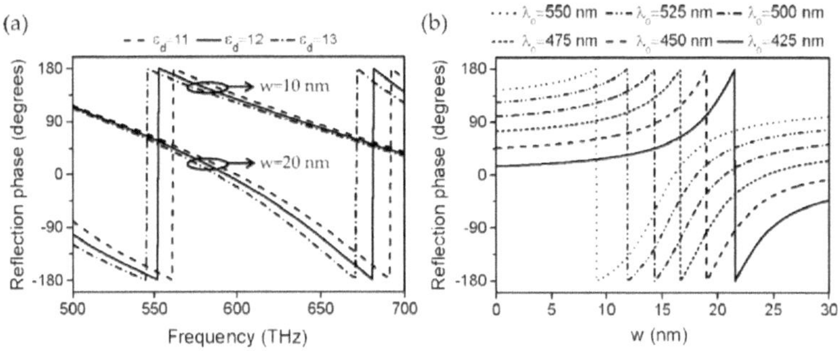

Figure 6.6 Tunability and intrinsic dispersion. (a) Reflection phase from a periodic metasurface with unit cells shown in Fig. 6.4a. Dielectric permittivity is swept from $\varepsilon_d = 11$ to $\varepsilon_d = 13$ to tune the frequency response. (b) Evolution of the reflection phase with the wavelength (reprinted (figure) with permission from Ref. [15]. Copyright (2014) by the American Physical Society).

for different values of ε_d. Two sample filling ratios are shown here to demonstrate the tunability of the operational frequency. We will discuss in Section 6.3.1 how such tunability may be incorporated in a beam deflecting surface to adjust the operational frequency of the surface.

The bandwidth of operation of a gradient metasurface implemented with hybrid nanoelements depends on the frequency dispersion of the constitutive materials, as well as the dispersive propagation effect in the substrate (or spacing layers between transmitting metasurfaces). Such frequency-dependent constituents anticipate a narrow-band behavior for gradient metasurfaces that are realized with these building blocks. However, and quite surprisingly, the proposed spatial phase modulators, exhibit an exceptional frequency response that makes them a suitable choice for a wide range of radiation engineering applications. This effect resides in the rather stable frequency dispersion of the effective impedance surface when a parallel nanoresonator is utilized. Inspecting Eq. 6.7, the equivalent local reactance of the surface can be written in the general format $Z_{\text{surface}} \propto \left(\omega(\varepsilon_p + \varepsilon_d)\right)^{-1}$. In reality, and especially at shorter wavelengths, both dielectric and plasmonic permittivities are decreasing toward higher frequencies and, thus, Z_{surface} exhibits suppressed frequency dispersion compared to its constitutive materials. The remaining phase variation in the surface impedance as well as the contributing propagation in the thin substrate create a relatively linear dispersion, as verified in Fig. 6.6b. Here, we are inspecting the variations in the reflection phase signature of the particle cluster in Fig. 6.5b, but now for various operating wavelengths. Interestingly, although the absolute scattering phase is frequency dependent, its gradient is more or less dispersionless. In other words, if we rely on the gradient of the scattering phase for a specific radiation pattering purpose, the operation is expected to be broadband. This is in fact the condition of interest for many applications, including beam steering, focusing, and integrated coupling. We note that this is not necessarily the case if we employ a series nanoresonator. In general, the nanoblock configuration must be chosen in view of the intrinsic material properties and the application of interest. As the proposed unit cell is simple and can be accurately and efficiently modeled with nanocircuit concepts, it is straightforward to predict the dispersion behavior of the surface and to select appropriate set of materials and geometrical combinations (i.e., parallel vs. series) to tailor the dispersion in the desired way.

Exploiting a resonant unit cell for the metasurface may rise concerns on the effects of ohmic loss in the response. This is particularly important at IR and optical frequencies as plasmonic metals typically show considerable losses over these regimes. Intrinsic material losses, however, only marginally affect the response since they are very much concentrated around the resonant point (e.g., in Fig. 6.5a). In addition, the interaction of the metasurface with the impinging wave is over a subwavelength thickness (e.g., $d = 40$ nm in the previous example) which furthermore reduces the effect of the surface resistance on the reflection efficiency. This is the reason behind very high reflectivity of the metasurface even around the resonance point, as plotted in Fig. 6.5b. Typically, less than 6% of the input power is absorbed in a gradient metasurface constructed with silver-based spatial phase modulators [15]. Employing even more lossy plasmonic metals does not drastically affect the performance. Figure 6.7a exemplifies the effect of higher intrinsic loss in the surface. Reflection coefficient is plotted here when we purposefully increase the collision frequency from the one of silver (low-loss metal) to moderate ($\gamma_{Ag} \times 2$) and high plasmonic losses ($\gamma_{Ag} \times 5$). Interestingly, the reflection phase is very stable to the surface absorption while the amplitude is also reasonably high even incorporating large plasmonic losses.

The second source of unwanted absorption in a reflecting metasurface is a nonideal back-plane mirror. Up until here, we assumed perfect reflection from the ground plane which is modeled with a perfect electric conductor (PEC) surface. A more realistic scenario is to employ a real metal, such as silver or gold to more accurately estimate the amount of absorption due to the penetration of the fields inside the metal. Figure 6.7b illustrates the effect of the back-plane material on the reflection from the silver-semiconductor metasurface in Fig. 6.5. Due to the low concentration of power around this surface (as the tangential component of the electric field vanishes on the metal) an imperfect ground only slightly increases the absorbed power (Fig. 6.7b). The reflection phase, on

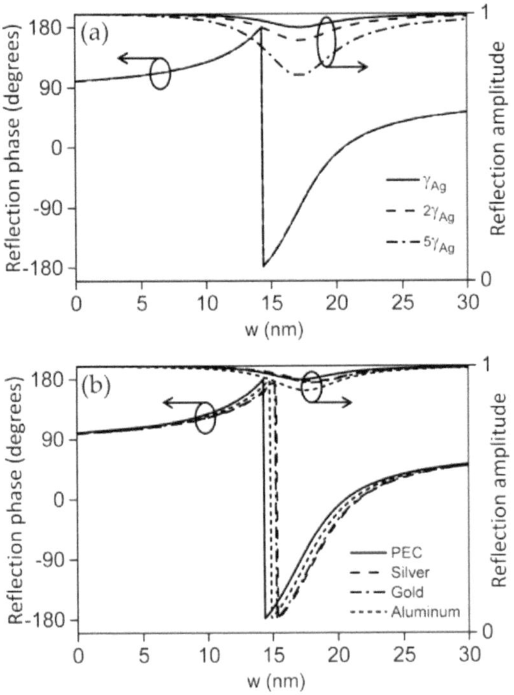

Figure 6.7 Effect of loss. Amplitude and phase of the reflection coefficient for the unit cell shown in Fig. 6.4a (a) with different amounts of intrinsic loss in the plasmonic portion and (b) replacing the back mirror with realistic optical metals.

the other hand, may shift down a little bit since the permittivity of the metallic plane is now a realistic finite value. It is important to underline that with the recent advances in film synthesis, it is in fact possible to realize plasmonic surfaces with absorption even less than conventional silver films [54]. In general, the total effect of loss in plasmonic metasurfaces is relatively insignificant for these effects.

6.2.3 *Polarization Control in Optical Lumped Resonators*

One of the rather challenging problems in the design of gradient metasurfaces is the one of providing a suitable control on their polarization response. We discussed earlier that to grant an acceptable control over the local scattering phase, some sort of resonance needs to be embedded on the surface constituents, for instance employing plasmonic nanoantennas or the proposed lumped nanoresonators. Quite remarkably, the artificial quasi-static resonance we create on the surface enables perfect control on the polarization response of the structure. We are capable of designing nanoresonators with strong polarization preference (i.e., $Y_{\text{eg},xx}$ and $Y_{\text{eg},yy}$ are drastically different) or, on the contrary, with the minimum sensitivity to the excitation axis, that is, $Y_{\text{eg},xx} = Y_{\text{eg},yy}$. Clearly, geometrically symmetric nanoresonators, such as plasmonic posts embedded in a dielectric substrate [15], will operate without any polarization preference (a sample gradient metasurfaces is shown in Fig. 6.15c). The 2D example of Fig. 6.4a, on the other hand, can be feasibly designed to operate under a specific polarization of the incident wave.

It is interesting to underline that this freedom in working with different polarizations is the result of exploiting lumped circuit elements. We can arbitrarily move around these circuit components on the surface to artificially create different effects for opposite polarizations (i.e., linearly polarized waves) or vice versa. This is not the case for plasmonic nanoantennas that naturally work only for the incident wave polarized along the antenna rod.

One exciting application for controllable polarization response is to create completely different effects toward opposite excitation directions. For instance, the unit cell proposed in Fig. 6.4a may be tailored to provide 0° to 360° phase variation under y-polarized illumination, while the same structure creates a constant scattering phase for an x-polarized wave. We designed such elements based on Eq. 6.7, and intentionally created very distinct phase signatures for parallel and series nanoresonators. The operation wavelength is $\lambda_0 = 500$ nm and silver-semiconductor nanoresonators are utilized in this case. The particle dimensions are set at $l = \sqrt{2}\lambda_0/8 = 88.4$ nm, $d = 50$ nm, and the metasurface is deposited on a 50 nm grounded quartz (SiO_2, $n = 1.55$) substrate. Figure 6.8a shows the reflection coefficient under TE (y-polarized) and TM (x-polarized) illuminations versus the width of the plasmonic portion in the cell. In the range between $w = 20$ nm and $w = 88$ nm, the resonance condition is hit under the TE illumination and a large span of scattering phase is therefore attainable. On the other hand, the

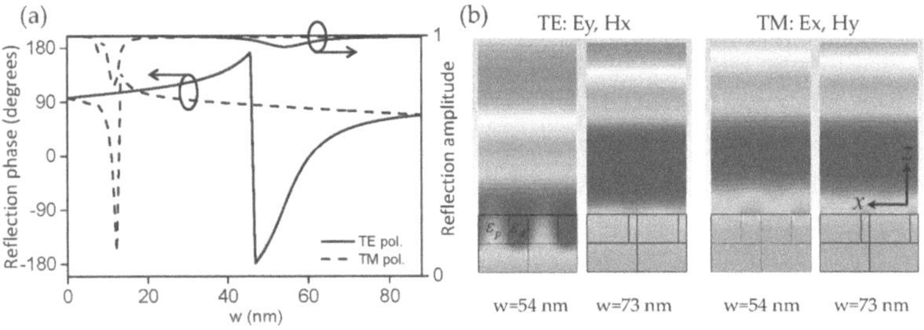

Figure 6.8 Polarization response. (a) Reflection coefficient from a grounded metasurface with the unit cell shown in Fig. 6.4a under two orthogonal excitations. (b) Effect of the geometry and incident polarization on the distribution of electric field over the metasurface. Two adjacent unit cells from the periodic metasurface are shown.

TM resonance is confined around $w = 15$ nm and, over the same region, less than 20° variation is created by the metasurface for this polarization. Different interactions of the input wave and the metasurface elements are also highlighted in Fig. 6.8b. Changing the filling ratio does not affect TM waves but it drastically modifies the TE reflection phase. The designed element can be used to realize interesting devices, such as ultrathin polarization beam splitters, which we will investigate in Section 6.3.3.

6.3 Beam Forming with Graded Metasurfaces

The fundamental building blocks proposed in the previous section provide us with the appropriate tool to manipulate the scattering phase over a surface. The general method outlined in Section 6.2 indicates that if it is possible to do so on a subwavelength scale, it is also possible to mold the outgoing wave at will. Throughout the current and the following sections we look at various devices employing this technique to imprint the desired pattern on the output. We must underline that in contrast to the metasurfaces examined in the previous section, we will not necessarily deal with periodic structures from now on. In general, the target scattering distribution may or may not impose a periodicity condition on the metasurface, and therefore the wave-shaping surface is in general nonperiodic (or as we refer to *gradient*). The reflection phase and amplitude curves plotted in Figs. 6.5 through 6.8 correspond to fully periodic surfaces (with the period $l << \lambda_0$). Subsequently, to realize gradient metasurfaces and satisfy Eq. 6.4, each element is individually selected based on the local phase variation predicted in these figures. Throughout this process we have been implicitly assuming that the mutual coupling between neighboring elements (which are now dissimilar) does not significantly affect their scattering properties. In other words, each nanoblock can individually provide a local scattering phase on its surface as if it is arranged in a periodic fashion. This is an acceptable assumption provided that we are working with well-defined, smoothly varying field distributions, as the following results indicate.

6.3.1 *Optical Reflectarrays and Transmitarrays*

In analogy to microwave reflectarrays and transmitarrays, gradient metasurfaces may be suitably designed to efficiently steer the direction of the outgoing wave (anomalous reflection or refraction), realizing arbitrary beam-steering functionality. This is somewhat similar to the operation of a blazed grating which, at one specific wavelength, is able to transfer the incident energy to one of its diffraction orders. Yet, distinct from conventional blazed gratings, here the desired outgoing beam is reconstructed on the surface with very high spatial resolution enabling near-field operation, steering at very steep angles, as well as broad bandwidth.

To realize anomalous reflection (or refraction) on a metasurface, the outgoing wave must follow a linear phase variation over the surface, separate from the phase profile of the excitation wave. As discussed through Fig. 6.3, for moderately small deflection angles the required surface properties can be attained with a simple ray approximation and a linear scattering phase $\angle R(x) = \alpha x$ must be enforced along the metasurface. Here, the relation between the incident angle θ_i and reflected angle θ_r is $\pm\alpha\lambda_0 = 2\pi\ (\sin\theta_\mathrm{r} - \sin\theta_\mathrm{i})$ and x denotes the bending axis toward the surface (a 2D setup is assumed). The scattering phase profile is plotted in Fig. 6.9b, assuming $\theta_\mathrm{i} = 0^\circ$ and $\theta_\mathrm{r} = -45^\circ$ (in accordance to Fig. 6.3b). To implement this surface, we quantize the phase profile into steps

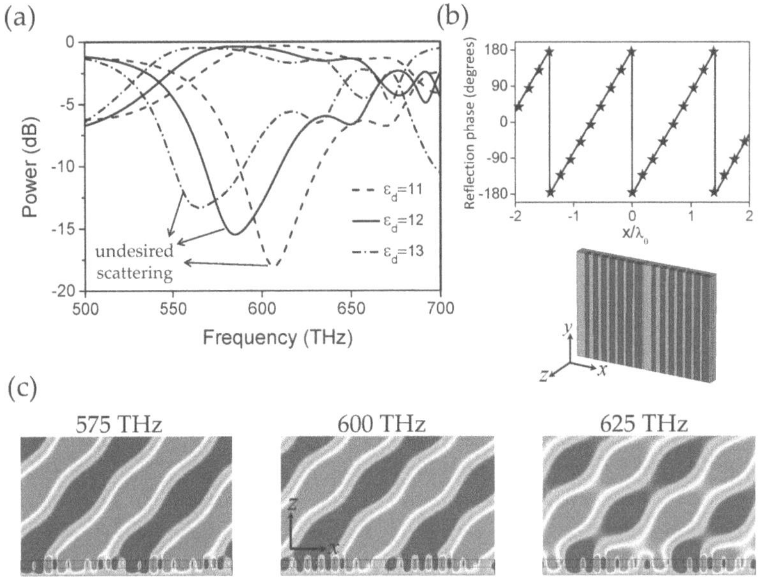

Figure 6.9 Tunable beam steering. (a) Normalized reflected power toward the desired direction when a TE-polarized plane wave illuminates the gradient metasurface shown in (b) toward the $-z$ direction. Permittivity of the dielectric portion of the surface nanoblocks is swept from $\varepsilon_\text{d} = 11$ to $\varepsilon_\text{d} = 13$ to tune the frequency response. (b) Linear phase function to bend a normal impinging wave by 45°. Quantized phases are shown by stars on the curve, and the corresponding metasurface is shown in the lower panel. (c) Distribution of the scattered field on the incident plane (xz) at three frequencies.

of lateral size $l = \sqrt{2}\lambda_0/8$, indicated by stars in Fig. 6.9b. We notice that due to the periodicity of the required phase pattern, the profile repeats itself over each 8 elements as well. In other words, the surface exhibits a *superlattice periodicity* of $L = |\lambda_0/(\sin\theta_\text{r} - \sin\theta_\text{i})| = \sqrt{2}\lambda_0$. This, however, is not an inherent requirement and we may realize the exact same scattering phase profile employing a fully nonperiodic surface. Here, to execute each phase step under TE illumination, we exploit the nanoblocks designed in Fig. 6.8, as shown in Fig. 6.9b, lower panel.

Figure 6.9 illustrates the wave bending performance of the designed metasurface. After imparting the linear phase profile on the scattered wave, a normally incident wave is successfully deflected toward -45° at 600 THz. It is important to underline the effect of subwavelength wave reconstruction in this technique. As demonstrated through the near-field plots in Fig. 6.9c, the outgoing plane wave is generated right above the surface elements with minimal undesired evanescent waves in the surface proximity. This is the direct consequence of satisfying exact boundary conditions of Eq. 6.1 over the surface. Following the superlattice periodicity of the implemented surface, the scattered wave may be also viewed as summation of multiple scattering orders, that is, $n = 0, \pm 1, \pm 2, \ldots$ Floquet modes, of a periodic structure. Enforcing a plane wave toward $\theta = \theta_r$ (similar to Eq. 6.6) as the ideal scattered wave is subsequently equivalent to entail the corresponding amplitude of all spatial orders to be zero, except for the $n = +1$ mode which is the plane wave toward the desired direction. In this regard the efficiency of the anomalous reflection (as opposed to specular reflection) can be defined as the ratio of the input power coupled to the first diffractive order (-45° in this example) to the total incident power. Figure 6.9a shows the normalized power coupled to the first diffractive order in comparison to all the other directions, versus excitation frequency. At the center design frequency 600 THz around 90% of the total power redirects toward -45°, 5% is scattered to unwanted spatial orders, and the rest get absorbed in the metasurface. We recall here that for simplicity of implementation, the loss/gain portions of the surface response are forced to zero from the ideal scenario in Fig. 6.3b, yet conversion efficiency above 90% is achieved. The performance of the surface is also very stable with frequency and sustains over 20% bandwidth (−3 dB bandwidth). We point out that a wave-squinting effect is naturally present owing to the dispersion of the Floquet modes. As a result, the reflection angle slightly varies over frequency, yet still the dominant portion of the input wave couples to the first diffraction order in this regime. The broadband operation of the surface is predictable if we inspect the nature of the process: an additional linear phase is imparted on the input, but, the absolute value of the phase is not a parameter in this process and the curve in Fig. 6.9b may have any arbitrary offset on the vertical axis. We discussed in Section 6.2.2 that the plasmonic-dielectric nanoblock maintains a stable phase profile over broad bandwidths and it merely adds a dispersive phase offset to the surface (Fig. 6.6b), which results in a broadband response for beam-steering applications. We underline again that the total wave manipulation process is accomplished here simply employing a single gradient metasurface with 50 nm $\approx \lambda/10$ thickness, and composed of two alternating materials.

There is a possibility to tune the frequency response of the surface by tailoring the properties of the nanoresonators. We looked at this option for the nanoblocks in Section 6.2.2 and here we examine this concept at the device level. As shown in Fig. 6.9a, maximum coupling range can be effectively tuned along the frequency by changing the permittivity of the dielectric portion of nanoresonators. In practice, the permittivity may be altered with temperature, optical tuning or via nonlinear effects. Tunable, or reconfigurable, beam-steering surfaces are envisioned based on this effect, with which we may efficiently control the frequency signature (i.e., color) of the steered beam with external bias. More details on reflective and transmitting beam-steering metasurfaces can be found in Refs. [15, 21].

To conclude this subsection, we investigate the effect of surface granularity on the performance of a gradient metasurface. Clearly, it is not possible to realize the required phase with infinite resolution and we always need to somehow quantize the phase pattern. Intuitively, more quantization levels appear to provide larger conversion efficiency to the desired scattering pattern. To gain a quantitative rule for the other designs, we study the effect of discretization on the performance of the beam-steering example. A gradient metasurface is designed to deflect a normally incident plane wave toward -45° at $\lambda_0 = 500$ nm. The superlattice periodicity (Fig. 6.9b) equals $\sqrt{2}\lambda_0 - 707$ nm which is now filled with $N = 2, 3, \ldots, 8$ different phase units (The previous example considered the case of $N = 8$). For each discretization number, the appropriate unit cell is designed employing silver-semiconductor resonators with $d = 50$ nm thickness, deposited on 50 nm quartz substrate. Figure 6.10 indicates the distribution of the power to different diffraction orders for these gradient metasurfaces. The incident power essentially couples to three propagating Floquet modes, where $n = +1$ indicated the desired direction of -45°.

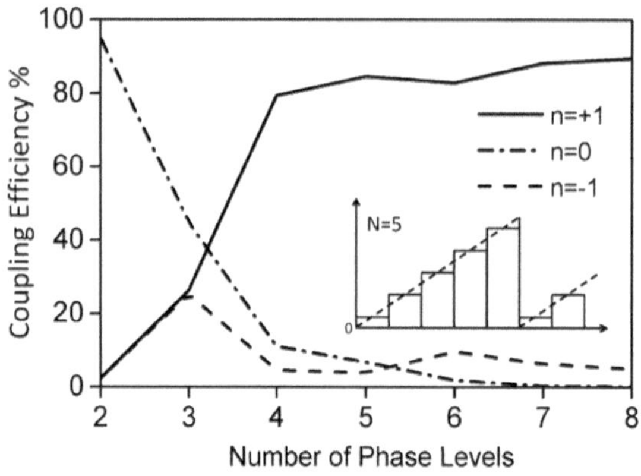

Figure 6.10 Discretization effect. Coupling efficiency to the propagating diffraction orders as a function of quantization levels for the phase profile shown in Fig. 6.9b. Inset illustrates the case of $N = 5$.

As expected, the efficiency of conversion to this mode increases as the discretization gets finer and finer. For $N = 2$ and 3, the surface exhibits odd, and even scattering symmetries and thus equal amount of power couples to $n = \pm 1$ modes. For $N > 3$ an artificial directional preference is obtained on the surface and therefore efficiency of coupling to $n = +1$ increases rapidly. Above $N = 4$, which represents a unit cell size of $l = 176$ nm $\approx \lambda_0/2.8$, we attain a reasonable performance from the structure. It must be noted that in this particular example, the metasurface exhibits a superlattice profile, which in turn creates a collective grating effect on top of the local phase management of the nanoresonators. In other applications, and to create a strong sense of local variation, a unit cell size of at least $\lambda_0/5$ is desirable. The minimum size of the surface element is a technological consideration related to the limitations of the considered fabrication methods; the composite nanorasonator block considered here can inherently provide a very large spatial resolution [15].

6.3.2 *Flat Lens*

Figure 6.11 shows another planar device based on gradient metasurfaces: a flat ultrathin lens. In the ray optics regime, flat

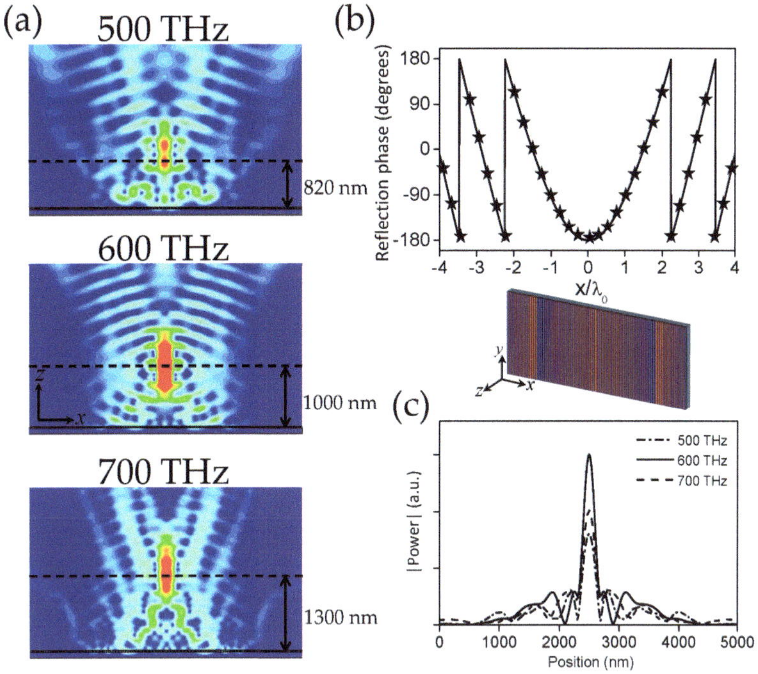

Figure 6.11 Broadband beam collimation. (a) Power density distribution of a flat mirror lens at three sample frequencies. The metasurface is illuminated along the $-z$ axis with a Gaussian beam profile and under TE polarization. (b) Quadratic phase function to focus the impinging wave at distance $2\lambda_0$ from the surface. Quantized phases are shown by stars, and the corresponding metasurface is shown in the lower panel. The numerical aperture of the lens is 0.9. (c) Power density along the focal plane, indicated by dashed lines in (a). Reprinted (figure) with permission from Ref. [15]. Copyright (2014) by the American Physical Society.

lenses impart a quadratic phase distribution (as shown in Fig. 6.11b) on the outgoing wave so that an incident normal plane wave efficiently converts to a spherical wave. Gradient metasurface-based lenses exhibit many advantages in terms of compact size, conformability, and the possibility of direct integration into optical systems. Through appropriately engineering the dispersion of unit cells, it is also possible to reduce the typical aberration effects present in conventional lenses [55]. These features make gradient metasurfaces an appealing choice to realize electromagnetically thin lenses over wide range of frequencies (see, for example, Refs. [15, 47, 55, 56]). The most interesting property of this technique, however resides in the ability of tailoring the outgoing wave front with high spatial resolution. This enables us to set the focal point arbitrarily close to the surface and realize lenses with very high numerical apertures.

We construct a mirror lens at $\lambda_0 = 500$ nm exploiting the nanoblocks characterized in Fig. 6.5, assuming a 2D variation on the surface and TE polarization of the incident wave. With the focal point at $f_l = 2\lambda_0$, the required phase distribution along the surface can be estimated as

$$\angle R(x) = 2\pi/\lambda_0 \left(\sqrt{x^2 + f_l^2} - f_l\right) \tag{6.8}$$

which is implemented in the form of a graded metasurface after proper discretization, as shown in Fig. 6.11b. Owing to the relative value of the scattered phase, we expect broadband focusing behavior from the structure, as demonstrated in Fig. 6.11a,c. While the surface is designed at 600 THz, it exhibits remarkable focusing properties over more than 30% fractional bandwidth.

6.3.3 *Polarization Beam Splitter*

As we have pointed out in the previous sections, quasi-static resonances imposed on the surface can be intentionally tailored to create a polarization-dependent response, or vice versa to effectively eliminate it. This effect suggests an appealing application of operating gradient metasurfaces as ultrathin polarizing beam splitters. Generally, a beam splitter is a device that divides the input beam into two separate beams with different polarization states (i.e., polarization beam splitters), wavelength (i.e., dichroic plates), or intensity. The core element of a conventional polarizing beam splitter is an electromagnetically thick layer of a birefringent material (such as Calcite) which intrinsically shows different responses for different polarizations of the incident beam. For instance, the device may be designed to reflect one polarization state while fully transmitting the orthogonal polarization or to create different refracting angles for the ordinary and extraordinary beams. Similar effects can be achieved employing a series of dielectric slabs aligned at their Brewster angle. More recently, a number of methods have been proposed to manipulate the propagation path of electromagnetic waves inside an artificial medium, depending on their polarization. Photonic crystals with intentionally engineered band structure and complex anisotropic TO-based beam splitters (see, for example, Refs. [57–59]) are among the most interesting recent proposals in this context. All these approaches, however, entail some sort of bulk birefringent properties to differentiate between two polarizations which inevitably results in electromagnetically large configurations.

An intriguing idea to efficiently split an unpolarized beam into two polarized components is to exploit a beam-steering gradient metasurface which is specifically tailored to operate solely for one polarization [11, 47, 49]. If so, the surface creates an anomalous reflection (or transmission) for one polarization and an ordinary specular reflection for the orthogonal polarization. With appropriate design, two beams can be tailored to have large divergence angles, comparable to those of a conventional Wollaston prism, but over an ultracompact profile and with very high efficiencies.

To verify the applicability of this technique for effective polarization splitting, a gradient metasurface is constructed at the visible range employing the intentionally designed surface blocks in Fig. 6.8. Imposing a linear phase profile of $\angle R_{\mathrm{TE}}(x) = \sqrt{2}\pi x/\lambda_0$ on the metasurface (as in Fig. 6.9b) enforces an anomalous reflection for the incident TE-polarized beam so as $\left(\sin\theta_{\mathrm{r}}^{\mathrm{TE}} - \sin\theta_{\mathrm{i}}^{\mathrm{TE}}\right) = \sqrt{2}/2$, while the TM-polarized beam will experience a specular reflection: $\theta_{\mathrm{r}}^{\mathrm{TM}} = \theta_{\mathrm{i}}^{\mathrm{TM}}$ since $\angle R_{\mathrm{TM}}(x)$ is approximately constant. Figure 6.12a shows the power distribution when this surface is illuminated at $\lambda_0 = 500$ nm with an oblique, circularly polarized Gaussian beam. Interacting with the subwavelength surface (50 nm thickness), the incident wave splits into two reflected beams which are verified as purely TE and TM in Fig. 6.12b,c [11]. We notice that there is a range of incident angles for which the gradient surface can efficiently operate as a polarizing beam splitter, quite distinct from conventional Brewster angle designs which require additional advanced alignments. Besides, it is worth to underline that such polarization-dependent radiation manipulation is simply achieved by alternating two common materials to form an extremely low-

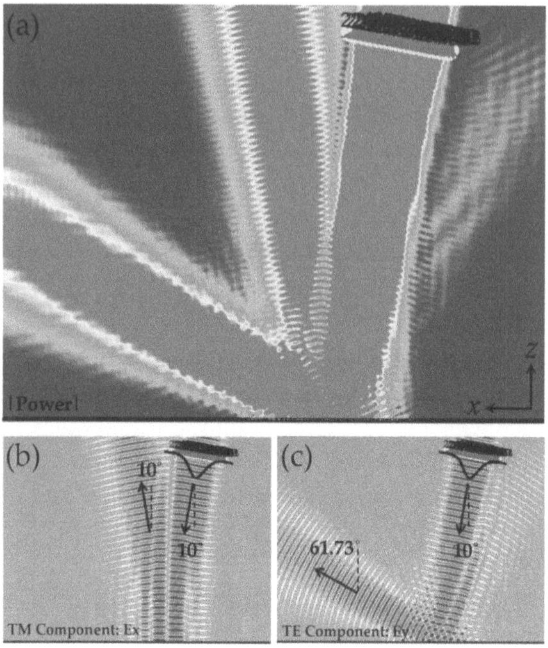

Figure 6.12 Polarization splitting. (a) Power density distribution of a planar polarization beam splitter illuminated under $\theta_i = 10^\circ$ with a Gaussian circularly polarized beam. (b) TM and (c) TE components of the total electric field distribution. The operation wavelength is 500 nm and surface profile is shown in Fig. 6.9b.

profile structure, which can be directly integrated into more complex optical elements.

6.4 Other Potential Applications

The flexibility offered by gradient metasurfaces to freely manipulate and mold electromagnetic wave fronts enables new opportunities for various flat optical elements with improved performances or peculiar functionalities. In the previous section we looked into a number of rather fundamental radiation control examples to efficiently redirect, focus, or split an optical beam with unprecedented resolution and very high efficiencies. Here, we discuss the potential impact of graded metasurfaces on integrated optics, energy harvesting, camouflaging, and optical computing with the intention of improving performance, reducing the size and complexity of the design and implementation, and enabling novel functionalities. The results presented in this section are based on our recent work presented in Refs. [11, 22, 23].

6.4.1 *Conformal Cloaking*

Wave reconstruction over a single textured surface can provide an exciting platform for ultrathin carpet cloaking [11, 22]. In the previous sections, we described the process of creating arbitrary scattering distributions by accurately engineering the profile of the surface to satisfy the required boundary conditions in Eq. 6.1. Through this process, the ultrathin metasurface is tailored to operate as a structure that scatters the incident wave in the predesignated manner. For instance, the flat lens presented in Section 6.3.2 operates analogous to a concave mirror or the reflectarray presented in Section 6.3.1 imitate a tilted mirror's functionality. Interestingly, we can likewise reverse this process and create a nonplanar gradient metasurface that scatters resembling a flat mirror. This indeed creates an ideal camouflage for any object put beneath such a surface so it can't be identified by an external observer who inspects the scattered wave from such surface.

Figure 6.13a illustrates a typical scattering distortion created due to the presence of an obstacle (a triangular PEC object in this example) on a flat conducting mirror. If the object was not there, the incident beam would simply reflect back from the ground plane maintaining its initial outline. The presence of the obstacle, however, distorts this field distribution and creates a large scattering shadow in the forward direction, make the object easily detectable. Theoretically, the field deformation is the result of a new boundary condition imposed by the object on its surface compared to the case where the object was absent. This is shown in Fig. 6.13b where we look at the phase and amplitude profile of the total field on a hypothetical boundary line above the surface of the obstacle. If somehow, the field distribution can be restored to its original values, the object will be effectively invisible. Such functionality is already

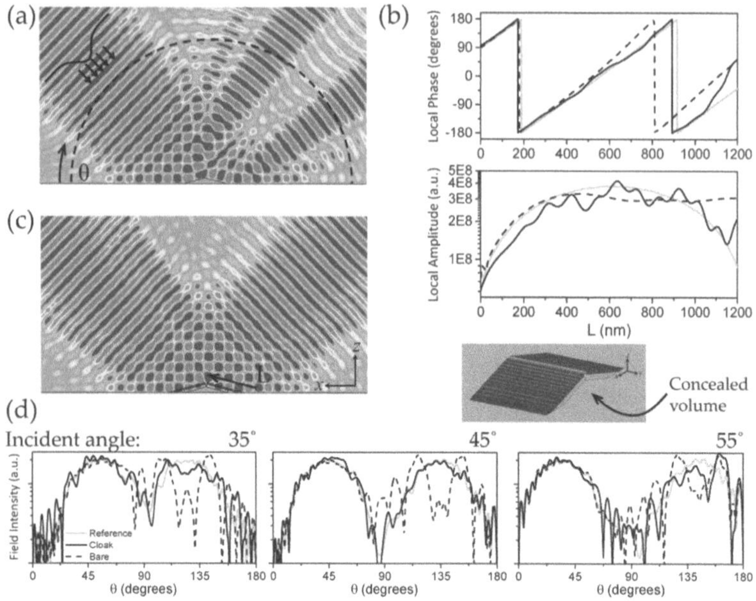

Figure 6.13 Conformal carpet cloaking. (a) Time snapshot of the electric field distribution when the uncloaked object is illuminated with an oblique Gaussian beam at a 45° angle. (b) Phase and amplitude of the total electric field on a hypothetical boundary line shown in (c). A flat mirror (gray), uncloaked object (dashed black), and cloaked object (solid black) are compared. The lower panel shows a 3D view of the setup showing the cloaking surface conformally covering the obstacle. (c) Same as (a) but with the metasurface cloak. (d) Total field intensity along a far-field half circle—dashed line in (a)—enclosing the structure. Three incident angles are shown demonstrating the scattering from the cloaked and uncloaked objects in comparison to a flat conducting mirror. All curves are at $\lambda_0 = 500$ nm. Adapted from Ref. [11].

assured with a gradient and inhomogeneous metasurface. We have shown that a single, ultrathin metamirror is sufficient to locally enforce an arbitrary value of phase or amplitude on its surface. Therefore, there is the possibility of wrapping a properly designed metasurface on the obstacle and impose the required boundary condition on its surface to effectively cloak the structure.

Figure 6.14 shows the exact reflection profile (solid lines) required on a triangular surface to optimally cloak it from an

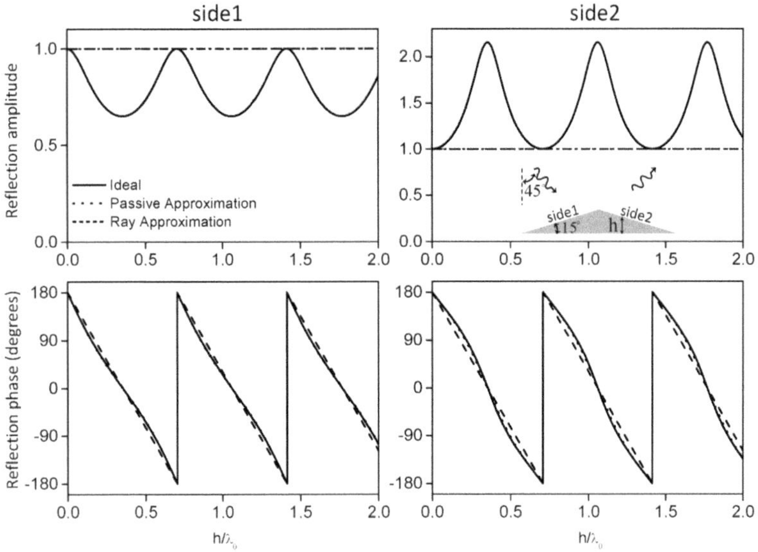

Figure 6.14 Comparison between exact and approximated local reflection coefficient for carpet cloaking. The metasurface is designed to cloak a triangular obstacle with side angles of 15° from an incident TE plane wave illuminating the structure with a 45° angle (shown in the inset). The required reflection phase and amplitude are shown for three cases: exact solution (solid lines), passive approximation of the exact solution (dotted lines), and ray approximation (dashed lines). The reflection coefficients are plotted along the edges of the triangle.

incident TE-polarized plane wave under 45° illumination. Triangular surface has 15° corner angles and the required local phase and amplitude of reflection are plotted versus the local height from the reference back plane (i.e., $h = z_{\text{local}}$) on both edges. Reflection coefficients are calculated for corresponding periodic metasurfaces under $45° \pm 15°$ illumination angles, respectively for the right and left edges. Similar to ideal wave deflection discussed in Fig. 6.3, ideal cloaking also requires active and passive surface elements. However, rather than alternating loss and gain portions over the surface, here, one edge is composed of lossy components, while the opposite edge must contain active elements. This is in perfect analogy with parity time induced invisibility [60, 61], where a combination of gain-loss materials with engineered symmetry can provide unidirectional invisibility. To simplify the cloaking configuration, here we implement a passive cloak in which the real part of the ideal surface admittances, which are required to obtain the exact desired scattering profile, are neglected (dotted lines). As discussed earlier, ray approximation is also applicable to estimate the required passive local phase on the surface. In this case, each surface element must provide a constant local phase of the reflection to compensate for the presence of the obstacle [22]. This approximation is shown with dashed lines in Fig. 6.14. For cloaking application, we design the metasurface based on the passive approximate approach and then the surface elements are optimized so that the near field is most close to the ideal cloaking response. We note that a 100% cloaking requires the exact lossy/gainy profile indicated by solid lines in Fig. 6.14 but a very good approximation with minimum residual unwanted scattering is attained with the passive surface as well [11].

The total electric field distribution when the obstacle is covered with such metasurface is shown in Fig. 6.13c. A 3D sketch of the setup is also shown in Fig. 6.13b, lower panel, indicating the PEC obstacle, the gradient cover, and the concealment volume beneath the surface to hide any arbitrary object. The original specular reflection of the incident beam (which is a Gaussian 5 μm wide optical beam at $\lambda_0 = 500$ nm) is restored employing a single 50 nm thick graded surface. This example is based on conjoined metal-dielectric nanorod resonators (see Fig. 6.4a) specifically tailored for TE illumination. We also extended this concept for unpolarized excitations and cloaking 3D objects on the surface [22], employing suitable 3D surface elements.

Inspecting Fig. 6.13b, the total field distribution on the hypothetical line above the structure is restored to the flat mirror condition (i.e., gray lines) after the designed metasurface is positioned around the obstacle. By reconstructing the field distribution in the near field, as expected, the far-field scattering pattern also resembles a flat conducting mirror, as shown in Fig. 6.13d for three different excitation angles. We have proved in Ref. [22] that our proposed surface cloak is quite stable to the angle of incidence. Indeed, as we design the cloak to satisfy the boundary condition for a specific direction of illumination, it maintains its performance over a reasonable range of incident angles, again owing to its ultrathin thickness. This is numerically illustrated in Fig. 6.13d as the cloaking performance is sustained over $\pm 10^\circ$ variations around the original design angle of 45°.

Achieving ideal transparency, both in the near field and the far field, with a simple textured surface offers a viable route to design controllable and reconfigurable cloaking blankets, which in principle can have extremely subwavelength thicknesses. This is in contrast to the initial TO-based proposals where optically thick, anisotropic, and inhomogeneous cloaking layers must be incorporated [62, 63], making them practically unattainable. Our proposed wave reconstruction technique entails a simple reactance engineering of a metasurface and is equally applicable to arbitrary shaped, electromagnetically large objects. We note that the concept is not limited to the visible and IR regimes where the hybrid nanoresonators initially introduced. With the choice of a suitable surface phase element, which may range from RF antennas to clusters of nanoparticles or biased graphene patches, the scattering pattern can be manipulated to effectively hide an object at any desired frequency. We envision a wide range of applications for ultrathin conformal cloaking blankets in camouflaging, switchable invisibility, and noise reduction in wireless communication systems [11, 22].

6.4.2 *Broadband Energy Harvesting*

Although the technique we discussed in this chapter originally aims at manipulating the radiation pattern and engineer the desired scattering signature, it can also provide many interesting implications for conceptually different applications. Here we show that gradient metasurfaces can replace the conventional mirrors in thin-film solar cells and significantly enhance their absorption efficiency over broad bandwidths.

Figure 6.15c illustrates the proposed configurations for 2D and 3D gradient metasurfaces operating as artificial ground planes in photovoltaic cells. The idea is to create a lateral standing wave profile inside the active layer in addition to the primary vertical standing waves to build up strong light concentration (i.e., local hot spots) and increase the collected photocurrent [11]. In this regard,

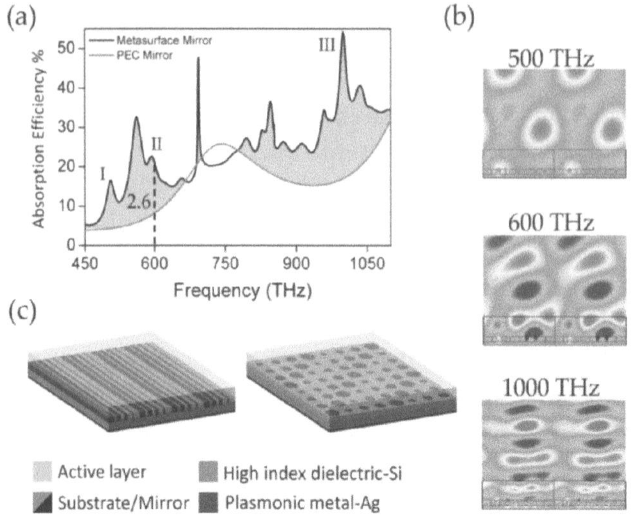

Figure 6.15 (a) Absorption efficiency of a 180 nm low-loss semiconductor layer when backed by a PEC (gray line) in comparison with a metasurface back reflector (black line). The solar cell is illuminated with a normal TE-polarized wave and the additional gained absorption is highlighted in gray. (b) Time snapshot of electric field distribution at three frequencies indicated in (a). Metasurface and active layer outlines are indicated in the plots. (c) Schematics of 2D and 3D gradient metasurfaces proposed as back mirrors in thin-film solar cells. Adapted from Ref. [11]. Copyright © 2015, copyright 2015 The Author(s) Published by the Royal Society. All rights reserved.

we design gradient metasurfaces to provide a constant momentum along the surface of the solar cell and therefore efficiently redirect the incident solar energy toward a new direction (which we set here at 45°). The artificially imposed surface momentum creates multiple reflections within the active layer and increases the effective propagation path length. Figure 6.15a demonstrates the absorption enhancement achieved by using a 2D graded metasurface mirror (as shown in Fig. 6.15c, left panel) at the back of a 180 nm semiconductor layer with initially poor absorption characteristics [11]. As predicted, the multiple artificial reflections created inside the active layer significantly boost the total absorption over an extremely broad bandwidth, following the linear dispersion of our proposed nanoblocks (see the discussions in Section 6.2.2).

Figure 6.15b highlights the created lateral standing wave profile at three sample frequencies. The surface is initially designed at 600 THz where an absorption enhancement of factor 2.6 (160%) is attained. It is also interesting to underline the effect of the produced broadband background absorptions on the original modes of the active layer. At the natural frequencies of the solar cell (which depend on the thickness and permittivity of the semiconductor [64, 65]), the incident wave can couple to in-plane propagative modes and efficiently get absorbed. This, in principle, is an extremely narrow-band effect (a few nanometers) in a conventional periodic solar cell. However, the gradient back mirror can noticeably broaden these peaks (which arise around 540 THz and 1000 THz in this example) and further boost the absorption enhancement.

6.4.3 *Nanoscale Signal Processing*

Inhomogeneous, locally designed metasurfaces can proficiently imprint a desired phase and amplitude map on an arbitrary input, with very high precisions. This opens up a possibility to also imprint an *operation* on the surface by implementing the operation's Green's function with a graded metasurface. The idea is based on implementing arbitrary functionalities in the *Fourier space domain* where the conventional convolution integral transforms to a simple multiplication operation. For instance, the first-order derivative entails a spatially inhomogeneous transfer function of $f(x) \propto jx$, which is straightforwardly accessible with a metasurface-based structure. This capability provides an interesting platform for optical analog computing systems to directly process an optical wave as it passes through a potentially compact stack of gradient metasurfaces [23]. This technique may lead to all optical ultrafast analog computing devices which are directly integrable to other optical systems, such as cameras and imaging devices.

6.5 Conclusions and Outlook

Here we reviewed our recent results and proposals on the concept of local wave manipulation and radiation patterning employing gradient metasurfaces. We focused on four main aspects: First, a general methodology based on equivalence principle is proposed to translate the scattering manipulation problem into an equivalent gradient metasurface design. Second, we discussed efficient implementation of gradient metasurfaces at optical and IR frequencies. In this regard we reviewed our proposed plasmonic-dielectric composite particles which may exhibit deeply subwavelength static resonances with ultrathin profiles, tailored employing a nanocircuit interpretation. This structure has been shown to be robust to intrinsic material losses, which are typically the bottleneck of optical designs, and with an interesting linear frequency dispersion that tracks the natural dispersion of wave phenomena, allowing broadband responses. We have shown that the proposed element can be potentially designed to be tunable and polarization selective. Third, we examined the performance of the proposed nanoelements to realize simple wave manipulation operations (e.g., beam steering and collimation), with notable performance in terms of bandwidth, efficiency, and tunability. Finally, we showed that gradient metasurfaces can perform complex radiation patterning and wave management operations, such as conformal cloaking and optical signal processing based on a suitable local scattering control. Our results may provide interesting possibilities for integrated, compact optical components with novel and enhanced performances which are compatible with the current 2D fabrication technology.

Acknowledgments

This work was supported by the Air Force Office of Scientific Research with grant no. FA9550-13-1-0204, the Office of Naval Research with grant no. N00014-10-1-0942, and the Welch Foundation with grant no. F-1802.

References

1. K. F. Braun (1967). Electrical oscillations and wireless telegraphy. In *Nobel Lectures, Physics 1901–1921* (Elsevier, Amsterdam), p. 239.

2. Wood, R. W. (1905). *Physical Optics* (Macmillan, New York, USA).
3. Chen, H., C. Chan, et al. (2010). Transformation optics and metamaterials, *Nat. Mater.*, **9**(5), 387–396.
4. García-Meca, C., Martínez, A., et al. (2011). Engineering antenna radiation patterns via quasi-conformal mappings, *Opt. Express,* **19**(24), 23743–23750.
5. Hansen, R. C. (2009). *Phased Array Antennas*, Vol. 213 (John Wiley & Sons, New York, USA).
6. O'Shea, D. C., Suleski, T. J., et al. (2004). *Diffractive Optics: Design, Fabrication, and Test* (Spie Press Bellingham, WA, USA).
7. Munk, B. A. (2000). *Frequency Selective Surfaces: Theory and Design* (Wiley Interscience, New York, USA).
8. Merlin, R. (2007). Radiationless electromagnetic interference: evanescent-field lenses and perfect focusing, *Science*, **317**(5840), 927–929.
9. Yu, N., P. Genevet, et al. (2011). Light propagation with phase discontinuities: generalized laws of reflection and refraction, *Science*, **334**(6054), 333–337.
10. Kildishev, A. V., Boltasseva, A., et al. (2013). Planar photonics with metasurfaces, *Science*, **339**(6125), 1232009.
11. Mohammadi Estakhri, N., Argyropoulos, C., and Alù, A. (2015). Graded metascreens to enable a new degree of nanoscale light management, *Philos. Trans. R. Soc. A*, **373**(2049), 20140351.
12. Alù, A., Silveirinha, M. G., et al. (2007). Epsilon-near-zero metamaterials and electromagnetic sources: tailoring the radiation phase pattern, *Phys. Rev. B*, **75**(15), 155410.
13. Argyropoulos, C., D'Aguanno, G., and Alù, A. (2014). Giant second harmonic generation efficiency and ideal phase matching with a double epsilon-near-zero cross-slit metamaterial, *Phys. Rev. B*, **89**(23), 235401.
14. Maier, S. A. (2007). *Plasmonics: Fundamentals and Applications* (Springer).
15. Mohammadi Estakhri, N., and Alù, A. (2014). Manipulating optical reflections using engineered nanoscale metasurfaces, *Phys. Rev. B*, **89**(23), 235419.
16. Engheta, N. (2007). Circuits with light at nanoscales: optical nanocircuits inspired by metamaterials, *Science*, **317**(5845), 1698–1702.
17. Pancharatnam, S. (1956). Generalized theory of interference, and its applications. Part I. Coherent pencils, *Proc. Indian Acad. Sci.*, **44**, 247–262.

18. Berry, M. V. (1987). The adiabatic phase and Pancharatnam's phase for polarized light, *J. Mod. Opt.*, **34**(11), 1401–1407.
19. Hasman, E., Kleiner, V., et al. (2003). Polarization dependent focusing lens by use of quantized Pancharatnam–Berry phase diffractive optics, *Appl. Phys. Lett.*, **82**(3), 328–330.
20. Kang, M., Feng, T., et al. (2012). Wave front engineering from an array of thin aperture antennas, *Opt. Express*, **20**(14), 15882–15890.
21. Monticone, F., Mohammadi Estakhri, N., et al. (2013). Full control of nanoscale optical transmission with a composite metascreen, *Phys. Rev. Lett.*, **110**(20), 203903.
22. Mohammadi Estakhri, N., and Alù, A. (2014). Ultrathin carpet cloak and wavefront reconstruction with graded metasurfaces, *IEEE Antennas Wireless Propag. Lett., Special Cluster Transformat. Electromagn.*, **13**, 1775–1778.
23. Silva, A., Monticone, F., et al. (2014). Performing mathematical operations with metamaterials, *Science*, **343**(6167), 160–163.
24. Pozar, D. (1996). Flat lens antenna concept using aperture coupled microstrip patches, *Electron. Lett.*, **32**(23), 2109–2111.
25. Pozar, D. M., Targonski, S. D., et al. (1997). Design of millimeter wave microstrip reflectarrays, *IEEE Trans. Antennas Propag.*, **45**(2), 287–296.
26. Encinar, J. A. (2001). Design of two-layer printed reflectarrays using patches of variable size, *IEEE Trans. Antennas Propag.*, **49**(10), 1403–1410.
27. M. R. Chaharmir et al. (2010). Design of broadband, single layer dual-band large reflectarray using multi open loop elements, *IEEE Trans. Antennas Propag.*, **58**, 2875–2883.
28. D. F. Sievenpiper et al. (2003). Two-dimensional beam steering using an electrically tunable impedance surface, *IEEE Trans. Antennas Propag.*, **51**, 2713–2722.
29. Al-Joumayly, M. A., and Behdad, N. (2011). Wideband planar microwave lenses using sub-wavelength spatial phase shifters, *IEEE Trans. Antennas Propag.*, **59**(12), 4542–4552.
30. Harrington, R. F. (2001). *Time- Harmonic Electromagnetic Fields* (John Wiley & Sons, New York, USA).
31. Kuester, E. F., Mohamed, M. A., Piket-May, M., and Holloway, C. L. (2003). Averaged transition conditions for electromagnetic fields at a metafilm, *IEEE Trans. Antennas Propag.*, **51**, 2641–2651.

32. Pfeiffer, C., and Grbic, A. (2014). Bianisotropic metasurfaces for optimal polarization control: analysis and synthesis, *Appl. Phys. Rev.*, **2**(4), 044011.

33. Pfeiffer, C., and Grbic, A. (2013). Metamaterial huygens' surfaces: tailoring wave fronts with reflectionless sheets, *Phys. Rev. Lett.*, **110**(19), 197401.

34. Tretyakov, S. (2003). *Analytical Modeling in Applied Electromagnetics* (Artech House).

35. Grbic, A., and Merlin, R. (2008). Near-field focusing plates and their design, *IEEE Trans. Antennas Propag.*, **56**(10), 3159–3165.

36. Zhou, J., Koschny, T., et al. (2005). Saturation of the magnetic response of split-ring resonators at optical frequencies, *Phys. Rev. Lett.*, **95**(22), 223902.

37. Zhao, Y., and Alù, A. (2011). Manipulating light polarization with ultrathin plasmonic metasurfaces, *Phys. Rev. B*, **84**(20), 205428.

38. Aydin, K., Bulu, I., et al. (2005). Investigation of magnetic resonances for different split-ring resonator parameters and designs, *New J. Phys.*, **7**(1), 168.

39. Yen, T.-J., Padilla, W., et al. (2004). Terahertz magnetic response from artificial materials, *Science*, **303**(5663), 1494–1496.

40. Kern, D. J., Werner, D. H., et al. (2005). Metaferrites: using electromagnetic bandgap structures to synthesize metamaterial ferrites, *IEEE Trans. Antennas Propag.*, **53**(4), 1382–1389.

41. Chen, P.-Y., and Alu, A. (2013). Terahertz metamaterial devices based on graphene nanostructures, *IEEE Trans. Terahertz Sci. Technol.*, **3**(6), 748–756.

42. Holloway, C. L., Mohamed, M. A., et al. (2005). Reflection and transmission properties of a metafilm: with an application to a controllable surface composed of resonant particles, *IEEE Trans. Electromagn. Compatib.*, **47**(4), 853–865.

43. Holloway, C. L., Kuester, E. F., et al. (2012). An overview of the theory and applications of metasurfaces: the two-dimensional equivalents of metamaterials, *IEEE Antennas Propag. Mag.*, **54**(2), 10–35.

44. Li, Y., Liang, B., et al. (2013). Reflected wavefront manipulation based on ultrathin planar acoustic metasurfaces, *Sci. Rep.*, **3**, 2546.

45. Sun, S., He, Q., et al. (2012). Gradient-index meta-surfaces as a bridge linking propagating waves and surface waves, *Nat. Mater.*, **11**(5), 426–431.

46. Grady, N. K., Heyes, J. E., et al. (2013). Terahertz metamaterials for linear polarization conversion and anomalous refraction, *Science*, **340**(6138), 1304–1307.
47. Pors, A., Nielsen, M. G., et al. (2013). Broadband focusing flat mirrors based on plasmonic gradient metasurfaces, *Nano Lett.*, **13**(2), 829–834.
48. Yang, Y., Wang, W., et al. (2014). Dielectric meta-reflectarray for broadband linear polarization conversion and optical vortex generation, *Nano Lett.*, **14**(3), 1394–1399.
49. Sun, S., Yang, K.-Y. et al. (2012). High-efficiency broadband anomalous reflection by gradient meta-surfaces, *Nano Lett.*, **12**(12), 6223–6229.
50. Alu, A., and Engheta, N. (2008). Tuning the scattering response of optical nanoantennas with nanocircuit loads, *Nat. Photonics.*, **2**(5), 307–310.
51. Engheta, N. (2002). An idea for thin subwavelength cavity resonators using metamaterials with negative permittivity and permeability, *IEEE Antennas Wireless Propag. Lett.*, **1**(1), 10–13.
52. Ordal, M. A., Bell, R. J., et al. (1985). Optical properties of fourteen metals in the infrared and far infrared: Al, Co, Cu, Au, Fe, Pb, Mo, Ni, Pd, Pt, Ag, Ti, V, and W, *Appl. Opt.*, **24**(24), 4493–4499.
53. Adachi, S. (1985). GaAs, AlAs, and AlxGa1- xAs: Material parameters for use in research and device applications, *J. Appl. Phys.*, **58**(3), R1–R29.
54. Wu, Y., Zhang, C., et al. (2014). Intrinsic optical properties and enhanced plasmonic response of epitaxial silver, *Adv. Mater.*, **26**(35), 6106–6110.
55. Aieta, F., Genevet, P., et al. (2012). Aberration-free ultrathin flat lenses and axicons at telecom wavelengths based on plasmonic metasurfaces, *Nano Lett.*, **12**(9), 4932–4936.
56. Li, X., Xiao, S., et al. (2012). Flat metasurfaces to focus electromagnetic waves in reflection geometry, *Opt. Lett.*, **37**(23), 4940–4942.
57. Zabelin, V., Dunbar, L., et al. (2007). Self-collimating photonic crystal polarization beam splitter, *Opt. Lett.*, **32**(5), 530–532.
58. Ao, X., Liu, L., et al. (2006). Polarization beam splitter based on a two-dimensional photonic crystal of pillar type, *Appl. Phys. Lett.*, **89**(17), 171115.
59. Kwon, D.-H., and Werner, D. H. (2008). Polarization splitter and polarization rotator designs based on transformation optics, *Opt. Express*, **16**(23), 18731–18738.
60. Lin, Z., Ramezani, H., et al. (2011). Unidirectional invisibility induced by P T-symmetric periodic structures, *Phys. Rev. Lett.*, **106**(21), 213901.
61. Zhu, X., Feng, L., et al. (2013). One-way invisible cloak using parity-time symmetric transformation optics, *Opt. Lett.*, **38**(15), 2821–2824.

62. Li, J., and Pendry, J. (2008). Hiding under the carpet: a new strategy for cloaking, *Phys. Rev. Lett.*, **101**(20), 203901.
63. Valentine, J., Li, J., et al. (2009). An optical cloak made of dielectrics, *Nat. Mater.*, **8**(7), 568–571.
64. Bermel, P., Luo, C., et al. (2007). Improving thin-film crystalline silicon solar cell efficiencies with photonic crystals, *Opt. Express*, **15**(25), 16986–17000.
65. Pala, R. A., White, J., et al. (2009). Design of plasmonic thin film solar cells with broadband absorption enhancements, *Adv. Mater.*, **21**(34), 3504–3509.

Chapter 7

RF/Optical Scattering Manipulation Using Metasurface Coatings and Plasmonic Loadings

Zhi Hao Jiang, Anastasios H. Panaretos, and Douglas H. Werner

Department of Electrical Engineering, The Pennsylvania State University, University Park, PA 16802, USA

dhw@psu.edu

This chapter presents the recent progress on scattering manipulation of objects using anisotropic metasurfaces in the microwave (MW) range and plasmonic loadings at optical wavelengths. Different from previously reported transformation optics-enabled coatings, they are achieved based on modifying the complex Mie scattering coefficients of an object. In particular, the nonvanishing radial response of an anisotropic metasurface is exploited to accomplish near-perfect cloaking and angle-tolerant illusion for objects beyond the quasi-static limit. In the second part of this chapter, it is demonstrated how plasmonic core-shell particles can provide a compact and robust solution toward the realization of

Transformation Wave Physics: Electromagnetics, Elastodynamics, and Thermodynamics
Edited by Mohamed Farhat, Pai-Yen Chen, Sebastien Guenneau, and Stefan Enoch

ISBN 978-981-4669-95-5 (Hardcover), 978-981-4669-96-2 (eBook)
www.panstanford.com

nanocircuit loads that offer unprecedented flexibility in tuning the response of a nanodipole-type antenna. In particular it is shown that plasmonic core-shell particles can be engineered such that they exhibit internal resonances which resemble the open- and short-circuit states of either a series or a parallel LC resonator. It is further demonstrated that such engineered internal resonating characteristics of core-shell particles are a consequence of their nanomixture characteristics, which can be exploited for nanodipole antenna tuning. Indicative examples are provided that demonstrate the tuning range that core-shell particles are capable of offering.

7.1 Introduction

Scattering is one of the most fundamental and widely used wave–matter interaction phenomena which has been exploited in various fields including radio-frequency (RF) and optical sensing, communications, and so on [1, 2]. Conventional methods for the manipulation of scattering signatures of objects include modifying the shapes, changing constitutive materials, and adding structured material coatings. Over the past decade, the rapid development of the field of metamaterials and plasmonics has provided scientists and engineers with unprecedented capabilities to manipulate and exploit the electromagnetic scattering properties of objects. These advancements have led to a variety of devices and components with novel scattering properties, including invisibility and transparency, optical illusion, directional beam bending, superscattering, and so on [3–7]. In this chapter, the theory and associated design methodology along with several full-wave numerical examples as well as experimental demonstrations of metasurface and plasmonic loadings for achieving comprehensive scattering manipulation in the microwave (MW) and optical regimes, respectively, will be presented. As an extension, the unique switching properties of plasmonic core-shell nanoparticles will also be exploited to accomplish resonating mode control of nanodipole antennas. A time dependence of $e^{j\omega t}$ is adopted throughout this chapter.

7.2 Metasurface Coatings for Cloaking and Illusion

In this section, we will first present the theory and full-wave numerical simulations of ultrathin anisotropic metasurface coatings to achieve cloaking, which is one of the most interesting effects. In general, two approaches have been employed. One technique is based on transformation optics (TO), which exploits the inhomogeneous and anisotropic properties of bulk metamaterials for bending electromagnetic waves [6–8] around a region to be hidden. However, the TO-enabled cloaking devices often require a thickness larger than that of the cloaked region [8, 9], which makes them impractical in many applications. In addition, the employed coordinate transformation prohibits the electromagnetic communication between regions inside and outside the cloak, causing undesired isolation when the cloaked object is a sensor or an antenna [10, 11]. A second method toward invisibility and transparency has also been introduced based on Mie scattering [12–16]. Single- and multilayer coatings with either low or negative values of permittivity have been previously studied. More recently, cloaks realized using a thin metasurface have been proposed [17–19]. This technique minimizes the dominant Mie scattering coefficient by engineering the surface impedance of the metasurface, thus accomplishing significant scattering reduction in the quasi-static limit. Two primary advantages of the metasurface cloaks are their low-profile property and the allowance of nonvanishing fields or currents to be induced in the cloaked object. These properties makes them more intriguing than coordinate transformation and bulk material coating based cloaks for a wide range of potential applications including scattering signature control, noninvasive probing, defense security, and so on.

So far, near-perfect metasurface cloaks have been achieved mainly in the quasi-static limit regime, where only the zeroth-order Mie coefficient of the object to be cloaked is nonzero. Under such circumstances, a scalar impedance surface [20] with a zero longitudinal response is enough to fulfill the design goal. However, as the size of the object increases, higher-order Mie scattering coefficients start to contribute to the total scattering cross section and therefore more degrees of freedom are desired to achieve cloaking. One method is to increase the number of layers, which will increase the overall volume of the device [15, 21]. Here we consider an alternative approach, while still maintaining only a single metasurface layer. By introducing a nonzero longitudinal response to the metasurface, the first and second Mie scattering coefficients can be suppressed simultaneously, thereby facilitating near-perfect scattering reduction for moderately sized objects [22].

In addition to cloaking, illusion optics is also another field of great interest, which involves the transformation of the scattering signature of an object into that of another [23]. It represents more comprehensive scattering manipulation, thus it is more general than cloaking. So far, TO has been used to create various illusions devices. Similar to TO-based cloaking designs, illusion coatings suffer from the same issues of bulky sizes and/or the isolation between regions inside and outside the coating [24–26]. Here, we will show that metasurface anisotropy can be exploited to achieve low-profile, ultrathin illusion coatings through matching of complex Mie scattering coefficients. In contrast to the normal incidence limitation that characterizes previously demonstrated illusion devices, the metasurface illusion effect with the proposed approach can be maintained within a reasonable field of view (FOV).

7.2.1 *Scattering from an Anisotropic Metasurface-Coated Cylinder*

As shown in Fig. 7.1, consider a transverse magnetic (TM_z)-polarized plane wave obliquely incident upon an infinitely long dielectric cylinder with radius a_1, permittivity ε_1, covered by a metasurface at a radius of a_c with surface electric and magnetic polarizability density tensors denoted as $\chi_{\mathbf{E}} = \text{diag}[\chi_{E\rho}, \chi_{E\phi}, \chi_{Ez}]$ and $\chi_{\mathbf{M}} = \text{diag}[\chi_{M\rho}, \chi_{M\phi}, \chi_{Mz}]$, respectively [27]. A thin slab of material with permittivity ε_c is used to fill the space between the inner dielectric cylinder and the metasurface coating. The case of a cloaked perfect electric conductor (PEC) cylinder can be readily analyzed following

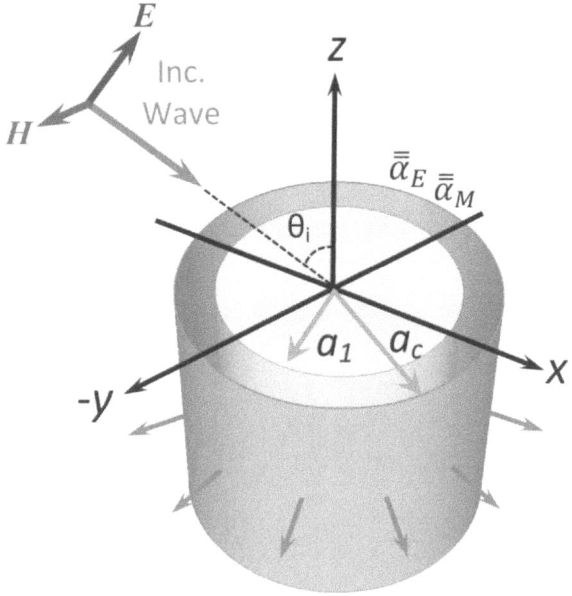

Figure 7.1 A metasurface-coated dielectric or conducting cylinder illuminated by an obliquely incident TM_z-polarized plane wave.

the same procedure, thus the associated analytical expressions will not be presented for succinctness. The metasurface can be realized by a periodic array of subwavelength-patterned metallic elements. Hence, the electromagnetic properties of the metasurface can be homogenized and approximated by the surface electric and magnetic polarizability density tensors, as indicated above. The tangential electric and magnetic fields at the surface of the inner coated cylinder and the outer metasurface satisfy the following conditions [28]:

$$\left(E_z^{a_c^+} - E_z^{a_c^-}\right) = -j\omega\varepsilon_0\chi_{M\varphi}\left(\frac{H_\varphi^{a_c^+} + H_\varphi^{a_c^-}}{2}\right) + \frac{\partial}{\partial z}\left(\chi_{E\rho}\frac{E_\rho^{a_c^+} + E_\rho^{a_c^-}}{2}\right) \tag{7.1}$$

$$\left(E_z^{a_1^+} - E_z^{a_1^-}\right) = 0 \tag{7.2}$$

$$\left(E_\varphi^{a_c^+} - E_\varphi^{a_c^-}\right) = -j\omega\varepsilon_0\chi_{Mz}\left(\frac{H_z^{a_c^+} + H_z^{a_c^-}}{2}\right)$$

$$-\frac{1}{\rho}\frac{\partial}{\partial\varphi}\left(\chi_{E\rho}\frac{E_\rho^{a_c^+} + E_\rho^{a_c^-}}{2}\right) \tag{7.3}$$

$$\left(E_\varphi^{a_1^+} - E_\varphi^{a_1^-}\right) = 0 \tag{7.4}$$

$$\left(H_z^{a_c^+} - H_z^{a_c^-}\right) = -j\omega\varepsilon_0\chi_{E\varphi}\left(\frac{E_\varphi^{a_c^+} + E_\varphi^{a_c^-}}{2}\right)$$

$$+\frac{\partial}{\partial z}\left(\chi_{M\rho}\frac{H_\rho^{a_c^+} + H_\rho^{a_c^-}}{2}\right) \tag{7.5}$$

$$\left(H_z^{a_1^+} - H_z^{a_1^-}\right) = 0 \tag{7.6}$$

$$\left(H_\varphi^{a_c^+} - H_\varphi^{a_c^-}\right) = j\omega\varepsilon_0\chi_{Ez}\left(\frac{E_z^{a_c^+} + E_z^{a_c^-}}{2}\right)$$

$$-\frac{1}{\rho}\frac{\partial}{\partial\varphi}\left(\chi_{M\rho}\frac{H_\rho^{a_c^+} + H_\rho^{a_c^-}}{2}\right) \tag{7.7}$$

$$\left(H_\varphi^{a_1^+} - H_\varphi^{a_1^-}\right) = 0. \tag{7.8}$$

By enforcing these boundary conditions, the nth-order TM_z and TE_z scattering coefficients $a_{n,\text{TM}}$ and $a_{n,\text{TE}}$ can be obtained. For a practical ultrathin metasurface comprised of an array of metallic patterns in a cylindrical configuration, the surface polarizability tensor parameters χ_{Mz}, $\chi_{M\phi}$, and $\chi_{E\rho}$ usually vanish provided that no magnetic materials are employed. At oblique angles of incidence, a nonvanishing z component of the magnetic field is a natural consequence of cross-polarization due to the asymmetry as seen by an incident wave, resulting in nonzero $a_{n,\text{TE}}$ $(n = 1, 2, \ldots)$ [29]. At normal incidence ($\theta_{\text{i}} = 90°$), the TE_z mode scattered field vanishes since there is no induced cross-polarization, such that the scattering coefficient $a_{n,\text{TM}}$ can be simplified as

$$a_{nTM} = \frac{\begin{vmatrix} J_n(k_1a_1) & J_n(k_0a_1) & Y_n(k_0a_1) & 0 \\ J'_n(k_1a_1)/\eta_1 & J'_n(k_0a_1)/\eta_0 & Y'_n(k_0a_1)/\eta_0 & 0 \\ 0 & J_n(k_0a_c) & Y_n(k_0a_c) & J_n(k_0a_c) \\ 0 & J'_n(k_0a_c)/\eta_0 - J_n(k_0a_c)/Z_s & Y'_n(k_0a_c)/\eta_0 - Y_n(k_0a_c)/Z_s & J'_n(k_0a_c)/\eta_0 \end{vmatrix}}{\begin{vmatrix} J_n(k_1a_1) & J_n(k_0a_1) & Y_n(k_0a_1) & 0 \\ J'_n(k_1a_1)/\eta_1 & J'_n(k_0a_1)/\eta_0 & Y'_n(k_0a_1)/\eta_0 & 0 \\ 0 & J_n(k_0a_c) & Y_n(k_0a_c) & H_n(k_0a_c) \\ 0 & J'_n(k_0a_c)/\eta_0 - J_n(k_0a_c)/Z_s & Y'_n(k_0a_c)/\eta_0 - Y_n(k_0a_c)/Z_s & H_n^{(2)\prime}(k_0a_c)/\eta_0 \end{vmatrix}}$$

$$Z_s = 1/\omega\varepsilon_0\chi_{Ez} + \frac{\omega\mu_0 a_c^2}{n^2\chi_{M\rho}^2} \tag{7.9}$$

where the primes in the terms J'_n, Y'_n, and $H_n^{(2)'}$ denote derivatives of the functions with respect to the arguments. By properly choosing the values of a_c, ε_c, $\chi_\mathbf{E}$, and $\chi_\mathbf{M}$, both the real and imaginary parts of $a_{n,\mathrm{TM}}$ and $a_{n,\mathrm{TE}}$ can be manipulated, thereby providing the capability to control the scattering signature of the coated object.

7.2.2 *Metasurface Cloaking beyond the Quasi-Static Limit*

To achieve the cloaking effect, the magnitudes of $a_{n,\mathrm{TM}}$ and $a_{n,\mathrm{TE}}$ must be suppressed, thereby reducing the total scattering width (SW). For a single metasurface at normal incidence in the TM_z mode case, we have two degrees of freedom available for controlling its electromagnetic properties, that is, χ_{Ez} and $\chi_{M\rho}$. The zeroth-order scattering coefficient $a_{0,\mathrm{TM}}$ can be suppressed by varying the value of χ_{Ez} but will not be affected by the value of $\chi_{M\rho}$. This is because the radial component of the magnetic field is zero for the fundamental zeroth-order cylindrical harmonic mode. The higher-order modes, however, can be controlled by both χ_{Ez} and $\chi_{M\rho}$. Hence, we are able to suppress the first two scattering coefficients of a cylinder with a moderate value of radius using the anisotropic metasurface, which cannot be achieved by a simple scalar impedance surface.

7.2.2.1 Metasurface cloaking for dielectric cylinders

We first demonstrate how a dielectric object can be cloaked by considering a dielectric cylinder with a relative permittivity $\varepsilon_{1\mathrm{r}} = 4$ and radius of $a_1 = 0.2\lambda_0$ coated by an annular metasurface positioned at a radius of $a_\mathrm{c} = 0.22\lambda_0$, where λ_0 is the free-space wavelength. The metasurface and the inner cylinder are separated by air. To achieve near-zero SW at the target frequency, which is chosen to be 2.5 GHz, the metasurface properties are designed to be $\chi_{Ez}/\lambda_0 = 0.97$ and $\chi_{M\rho}/\lambda_0 = -2.3$. It has been shown in the literature that an inductive impedance surface is capable of cloaking a dielectric cylinder in the quasi-static limit [17, 18]. For example, when the radius of a cylinder comprised of the same material is $0.1\lambda_0$ and the radius of the impedance surface coating is $0.11\lambda_0$, the required χ_{Ez}/λ_0 value to achieve near-perfect cloaking is −0.23, that is, a surface reactance $Z_\mathrm{s} = j261\Omega$, which is inductive. However,

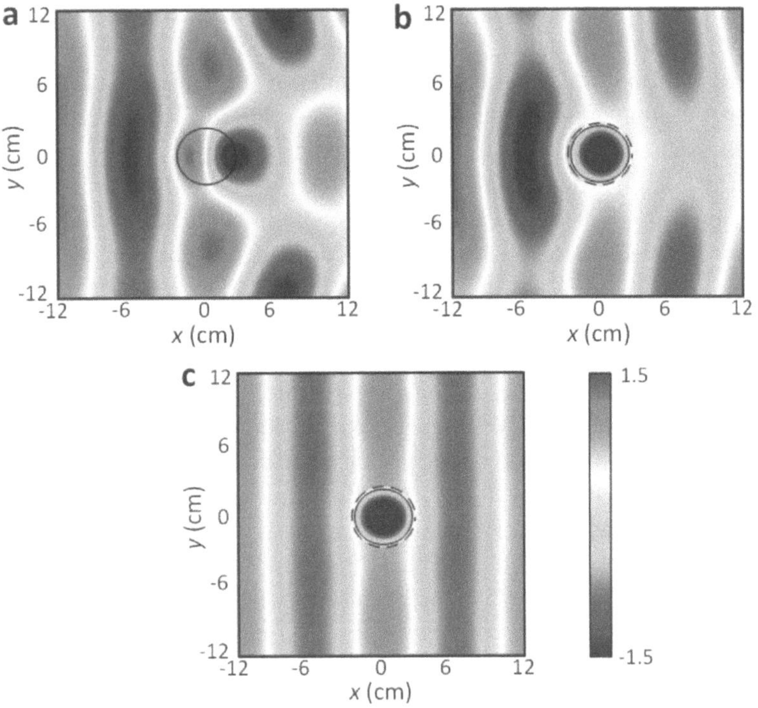

Figure 7.2 Snapshots of the analytically calculated total electric field distribution for (a) a bare dielectric cylinder with $a_1 = 0.2\lambda_0$, $\varepsilon_{1r} = 4$, and $\mu_{1r} = 1$, (b) the same dielectric cylinder coated by an annular scalar impedance surface with parameters $a_c = 0.22\lambda_0$ and $\chi_{Ez}/\lambda_0 = 0.964$, and (c) the same dielectric cylinder coated by an annular anisotropic metasurface with parameters $a_c = 0.22\lambda_0$, $\chi_{Ez}/\lambda_0 = 0.964$, and $\chi_{M\rho}/\lambda_0 = -2.3$.

as the radius of the cylinder increases, a capacitive metasurface is desired because the reactive properties of the cylinder change due to increased wave propagation retardation.

Figure 7.2a–c shows the analytically calculated distributions of the total electric field in the xy plane, with a plane wave polarized in the z direction illuminating from the $-x$ direction. For the dielectric cylinder alone, standing wave patterns can be observed due to scattering. The waves passing through the cylinder in the $+x$ direction experience a phase lag compared to waves propagating in the air region outside the cylinder. With a coating comprised of a scalar impedance surface, only the zeroth-order Mie scattering mode can be suppressed, so that the fields in the top and bottom regions of the cylinder are less disturbed. But the remaining presence of a strong first-order Mie scattering mode causes evident reflection in the $-x$ direction and field nulls in the $+x$ direction (see Fig. 7.2b). However, when the anisotropic metasurface is employed, significant improvement is achieved as shown in Fig. 7.2c. The characteristics of the incident plane wave are well maintained as it passes through the coated dielectric cylinder region, with almost no back reflection or forward scattering. The anisotropic metasurface achieves 98.7% SW reduction while the conventional scalar impedance surface yields only 66.4% reduction. Different from the TO-based cloaks, in this case, the electric field is able to penetrate through the metasurface and propagate inside the area enclosed by the coating. More interestingly, the field in the middle of the dielectric cylinder is enhanced, making the "cloaked object" capable of detecting incoming signals while at the same time being "less visible" to an outside detector.

7.2.2.2 Metasurface cloaking for conducting cylinders

In the next example, we consider the metasurface cloaking of a conducting cylinder, which is of more interest since the hidden object is impenetrable in this case. Here a PEC cylinder was chosen with a radius of $a_1 = 0.167\lambda_0$ while the annular anisotropic metasurface coating is placed at a radius of $a_c = 0.217\lambda_0$. The surface polarizability parameters of the metasurface required in order to achieve near-perfect cloaking are $\chi_{Ez}/\lambda_0 = 0.57$ and $\chi_{M\rho}/\lambda_0 = -0.34$, resulting in a SW reduction of 98.6%. Figure 7.3a–c reports the analytically calculated distributions of the total electric field in the xy plane, with a plane wave polarized in the z direction illuminating from the $-x$ direction. It can be observed that the bare PEC cylinder causes standing wave patterns due to scattering. With a coating comprised of a scalar impedance surface that has an optimum isotropic surface polarizability $\chi_{Ez}/\lambda_0 = 0.43$, the zeroth-order cylindrical harmonic scattering is suppressed, so that the fields in the top and bottom regions of the cylinder are less disturbed. However, the remaining presence of a strong first-

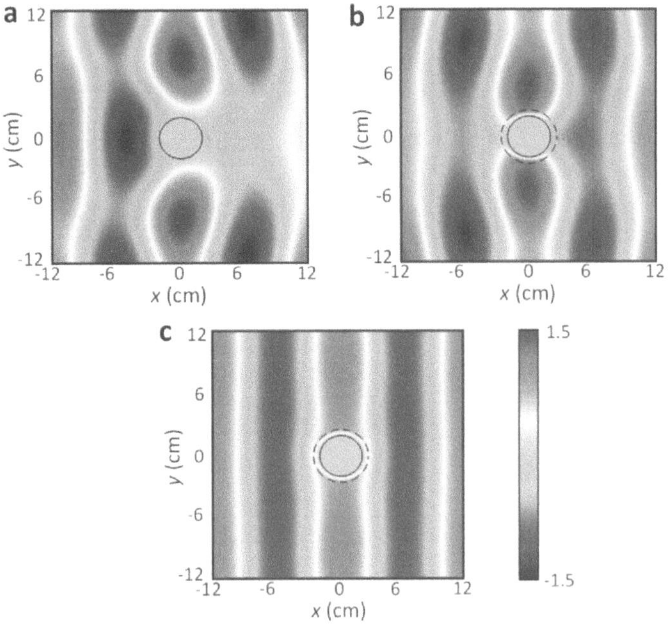

Figure 7.3 Snapshots of the analytically calculated total electric field distribution for (a) a bare PEC cylinder with $a_1 = 0.167\lambda_0$, (b) the same PEC cylinder coated by an annular scalar impedance surface with parameters $a_c = 0.217\lambda_0$ and $\chi_{Ez}/\lambda_0 = 0.43$, and (c) the same PEC cylinder coated by an annular anisotropic metasurface with parameters $a_c = 0.22\lambda_0$, $\chi_{Ez}/\lambda_0 = 0.57$, and $\chi_{M\rho}/\lambda_0 = -0.34$.

order cylindrical harmonic scattering mode, which produces dipolar scattered fields, causes evident reflection in the $-x$ direction and field nulls in the $+x$ direction (see Fig. 7.3b). When the anisotropic metasurface coating is employed, significant scattering reduction is achieved. The scattering reduction using the scalar impedance surface is only 55.8% while the anisotropic metasurface achieves a much improved 98.6% reduction. Similar to the dielectric cylinder metasurface cloak case, the waves are able to propagate inside the region between the coating and the cylinder. Nonzero tangential magnetic fields exist on the surface of the PEC cylinder, which leads to a nonzero induced current that can be utilized for noninvasive sensing applications.

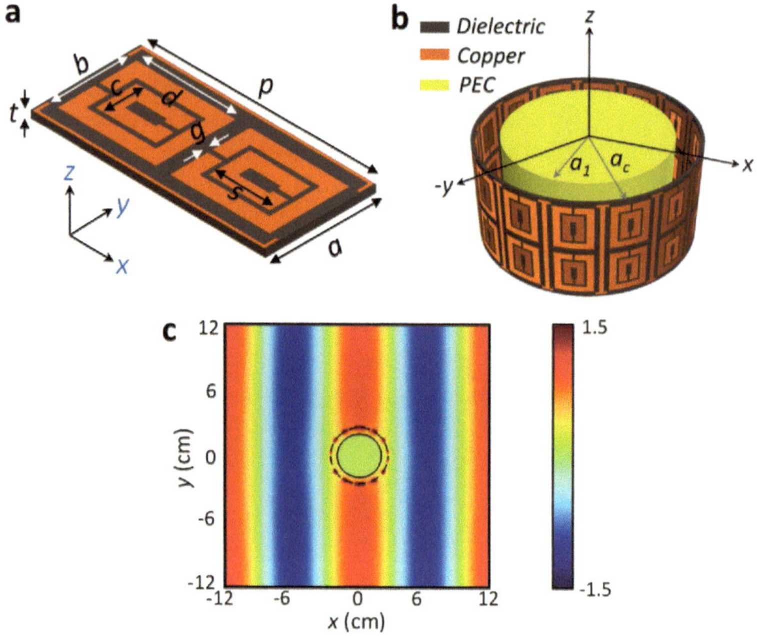

Figure 7.4 (a) Unit cell of the optimized anisotropic metasurface. (b) The configuration of a PEC cylinder covered by an anisotropic metasurface. The dimensions are $a = 12$, $p = 25.2$, $b = 8.4$, $c = 4.8$, $d = 10.8$, $s = 6$, $g = 0.48$, $t = 0.72$, $a_1 = 20$, and $a_c = 26$ (all in millimeters). The substrate material of the coating is Rogers RT/duroid 5880 ($\varepsilon_r = 2.2$ and $\delta_{tan} = 0.009$). (c) Snapshots of the total electric field distribution using full-wave simulations for the PEC cylinder coated by the designed anisotropic metasurface.

To realize this design using a practical metasurface, a composite broadside-coupled split-ring resonator (SRR) [30, 31] and an I-shaped short dipole (ISD) [32, 33] composite array was used to provide the desired surface polarizability properties. The metasurface unit cell is shown in Fig. 7.4a, which has a width of 11.64 mm so that 14 unit cells will exactly fit into the coating circumference as displayed in Fig. 7.4b. The thickness of the substrate layer is only $0.006\lambda_0$, which can be considered as a sheet with essentially no electrical thickness, that is, the retardation effect can be ignored. The metallic patterns of the metasurface are printed on both sides of the dielectric substrate. During the design of the metasurface, the high-frequency structure simulator (HFSS) finite element solver was used to perform the full-wave unit cell plane-wave scattering simulations at different incident angles in order to retrieve the effective surface polarizability tensor parameters using an anisotropic inversion technique [32]. Due to the subwavelength size of the unit cell in the φ direction, the metasurface can be treated as locally flat such that its electromagnetic properties are well approximated by a planar configuration [32]. Since the SRR provide negligible electric surface polarizability compared to that produced by the ISDs, the two components are designed separately. The array of ISDs can be modeled as a series LC resonator. The length of the end loads and the separation distance of the end loads between adjacent cells controls the capacitance, while width and length of the central section of the I-shaped element determine the inductance. These four parameters are first tuned to achieve the desired χ_{Ez}/λ_0 value. The SRRs are then designed to obtain the desired $\chi_{M\rho}/\lambda_0$ value by tuning the radius and gap size of the SRRs. Finally, the ISDs and SRRs are combined together and fine-tuned to account for the weak mutual coupling between them. The discrete metasurface-coated PEC cylinder was simulated using HFSS in a parallel plate waveguide with a height the same as that of one unit cell. A z-polarized plane wave is illuminating from the $-x$ direction. Fig. 7.4c reports the distributions of the total electric field in the xy plane obtained from the full-wave simulation. The full-wave simulation result achieves an excellent agreement with that obtained from the analytical solution as displayed in Fig. 7.3c, confirming the validity of the proposed metasurface concept and the analytical formulation. The good correspondence also indicates that the metasurface indeed behaves effectively as a homogenous quasi-2D coating with predesigned surface polarizability properties, except in the extreme close proximity of the metasurface.

Different from the plasmonic cloak, which is based on the induced dipole moment cancellation between the core and shell with permittivity values of opposite signs, the metasurface cloak is achieved by induced current interference cancellation. Figure 7.5a–c shows the magnitudes and phases of the average induced currents on the outer metasurface coating and the surface of the inner PEC cylinder corresponding to the magnetic field of the nth ($n = 0, 1, 2$)-

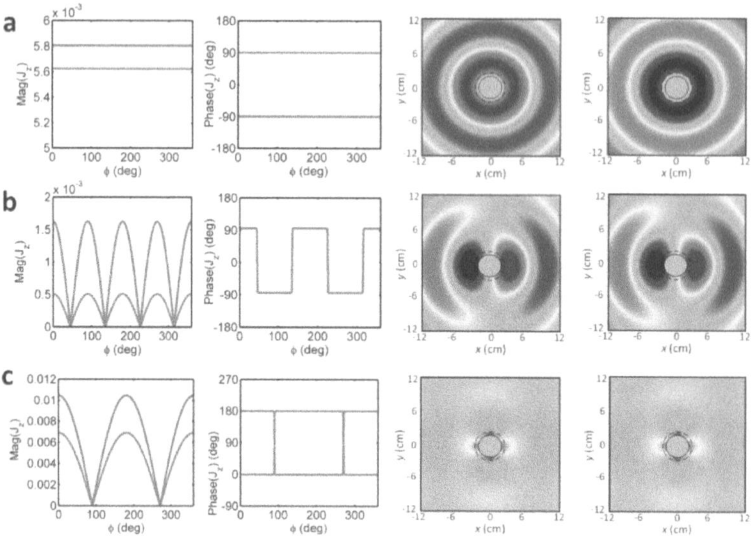

Figure 7.5 Induced current amplitudes and phases on the anisotropic metasurface (dashed blue) and the surface of the PEC cylinder (solid red) for (a) the zeroth-, (b) first-, and (c) second-order scattering modes. Corresponding snapshots of the distributions of the electric field radiated from the two induced current sheets are also plotted. Left column: E-field radiated from induced currents on the surface of the PEC cylinder. Right column: E-field radiated from induced currents on the anisotropic metasurface.

order scattering modes. It can be seen from these plots that for the fundamental zeroth- and first-order scattering modes, the average induced currents on the metasurface are out of phase and slightly weaker when compared to those on the surface of the PEC cylinder. Such a scattering problem can be understood as a two-step process: the incident wave induces currents on the scatterers followed by the reradiation of these induced currents to form the scattered fields. To cancel out the scattered fields outside the metasurface, the response of the two induced current sheets need to be almost exactly out of phase since the separation distance between them is negligible compared to the operating wavelength. Due to the logarithmic drop in the magnitude of the zeroth-order Hankel function for small arguments, the induced currents on the surface of the PEC cylinder must be stronger than those on the metasurface. For the second-order scattering mode, the induced currents are in phase such that the quadrupolar scattering mode is not cancelled. These currents are too weak to cause noticeable scattering. The surface currents induced on a bare PEC cylinder are also included in Fig. 7.5a–c for the purpose of comparison.

The radiated fields from the current sheets belonging to different scattering modes can be expressed using the free space Green's function in a cylindrical coordinate system. The snapshots of the distributions of the electric field radiated from the two induced current sheets individually are also shown in Fig. 7.5a–c. The scattered (or radiated) fields for the first three modes have omnidirectional, bidirectional, and quad-directional patterns, respectively. The field patterns for the first two modes are out of phase, thus are cancelled out, whereas the field patterns for the third mode are superimposed in phase. Above all, the examination of the averaged induced currents and their resulting radiated fields for different scattering modes clearly explain the physical mechanism of the observed scattering reduction phenomenon.

7.2.3 *Angle-Tolerant Metasurface Illusion*

As an extension of cloaking, which only involves the suppression of the magnitudes of the Mie scattering coefficients, the scattering signature of an object can be transformed into that of another preselected object by tailoring both the real and imaginary parts of the nonvanishing Mie scattering coefficients, even within a certain range of the incident angle. Here, as a proof-of-concept example, we demonstrate how a metasurface can be designed to greatly increase the scattering strength of an object. Particularly, the scattering perception of a dielectric cylinder will be altered to appear like that of a PEC cylinder, which produces much stronger scattered fields.

The dielectric cylinder to be coated has a radius of $a_1 = 0.106\lambda_0$ and a relative permittivity of 2.1, while the targeted conducting cylinder is comprised of PEC with a radius of $a_t = 0.16\lambda_0$ at 2.5 GHz. To mimic the scattering signature of the PEC cylinder within a FOV up to 20° off normal, the simplest way is to shield the dielectric cylinder with a copper sheet placed at a radius of a_t. However, this completely blocks any signal from transmitting into or out of the region inside the shielding. In contrast, with an anisotropic metasurface, the illusion effect can be achieved with a coating that has a lower profile. Simultaneously, communication between the interior of the metasurface coating and the outside world is permitted. This is significantly different from the TO-enabled illusion devices where the inhomogeneous and anisotropic material coating has a much larger profile than the targeted object. The optimized parameters of the metasurface coating were found to be $a_c = 0.111\lambda_0$, $\chi_{Ez}/\lambda_0 = 0.55$ and $\chi_{M\rho}/\lambda_0 = 0.13$.

Figure 7.6a–c compares the analytically calculated distributions of the total electric field in the *xy* plane for the three structures under

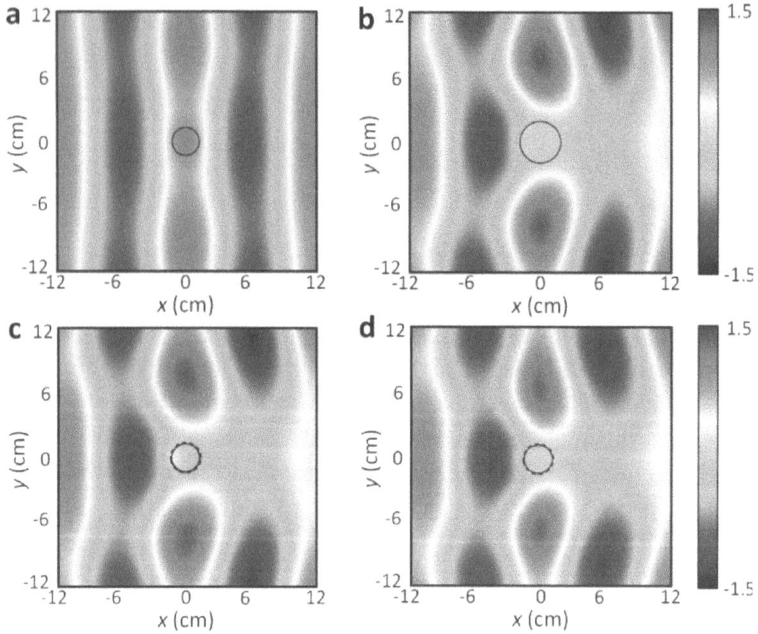

Figure 7.6 Snapshots of the analytically calculated total electric field distribution for (a) a bare dielectric cylinder with $a_1 = 0.106\lambda_0$, $\varepsilon_{1r} = 2.1$, and $\mu_{1r} = 1$, (b) the targeted PEC cylinder with $a_1 = 0.16\lambda_0$, (c) the same dielectric cylinder coated by an annular homogeneous anisotropic metasurface with parameters $a_c = 0.111\lambda_0$, $\chi_{Ez}/\lambda_0 = 0.55$, and $\chi_{M\rho}/\lambda_0 = 0.13$, and (d) the same dielectric cylinder coated by an annular discrete anisotropic metasurface comprised by an SRR-ISD composite array.

plane-wave excitation at normal incidence ($\theta_i = 90°$): a bare PEC cylinder, a bare dielectric cylinder, and the dielectric cylinder coated by the custom designed metasurface. A plane wave polarized in the z direction is assumed to be illuminating from the $-x$ direction. It can be seen that the metasurface-coated dielectric cylinder produces a field distribution almost the same as that of the PEC cylinder both qualitatively and quantitatively. Physically, the surface currents induced on the metasurface are engineered to properly interfere with the fields induced inside the inner dielectric cylinder, giving rise to the altered scattered fields. The far-zone bistatic SW as a function of φ for both normal incidence and oblique incidence with $\theta_i = 70°$ are displayed in Fig. 7.7a,b, which were collected at an elevation

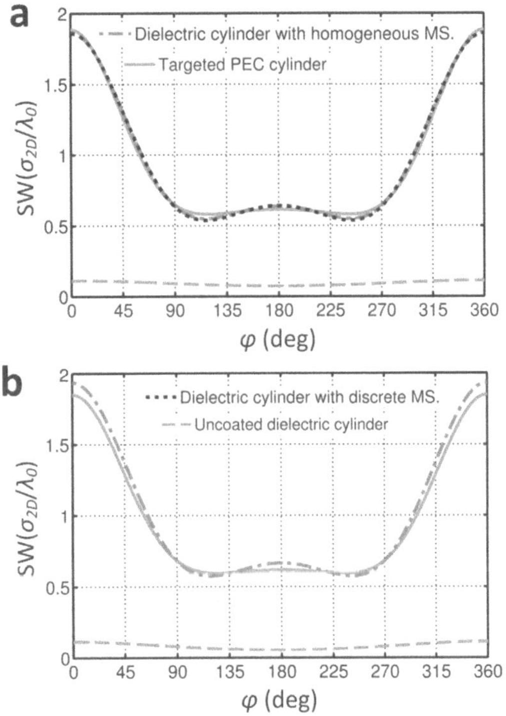

Figure 7.7 The analytically calculated and full-wave simulated far-field SW patterns of a targeted copper cylinder, an uncoated dielectric cylinder, and a dielectric cylinder with a homogenous and discrete metasurface (MS) coating for normal and $\pm 20°$ oblique incidence angles.

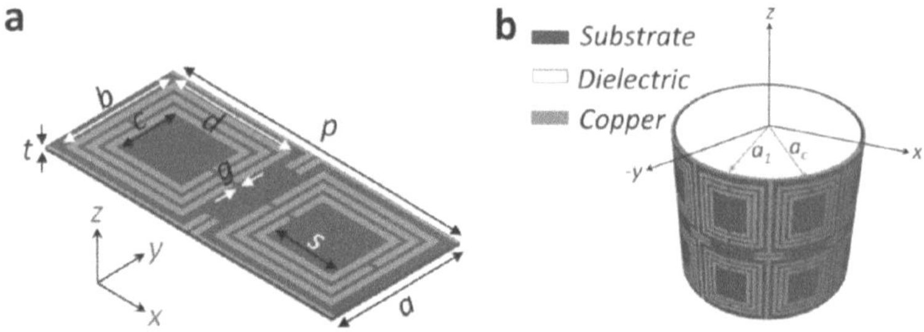

Figure 7.8 (a) Unit cell of the optimized anisotropic metasurface. (b) The configuration of the dielectric cylinder covered by an anisotropic metasurface. The dimensions are $a = 10.47$, $p = 23.47$, $b = 8.9$, $c = 4.7$, $d = 9.6$, $s = 5.4$, $g = 0.3$, $t = 0.25$, $a_1 = 12.7$, and $a_c = 13.34$ (all in millimeters). The substrate material of the coating is Rogers RT/duroid 5880 ($\varepsilon_r = 2.2$ and $\delta_{tan} = 0.009$).

angle of $180° - \theta_i$. It can be seen that the SW patterns produced by the dielectric cylinder coated with the metasurface agree well with those of the PEC cylinder, with an average deviation of only 2.25% at normal incidence and 4.66% at $\pm 20°$ oblique incidence angles. At larger incidence angles of $\pm 30°$, the deviation slightly increases to 8.89%.

Similar to the cloaking example, a composite SSR and ISD array was used to realize the desired surface polarizability properties of the metasurface. The unit cell is depicted in Fig. 7.8a, which has a width of 10.47 mm so that 8 unit cells will exactly fit into the coating circumference as displayed in Fig. 7.8b. The thickness of the substrate layer is merely $0.002\lambda_0$. The metallic patterns of the metasurface are printed only on the outer side of the dielectric substrate. The discrete metasurface-coated dielectric cylinder was simulated using HFSS in a parallel plate waveguide with a height the same as that of one unit cell. A z-polarized plane wave is illuminating from the $-x$ direction. Figure 7.6d reports the distribution of the total electric field in the xy plane obtained from the full-wave simulations. The full-wave simulation results are in excellent agreement with those obtained from the analytical solution, confirming the validity of the proposed metasurface concept. Due to the fact that the unit cells of the metasurface operate at frequencies away from the resonance and only a monolayer is employed, the resulting illusion coating has negligible absorption loss. In Fig. 7.7a, it is shown that the bistatic SW pattern of the discrete metasurface-coated, infinitely long dielectric cylinder at normal incidence well mimics that of the targeted PEC cylinder, with an average deviation of only 2.65%.

To validate the concept, experiments were carried out to show that the metasurface can transform the scattering signature of a finite-length dielectric cylinder into that of a finite-length conducting cylinder by using a composite spiral resonator and ISD array (see Fig. 7.9a). The spiral resonator was chosen due to it ability to provide magnetic out-of-plane response with a more compact size than an SRR. In particular, a $2.34\lambda_0$ long Teflon cylinder is employed. The Teflon has a relative permittivity of 2.1 and a radius of $0.106\lambda_0$. The detailed unit cell geometry is illustrated in the inset of Fig. 7.9a. The metasurface coating was fabricated and wrapped around the Teflon cylinder, as shown in Fig. 7.9b. Four layers of foam, each with a 1 mm thickness, were used as the spacer in between the metasurface and the Teflon cylinder. The far-field SW patterns of the Teflon cylinder with the metasurface coating in the horizontal (xy) plane were measured at normal incidence in an anechoic chamber using two horn antennas. One of the antennas was used as the transmitter while the other was used as the receiver. Due to the finite size of the horn antennas, the scattering information within the $160^\circ \leq \varphi \leq 200^\circ$ angular range could not be obtained. Figure 7.9c shows the simulated far-field SW patterns of the targeted PEC cylinder, the Teflon cylinder with and without the metasurface coating, and the measured far-field SW pattern of the Teflon cylinder with the metasurface coating. It can be seen that very good agreement between simulation and measurement is achieved, showing greatly enhanced scattering due to the metasurface coating. With the metasurface coating, the measured SW pattern exhibits an average deviation of 3.58% from the simulated result of the targeted PEC cylinder. This experiment validates that the anisotropic metasurface is able to manipulate the scattering perception of a finite-length dielectric object by electromagnetically transforming it into that of a conducting cylinder.

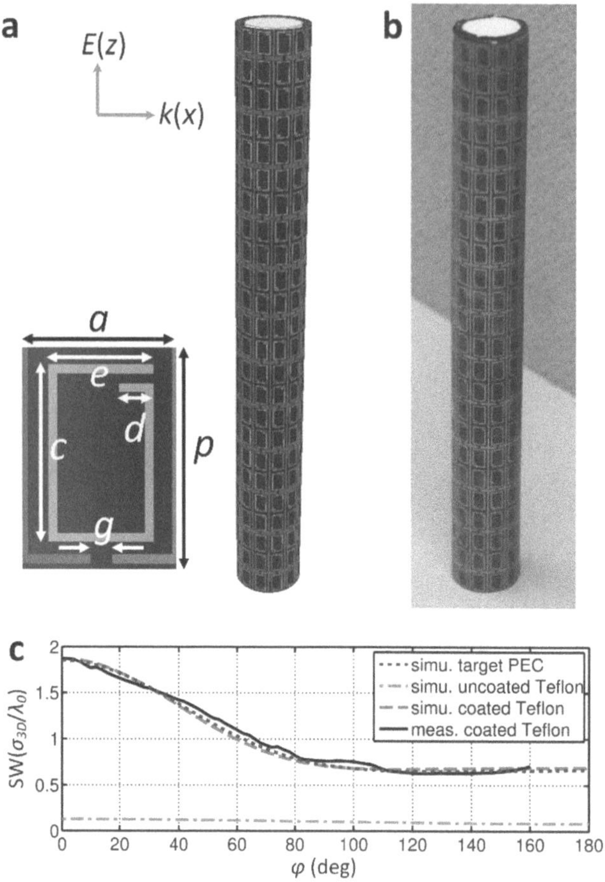

Figure 7.9 (a) Configuration of a finite-length Teflon cylinder coated by an anisotropic metasurface. The dimensions are $a = 8.74$, $p = 11.72$, $c = 9.5$, $d = 1.5$, $e = 6$, and $g = 0.3$ (all in millimeters). The substrate material of the coating is Rogers RT/duroid 5880 ($\varepsilon_r = 2.2$ and $\delta_{\tan} = 0.009$) with a thickness of 0.25 mm. (b) A photograph of the fabricated metasurface surrounding a Teflon cylinder. (c) The simulated far-field SW patterns of the targeted PEC cylinder, the Teflon cylinder with and without the metasurface coating, and the measured far-field SW pattern of the Teflon cylinder with the metasurface coating.

7.3 Optical Plasmonic Core-Shell Particles Exhibiting Zero-Impedance and Zero-Admittance Properties

The realization of electric circuit functionality at the nanoscale is not trivial. This is primarily due to the fact that circuit components are not available at optical frequencies, which is mainly because there is no standard methodology to achieve a response equivalent to that of a capacitor or an inductor. At the same time it would be highly desirable to be able to introduce circuit functionality in optical devices simply because this would permit their response to be custom engineered. In addition to that, the principles of operation for many optical devices heavily rely on their RF and MW counterparts. Consequently, all the methodologies and techniques developed for networks comprised of conventional electronic circuits could be implemented at optical frequencies, if nanocircuit elements were available. Toward this end plasmonic core-shell particles are proposed as a compact and robust solution to mimic the electric response of a general RLC circuit.

In particular we propose plasmonic core-shell particles as a means to devise nanostructures that exhibit the response of either a series or a parallel resonating LC circuit. In the same context it has been demonstrated [34–36] that when the faces of a Janus particle have permittivity opposite in sign, the particle exhibits an internal resonance. The character of this resonance depends on the polarization of the excitation field: it resembles the resonance of either a series or a parallel LC circuit, when the electric field is vertical or parallel to the plane that separates the two faces. The key element for the realization of these properties is the combination of the two opposite in sign permittivity faces. However, at optical frequencies such material contrast can be also realized using plasmonic core-shell particles in the case where the core/shell material is some noble metal, and correspondingly the constitution of the shell/core is some material characterized by a conventional positive permittivity. Consequently it is expected that such structures can be good candidates for the development of series and parallel LC nanoresonators.

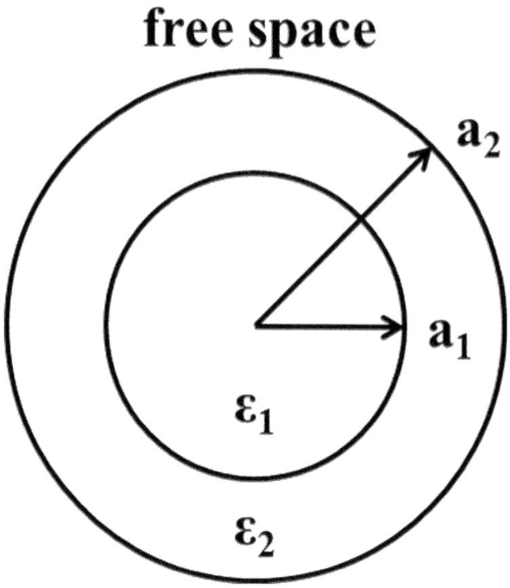

Figure 7.10 Geometry of a radially inhomogeneous stratified dielectric sphere.

To characterize a device, and in this particular case a core-shell particle, in terms of its electric response one needs to first introduce a suitable impedance definition for the structure under study. For this reason we treat the core-shell particle as a radially inhomogeneous spherical transmission line [37–39]. The advantage that this approach offers is that the TM input admittance of a radially inhomogeneous spherical transmission line can be employed to characterize the admittance response of the particle under study. Consequently, the investigation of the possibility that a core-shell particle can function as an LC resonator reduces to the simple task of setting either its admittance, or its impedance, equal to zero.

It should be noted here that for the particles examined in this study it is assumed that they are purely dielectric, and that only the first dipolar mode is sufficient to describe their electromagnetic response. Based on this assumption the necessary conditions are derived so that a core-shell particle can exhibit either a short- or an open-circuit state. A further investigation of these conditions reveals that the aforementioned states can be completely defined as a combination of the shell and core material, as well as the particle's volume fraction. According to our findings the short- and open-circuit states of a core-shell particle are associated with the dipolar response of the core, with respect to the material of the shell. A more intuitive interpretation of these phenomena can be arrived at after examining the low-frequency behavior of a core-shell particle's admittance. First, it is demonstrated that in this frequency limit, the dielectric material properties of a homogeneous dielectric particle are directly mapped to its TM admittance response. The same analysis is performed in the case of a core-shell particle and it is demonstrated that its admittance corresponds to an effective homogeneous sphere whose material properties are given by the effective permittivity of the Clausius–Mossotti (CM) mixing rule: the filler and the host of this mixture is the core and the shell material, respectively. This outcome is of paramount importance because it reveals that the zero admittance and the zero impedance states of the particle are essentially associated with an effective dielectric permittivity that has zero and infinite value, respectively.

7.3.1 *Elements of Radially Inhomogeneous Spherical Transmission Line Theory, and Impedance Characterization of a Core-Shell Particle*

Let us consider the two layer spherical structure shown in Fig. 7.10. The exterior (shell) and interior (core) layers are characterized by relative permittivity values of ε_2 and ε_1 respectively, while the surrounding medium is considered as free space. The wavenumbers in the core, the shell, and the surrounding medium are defined as $k_1 = \omega/c_0\sqrt{\varepsilon_1}$, $k_2 = \omega/c_0\sqrt{\varepsilon_2}$, and $k_3 = \omega/c_0$, respectively, while c_0 is the speed of light. Additionally, the refractive index of the core and the shell are given by $m_1 = k_1/k_3$ and $m_2 = k_2/k_3$. Finally, the electrical length of the particle and the core with respect to the wavenumber of the surrounding medium is $\nu \equiv k_3 a_2$ and $\alpha \equiv k_3 a_1$, respectively.

According to spherical transmission line theory the TM input admittance of mode n for this structure is defined as

$$Y_{2,n}^{(e)} = Y_{0,n}^{(e)}(m_2\nu)\frac{1-\dfrac{\chi_n(m_2\nu)}{\psi_n(m_2\nu)}\dfrac{\psi_n(m_2\alpha)}{\chi_n(m_2\alpha)}\dfrac{\dfrac{1}{Y_{1,n}^{(e)}}-\dfrac{1}{Y_{0,n}^{(e)}(m_2\alpha)}}{\dfrac{1}{Y_{1,n}^{(e)}}+\dfrac{1}{Y_{ext,n}^{(e)}(m_2\alpha)}}}{1-\dfrac{\chi_n'(m_2\nu)}{\psi_n'(m_2\nu)}\dfrac{\psi_n'(m_2\alpha)}{\chi_n'(m_2\alpha)}\dfrac{-Y_{0,n}^{(e)}(m_2\alpha)+Y_{1,n}^{(e)}}{Y_{ext,n}^{(e)}(m_2\alpha)+Y_{1,n}^{(e)}}} \tag{7.10}$$

where the (e) superscript denotes quantities associated with the TM response. The various quantities in the previous expression are defined as follows: the input TM admittance of the core normalized to the admittance of free space $Y_0 = \sqrt{\varepsilon_0\mu_0^{-1}}$is defined as

$$Y_{1,n}^{(e)} = jm_1\frac{\psi_n(m_1\alpha)}{\psi_n'(m_1\alpha)} \tag{7.11}$$

Furthermore, the fringing fields in the core exterior, or the field in the shell due to scattering by the core, are characterized by the TM admittance

$$Y_{ext,n}^{(e)}(m_2\alpha) = \frac{m_2}{j}\frac{\chi_n(m_2\alpha)}{\chi_n'(m_2\alpha)} \tag{7.12}$$

Finally, we define the characteristic TM admittance of a spherical transmission line with a refractive index m and a radius$r = r_0$ as

$$Y_{0,n}^{(e)}(mk_3r_0) = jm\frac{\psi_n(mk_3r_0)}{\psi_n'(mk_3r_0)} \tag{7.13}$$

Note here that the primed quantities in the previous expressions indicate first-order derivatives. Also in the preceding expressions the Riccati–Bessel functions are defined as

$$\psi_n(\rho) = \sqrt{\pi\rho/2}J_{n+1/2}(\rho) \tag{7.14}$$

and

$$\chi_n(\rho) = \sqrt{\pi\rho/2}H_{n+1/2}^{(2)}(\rho) \tag{7.15}$$

where $J_{n+1/2}(\bullet)$ and $H_{n+1/2}^{(2)}(\bullet)$ are the half-order Bessel and Hankel functions of the first and second kind, respectively.

7.3.2 *Zero-Impedance and Zero-Admittance Conditions on the Surface of a Core-Shell Particle*

Having established a compact framework for the characterization of the particle's impedance properties, we next examine the possibility of realizing zero-admittance and zero-impedance conditions on the surface of an electrically small dielectric core-shell particle, given that its electromagnetic response can be fully described by its dipolar mode. Consequently, for the remainder of this analysis, when it is not specified, it should be assumed that $n = 1$.

Now, from Eq. 7.10 it can be concluded that the particle's input admittance attains very large values when the denominator goes to zero, or more specifically when

$$\frac{Y_{1,1}^{(e)} - Y_{0,1}^{(e)}(m_2\alpha)}{Y_{1,1}^{(e)} + Y_{ext,1}^{(e)}(m_2\alpha)} = \frac{\psi_1'(m_2\nu)\,\chi_1'(m_2\alpha)}{\chi_1'(m_2\nu)\,\psi_1'(m_2\alpha)} \tag{7.16}$$

Similarly, the input admittance of the particle goes to zero when the numerator in Eq. 7.10 is equal to zero which results in the following condition:

$$\frac{Y_{1,1}^{(e)} - Y_{0,1}^{(e)}(m_2\alpha)}{Y_{1,1}^{(e)} + Y_{ext,1}^{(e)}(m_2\alpha)} = -\frac{Y_{0,1}^{(e)}(m_2\alpha)}{Y_{ext,1}^{(e)}(m_2\alpha)}\frac{\psi_1(m_2\nu)\,\chi_1(m_2\alpha)}{\chi_1(m_2\nu)\,\psi_1(m_2\alpha)} \tag{7.17}$$

The expressions in Eq. 7.16 and Eq. 7.17 constitute the necessary conditions such that a nonmagnetic two-layer spherical particle can exhibit either zero input impedance or zero input admittance. A better insight into the behavior of these conditions is obtained if we Taylor expand them and keep only the first terms in the series. Specifically we get:

$$\frac{\varepsilon_1 - \varepsilon_2}{\varepsilon_1 + 2\varepsilon_2} = \frac{1}{p} \tag{7.18}$$

and

$$\frac{\varepsilon_1 - \varepsilon_2}{\varepsilon_1 + 2\varepsilon_2} = -\frac{1}{2p} \tag{7.19}$$

where $p \equiv (a_1/a_2)^3$ is the volume fraction of the particle. Evidently, by judiciously choosing the material constitution of the core and shell material, as well as the particle's volume fraction we can devise nanoparticles that exhibit an internal resonance that resembles either a series or a parallel resonant LC circuit.

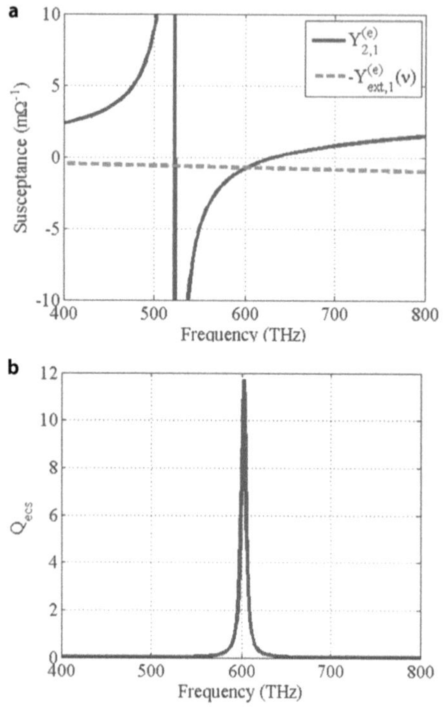

Figure 7.11 (a) Core-shell particle input susceptance along with minus the susceptance of the fringing field. (b) Core-shell particle extinction efficiency.

In Fig. 7.11a, the input reactance of a core-shell particle with an outer radius of 20 nm and an inner radius of 10 nm is plotted as a function of frequency. The material constitution of the core is assumed to be silver modeled as a Drude dielectric whose permittivity, after fitting experimental data [40], is given by

$$\varepsilon_r^{\mathrm{Ag}} = \varepsilon_\infty + \frac{f_p^2}{j\nu_L f - f^2} \tag{7.20}$$

where $\varepsilon_\infty = 5$, $\nu_L = 5.13$ THz and $f_p = 2213$ THz. The material of the shell is set to a nondispersive dielectric with permittivity $\varepsilon_r = 5$. Note that for this particular particle the first mode suffices to describe its electromagnetic response therefore we examine only the behavior of its susceptance corresponding to the first dipolar mode, or equivalently we set $n = 1$.

Notice that the input susceptance of the particle under study is equal to zero at 632 THz while at 522 THz the same quantity exhibits large values. This is in accordance with the response expected by the conditions in Eqs. 7.18 and 7.19. As a matter of fact given that $p = 1/8$ and $\varepsilon_2 = 5$, these two conditions are satisfied when $\varepsilon_1 = -7$ and $\varepsilon_1 \approx -12.14$, respectively. Indeed, the real part of the relative permittivity of silver at these two frequencies is $\mathrm{Re}\{\varepsilon_r^{\mathrm{Ag}}$ (632 THz)$\} = -7.25$ and $\mathrm{Re}\{\varepsilon_r^{\mathrm{Ag}}$ (522 THz)$\} = -12.95$. In the same plot we have also included minus the susceptance of the fringing field surrounding the particle for the same mode. The TM admittance of this field is given by

$$Y_{\mathrm{ext},1}^{(e)}(\nu) = \frac{1}{j}\frac{\chi_1(\nu)}{\chi_1'(\nu)} \tag{7.21}$$

Notice that at the two frequencies which correspond to the zero reactance and zero susceptance states, the particle does not exhibit any extreme scattering response since the resonance condition

$$\mathrm{Im}\left\{Y_{2,1}^{(e)} + Y_{\mathrm{ext},1}^{(e)}(\nu)\right\} = 0 \tag{7.22}$$

is satisfied, in this case around 600 THz. This is further verified by observing that the peak in the particle's extinction efficiency occurs at the same frequency, as seen in Fig. 7.11b. It is further evident that there is no strong interaction between the particle and the exterior space when its input impedance becomes zero and very large, respectively.

7.3.3 *Material Interpretation of the Core-Shell's Response*

From Eqs. 7.16 and 7.17 it becomes evident that the internal resonances of a core-shell particle are a consequence of a specific material combination between the core and shell, or equivalently the dipole moments of the two layers interact in such a way that the particle exhibits the aforementioned internal resonant characteristics. The transmission line representation, however, of a core-shell particle permits an alternative material based interpretation of this phenomenon. This interpretation serves also as the basis for the loading scheme described in the next section.

First, let us consider a homogeneous dielectric sphere with relative permittivity ε_2 and radius a_2. The electrical length of the particle is assumed to be such that the first dipolar mode suffices to describe its electromagnetic response. The input admittance of this object can be deduced from Eq. 7.10 if the material of the core is set equal to that of the shell. In particular, this yields

$$Y_{2,1}^{(e)} = Y_{0,1}^{(e)}(m_2 \nu) \tag{7.23}$$

If we Taylor expand the preceding expression and retain only the first term in the series, the input admittance of the particle becomes:

$$Y_{0,1}^{(e)}(m_2 \nu)\, Y_0 \approx \frac{j\omega\varepsilon_0 a_2}{2}\varepsilon_2 \tag{7.24}$$

This last outcome is of paramount importance regarding a dielectric particle's electromagnetic response, since it reveals that for an electrically small homogeneous dielectric sphere its material properties are directly mapped to its impedance characteristics.

Given the previous result, let us now perform a similar analysis for the input admittance of an electrically small core-shell particle. After taking the Taylor expansion of Eq. 7.10, for $n = 1$, we obtain

$$Y_{2,1}^{(e)} Y_0 \approx \frac{j\omega\varepsilon_0 a_2}{2}\varepsilon_2 \frac{1 + 2p\dfrac{\varepsilon_1 - \varepsilon_2}{\varepsilon_1 + 2\varepsilon_2}}{1 - p\dfrac{\varepsilon_1 - \varepsilon_2}{\varepsilon_1 + 2\varepsilon_2}} \tag{7.25}$$

By direct comparison between the two previous expressions one can conclude that for a core-shell particle whose electromagnetic response can be described by the first TM mode, its input admittance corresponds to that of a homogeneous sphere with relative dielectric permittivity given by

$$\varepsilon_r = \varepsilon_2 \frac{1 + 2p\dfrac{\varepsilon_1 - \varepsilon_2}{\varepsilon_1 + 2\varepsilon_2}}{1 - p\dfrac{\varepsilon_1 - \varepsilon_2}{\varepsilon_1 + 2\varepsilon_2}} \tag{7.26}$$

Note that this is the effective permittivity of a binary mixture governed by the CM mixing rule, where the host and the filler of the mixture is the shell and the core material, respectively. Note also that when the conditions in Eqs. 7.18 and 7.19 are satisfied, Eq. 7.26 yields an effective particle permittivity equal to infinity and zero, respectively. However, due to the direct mapping of the particle's material to its input admittance, these extreme permittivity values are translated to extreme admittance responses.

7.4 Tunable Optical Nanoantenna Loaded by Plasmonic Core-Shell Particles

It was demonstrated in the previous section that a core-shell particle exhibits internal mixture properties. Herein we further exploit this property and we devise core-shell-based, tunable nanoantenna loads. Recently, there has been a great research effort toward the theoretical and experimental investigation of optical nanoantennas [41–46]. Different nanoantenna designs have been proposed, where in most cases they heavily rely on concepts developed for their RF counterparts [47, 48]. Of particular importance for the development of optical nanoantenna theory are the findings of Alù et al. [49, 50] where it has been demonstrated that it is possible to define the impedance of an optical nanodipole, and additionally this impedance can be tuned by appropriately filling the antenna feeding gap with different materials. Essentially, the gap defined by the nanodipole arms can be considered as a cylindrical capacitor, therefore by appropriately choosing the material constitution within it, we can achieve either a capacitive or an inductive equivalent admittance response [50, 51]. The aforementioned nanodipole loading scenario is the optical analogue of what is common practice among RF engineers, that is the incorporation of lumped elements into antenna structures. A typical example is wire antennas, operating in the MHz range, judiciously loaded along their length with combinations of discrete LC (inductor/capacitor) loads. The latter effectively modify the current distribution along the antenna, and thus additional control is achieved over their radiation and scattering properties [52–58]. It becomes evident that in the case of nanodipole antennas the tuning options depend on the material variety that can be used to fill its loading gap. The fact that electrically small core-shell particles are characterized by an internal mixture structure makes them ideal candidates for the realization of nanodipole loading elements. Essentially, by judiciously choosing the particle's material constitution as well as its volume fraction, we can engineer its effective material properties so that they mimic some desired circuit functionality, and subsequently utilize them as nanoantenna loads.

More precisely our approach relies on the observation that the gap volume defined by the nanodipole arms can be considered as a mixing volume. First, in this volume we dilute a homogeneous dielectric sphere and we derive the mixture's effective permittivity properties. This first step is of great importance because it demonstrates a compact way to devise nanodipole loads using readily available spherical particles. In addition to that the corresponding analysis relies solely on quasi-static assumptions; therefore the mixture's effective dielectric properties can be directly mapped to an admittance response of a circuit. In the previous step it was assumed that the loading sphere has a fixed size therefore the only tuning option comes from changing its material constitution. Additional degrees of freedom are introduced in this loading scheme if the homogeneous dielectric sphere is substituted by a core-shell particle. Consequently, we devise a loading scheme comprised by two successive mixing scenarios, since as demonstrated in the previous section a core-shell particle can be treated as an effective homogeneous sphere where its permittivity is the outcome of mixing the core into the shell material, according to the CM mixing rule.

7.4.1 *Nanodipole Geometry and Response*

Let us consider the nanodipole antenna illustrated in Fig. 7.12. The structure is immersed in free space and it consists of two cylindrically shaped arms comprised of silver, separated by a 25 nm long gap. The length and radius of each arm is 30 nm and 15 nm, respectively. Hemispherical caps with a radius of 15 nm are placed at the end of each dipole arm. In this study silver is modeled according to the Drude model presented in the previous section. Our objective is to control the nanoantenna's dipole mode, therefore the structure is illuminated by a plane wave polarized parallel to the antennas' axis of rotational symmetry, as indicated in Fig. 7.12. The structure's extinction cross section (ECS) for this illumination scenario is shown plotted in Fig. 7.13a, and it exhibits one resonance peak at 680 THz. This corresponds to a structure mode where the two arms have been simultaneously excited. Similar to its RF counterpart one way to tune the resonance frequency of this structure's ECS is by changing the material constitution of the gap volume defined by the two antenna arms. Evidently, this modification alters the displacement current

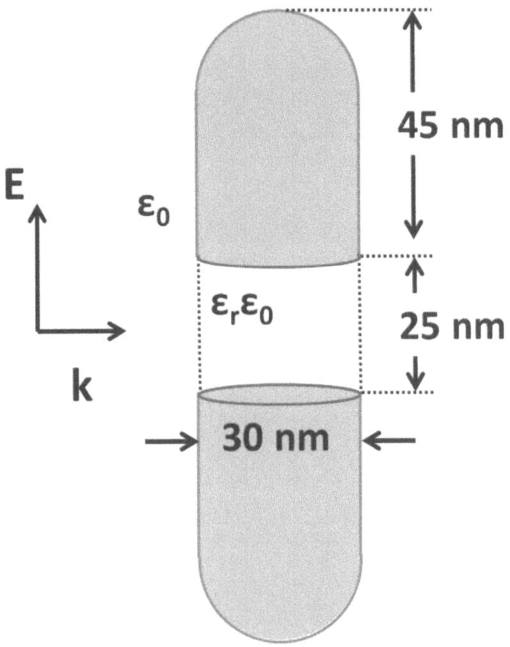

Figure 7.12 Nanodipole geometry.

distribution across the nanostructure, and as a consequence the radiation characteristics of the antenna change as well.

The response of the nanoantenna under different loading scenarios can be studied in its receiving mode, or by examining its scattering signature under plane-wave excitation. However, it is more practical, more intuitive, and computationally less demanding to be able to quantify the antenna's response in its transmitting mode. This requires the extraction of an equivalent circuit that can fully capture the impedance interactions between the antenna and the load of interest. For this reason it is necessary to derive the antenna's intrinsic (unloaded) impedance, as well as to devise an appropriate circuit representation for the loading volume's admittance response when filled by different materials.

First, the extraction of the antenna's intrinsic impedance is realized using a procedure similar to the one described in Ref. [59]. According to this procedure the structure is modeled using the

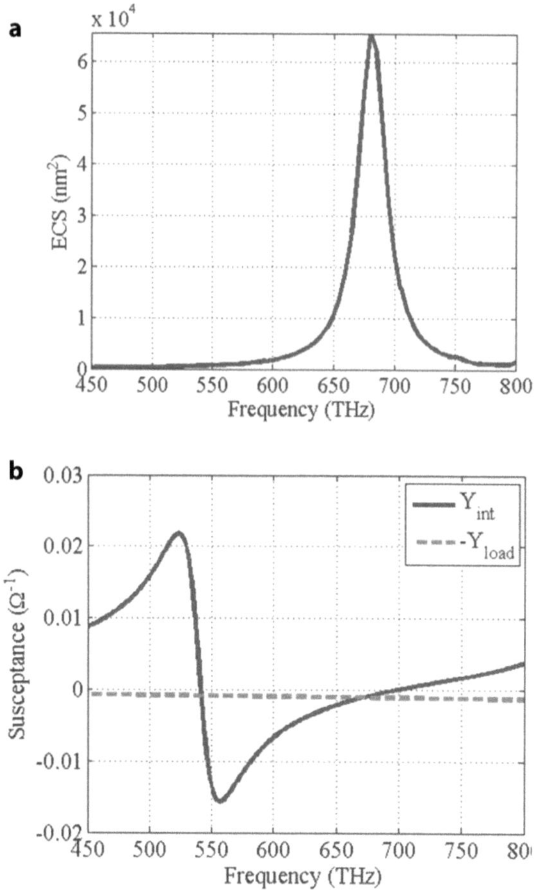

Figure 7.13 (a) Nanodipole extinction cross section. (b) Nanodipole's intrinsic susceptance along with minus the susceptance of the free-space filled volume computed using Eq. 7.27.

body-of-revolution solver of COMSOL. The gap is excluded from the computational domain, while along the boundary of the gap (i.e., the top, bottom, and right edges), a uniform and constant z-directed excitation electric field distribution is defined, whereas no excitation is defined along the left edge since it is collocated with the symmetry axis. Then the effect of the free space filled gap is treated as a load that tunes accordingly the response of the antenna. Due to the fact that we assume the electric field within the gap is constant and uniform, the effect of any load may be expressed by the admittance of a cylindrical capacitor:

$$Y_{\text{load}} = j\omega\varepsilon_0\varepsilon_r\frac{\pi a^2}{h} \tag{7.27}$$

where a and h are the radius and the height of the nanodipole gap, respectively, while ε_r corresponds to the relative permittivity of the material that it is filled with. Fig. 7.13b summarizes the effect of the gap's admittance on the nanodipole's intrinsic admittance. The latter is displayed in the figure by the solid line. The dashed line corresponds to minus the susceptance of the gap, as given by Eq. 7.27, when $\varepsilon_r = 1$. Notice that the frequency where the resonance condition $\text{Im}\{Y_{\text{int}} + Y_{\text{load}}\} = 0$ is satisfied corresponds to the structure's extinction efficiency peak, as predicted in Fig. 7.13a.

There are two main outcomes of the previous numerical study: First we have demonstrated that the proposed methodology for the extraction of this antenna's intrinsic impedance, in conjunction with the simplified capacitor admittance model to quantify the loading effects of different gap filling materials, can accurately enough capture the scattering properties of the device under study. Second, in this particular case the free-space filled gap, due to its positive dielectric permittivity, corresponds to a purely capacitive load. Consequently, it is expected that greater tuning flexibility can be achieved if the nanoantenna is loaded by a general Lorentzian-type dielectric. As a matter of fact Eq. 7.27 dictates that there is a direct mapping of the load material constitution to its admittance. Therefore, the loading of the antenna with a general Lorentzian-type dielectric is equivalent to a loading scheme that utilizes a general RLC circuit element.

The question now becomes how to devise a load for the nanodipole under study that exhibits the desired Lorentzian response. Toward this end we recall that a Lorentzian response can be achieved in the case of a mixture where, for instance, the host is a nondispersive dielectric and the filler a dispersive Drude-type dielectric [60]. This scenario can be easily adapted to the geometry under study if we assume that the mixing volume is the antenna gap, and within this gap a homogeneous Drude, or any dispersive dielectric, sphere is diluted.

7.4.2 *Loading the Gap Volume with a Homogeneous Dielectric Sphere*

In this section we examine the effective permittivity properties of the gap volume in the case where at its center there exists a 10 nm radius homogeneous dielectric sphere. This system is treated as a mixture, where the dielectric sphere is diluted in the free space host at a volume fraction

$$p = (4\pi/3) \times \left[10^3/(\pi 15^2 \times 25)\right] \approx 0.237. \tag{7.28}$$

To derive an effective medium model for this mixing scenario, we employ the partially filled capacitor representation for the mixture's effective permittivity [61–63]. This methodology dictates that the mixing rule that characterizes the nanoantenna gap loaded by the 10 nm spherical homogeneous dielectric particle is given by

$$\varepsilon_e - \varepsilon_h = -\varepsilon_h \sum_{i=1}^{3} \frac{w_i h_i}{u - (1 - h_i)} \tag{7.29}$$

with

$$\{w_1, w_2, w_3\} \approx \{0.2911, 0.0205, 0.0007\}$$

and

$$\{h_1, h_2, h_3\} \approx \{0.7718, 0.5956, 0.2681\}.$$

Also in the preceding expression $u \equiv \varepsilon_h/(\varepsilon_h - \varepsilon_f)$, where ε_h is the host material relative permittivity and ε_f is the relative permittivity of the spherical dielectric load. The details of the derivation for the expression given in Eq. 7.29 are as follows: the system of the cylindrical gap volume with the homogeneous dielectric sphere at its center is modeled using a quasi-static solver. A z-directed uniform and constant electric field distribution is established within this loaded cylindrical capacitor. We first set $\varepsilon_h = 1$, and then discretely vary the relative permittivity of the filler ε_f from 1 to 20,000; for each permittivity value a simulation is performed where an estimate of the system's relative effective permittivity is obtained using the following volume averaging formula:

$$\varepsilon_e = 1 + \iiint_V P_z dr^3 \left(\varepsilon_0 \iiint_V E_z dr^3\right)^{-1} \tag{7.30}$$

where P_z represents the z component of the dielectric material's polarization density. Note that the integrals are evaluated over the volume of the gap. At the end of this process a set of numerically derived data is generated which describes the variation of the cylindrical capacitor's relative effective permittivity as a function of the filler's relative permittivity, $\varepsilon_e = \varepsilon_e\left(\varepsilon_f\right)$. The next step requires mapping the numerically derived relative effective permittivity data onto a partially filled capacitor model. This corresponds to an N-term expansion such as the one shown in Eq. 7.29, where the number of terms as well as the poles and zeroes need to be determined. This was achieved via the covariance matrix adaptation evolution strategy (CMA-ES), which is a robust and efficient global optimization scheme [64, 65]. The outcome of this optimization procedure is the relative effective permittivity expression shown in Eq. 7.29.

To verify the applicability and accuracy of this effective medium model we examine the nanoantenna's response in the case of a load comprised of a homogenous silver sphere. In particular, Fig. 7.14a displays the nanoantenna's intrinsic input susceptance, along with minus the susceptance of the homogeneous silver sphere load. To derive the admittance of this load, we first substitute ε_{f} into Eq. 7.29 with the relative permittivity of silver as given by Eq. 7.20. The resulting relative effective permittivity is subsequently inserted into Eq. 7.27 and the admittance of the load under consideration is derived. According to the admittance behavior depicted in Fig. 7.14a the loaded nanoantenna for the frequency range of interest should exhibit a resonance around 650 THz. This was validated by computing the structure's ECS via full wave simulations, which is shown in Fig. 7.14b. Evidently, there is very good agreement between the frequencies where the antenna's total susceptance goes to zero and the frequency at which the ECS peak occurs.

7.4.3 *Loading the Gap Volume with a Plasmonic Core-Shell Particle*

Now, given that the spherical particles we consider in this study have a fixed radius of 10 nm, the only tunability that can be achieved is from the different material constitution used for the filler dielectric

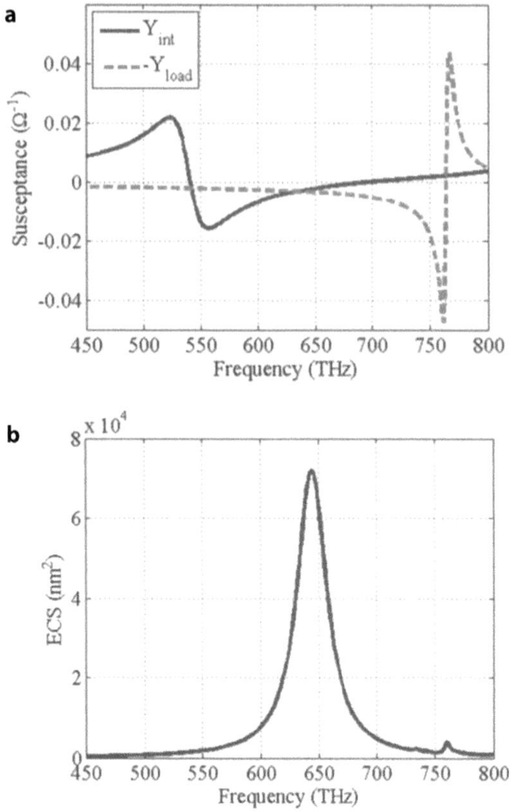

Figure 7.14 (a) Nanodipole's intrinsic susceptance along with minus the susceptance of the homogeneous silver sphere load. (b) ECS of the homogeneous silver sphere–loaded nanodipole.

sphere. In particular, it is desirable to control the line shape and the resonance frequency of the Lorentzian response of the mixture shown with the dashed line in Fig. 7.14a. A wider range of filler material, and thus tuning options, can be achieved if we substitute the homogeneous dielectric sphere with a core-shell sphere. In the previous section it was demonstrated that in the low-frequency limit a dielectric core-shell sphere exhibits the input admittance of an effective homogeneous dielectric sphere. The relative permittivity properties of this sphere are obtained after applying the CM mixing rule to a mixture where the core material is diluted in the shell material, at a volume fraction equal to that of the particle's volume fraction. Consequently, by appropriately choosing the core and shell material constitution as well as the particle's volume fraction we can realize a wide variety of loading material properties for the nanoantenna under study.

To demonstrate how core-shell-type particles can accommodate the need to devise more flexible material constitutions for the loading sphere we examine the following core-shell particle: the shell and core radius are $a_2 = 10$ nm and $a_1 = 7$ nm, respectively. The core material is defined as a dielectric with relative permittivity $\varepsilon_1 = 3$, while the shell material is assumed to be silver. First, the effective material properties of the core-shell particle were derived using Eq. 7.26, and then this relative effective permittivity is substituted into Eq. 7.29 in order to derive the equivalent relative effective permittivity of the nanoantenna load. Finally using Eq. 7.27, the equivalent admittance of the load was computed. This admittance along with the nanodipole's intrinsic impedance is depicted in Fig. 7.15a. The admittance model predicts two frequency points where the structure may exhibit a resonance. This prediction is supported by the antenna's ECS response shown in Fig. 7.15b, where the frequencies where the ECS peaks occur have been accurately predicted by the previously introduced equivalent circuit methodology.

As a final comment it should be noted that both load admittances as shown in Figs. 7.14a and 7.15a are characterized by a Lorentzian response for the frequency range of interest. As mentioned previously this is the typical characteristic of a mixture comprised by a nondispersive host and a Drude, or a general Lorentzian dielectric filler. It becomes also evident that the manipulation of this resonance is the key toward the realization of a wide tuning range for the nanodipole under study. The proposed tuning mechanism controls the characteristics of this Lorentzian via the combined effect of two consecutive mixing rules: the first mixing occurs within the core-shell particle which is characterized by an effective homogeneous relative permittivity that would not be feasible to obtain using some homogeneous bulk material. The second mixing rule corresponds to the dilution of this effective homogeneous particle into the gap volume defined by the nanoantenna arms.

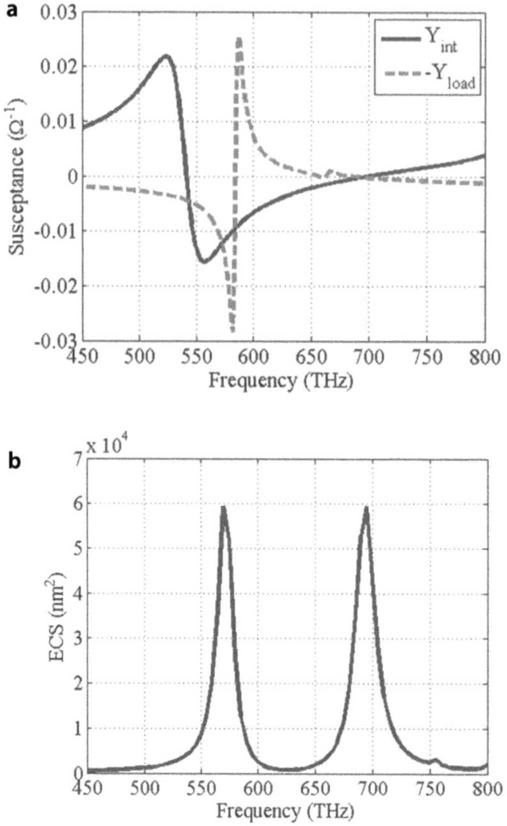

Figure 7.15 (a) Nanodipole's intrinsic susceptance along with minus the susceptance of the core-shell particle load. (b) ECS of the core-shell-loaded nanodipole.

7.5 Conclusion

In summary, we have presented novel scattering control methodologies both in the MW regime and at optical wavelengths based on the Mie scattering theory. The anisotropy of metasurfaces has been exploited to successfully achieve near-perfect cloaking and illusion effects for objects with practical sizes beyond the quasi-static limit. A spherical transmission line approach has been adopted to provide the mathematical framework in order to design open-

and short-circuit nanoswitches based on core-shell particles. These unique properties of core-shell particles were further coupled to nanoantennas, which enables the creation of tunable resonant-mode nanodipoles.

Acknowledgments

This work was partially supported by the National Science Foundation's Material Research Science and Engineering Center (MRSEC) under grant no. DMR-0820404.

References

1. Tsang, L., Kong, J. A., and Ding, K.-H. (2000). *Scattering of Electromagnetic Waves: Theories and Applications* (John Wiley & Sons, USA).
2. Bohren, C. F., and Huffman, D. R. (1983). *Absorption and Scattering of Light by Small Particles* (Wiley, USA).
3. Engheta, N., and Ziolkowski, R. (2006). *Metamaterials: Physics and Engineering Explorations* (Wiley-IEEE Press, USA).
4. Cui, T. J., Smith, D. R., and Liu, R. (2009). *Metamaterials: Theory, Design, and Applications* (Springer, UK).
5. Chen, P. Y., Soric, J., and Alù, A. (2012). Invisibility and cloaking based on scattering cancellation, *Adv. Mater.*, **24**, 281–304.
6. Werner, D. H., and Kwon, D.-H. (2014). *Transformation Electromagnetics and Metamaterials: Fundamental Principles and Applications* (Springer, UK).
7. Kwon, D.-H., and Werner, D. H. (2010). Transformation electromagnetics: an overview of the theory and its application, *IEEE Antennas Propag. Mag.*, **52**, 24–46.
8. Pendry, J. B., Schurig, D., and Smith, D. R. (2006). Controlling electromagnetic fields, *Science*, **312**, 1780–1782.
9. Schurig, D., Mock, J. J., Justice, B. J., Cummer, S. A., Pendry, J. B., Starr, A. F., and Smith, D. R. (2006). Electromagnetic metamaterial cloaking at microwave frequencies, *Science*, **314**, 977–980.
10. Alù, A., and Engheta, N. (2009). Cloaking a sensor, *Phys. Rev. Lett.*, **102**, 233901(1)–(4).

11. Greenleaf, A., Kurylev, Y., Lassas, M., and Uhlmann, G. (2011). Cloaking a sensor via transformation optics, *Phys. Rev. E*, **83**, 016603(1)-(6).
12. Kerker, M. (1975). Invisible bodies, *J. Opt. Soc. Am.*, **65**, 376–379.
13. Alù, A., and Engheta, N. (2005). Achieving transparency with plasmonic and metamaterial coatings, *Phys. Rev. E*, **72**, 016623(1)-(9).
14. Silveirinha, M. G., Alù, A., and Engheta, N. (2007). Parallel-plate metamaterials for cloaking structures, *Phys. Rev. E*, **75**, 036603(1)-(16).
15. Qiu, C.-W., Xu, X., and Feng, Y. (2009). Spherical cloaking with homogeneous isotropic multilayered structures, *Phys. Rev. E*, **79**, 047602(1)-(4).
16. Rainwater, D., Kerkhoff, A., Soric, J. C., Moreno, G., and Alù, A. (2012). Experimental verification of three-dimensional plasmonic cloaking in free-space, *New J. Phys.*, **14**, 013054(1)-(13).
17. Alù, A. (2009). Mantle cloak: invisibility induced by a surface, *Phys. Rev. B*, **80**, 245115(1)-(4).
18. Chen, P. Y., and Alù, A. (2011). Mantle cloaking using thin patterned metasurfaces, *Phys. Rev. B*, **84**, 205110(1)-(13).
19. Soric, J. C., Chen, P. Y., Kerkhoff, A., Rainwater, D., Melin, K., and Alù, A. (2013). Demonstration of an ultralow profile cloak for scattering suppression of a finite-length rod in free space, *New J. Phys.*, **15**, 033037(1)-(18).
20. Munk, B. A. (2000). *Frequency Selective Surface: Theory and Design* (John Wiley & Sons, USA).
21. Wang, X., and Semouchkina, E. (2013). A route for efficient non-resonance cloaking by using multilayer dielectric coating, *Appl. Phys. Lett.*, **102**, 113506(1)-(4).
22. Jiang, Z. H., and Werner, D. H. (2013). Exploiting metasurface anisotropy for achieving near-perfect low-profile cloaks beyond the quasi-static limit, *J. Phys. D Appl. Phys.*, **46**, 505306(1)-(11).
23. Lai, Y., Chen, H. Y., Han, D. Z., Xiao, J. J., Zhang, Z.-Q., and Chan, C. T. (2009). Illusion optics: the optical transformation of an object into another object, *Phys. Rev. Lett.*, **102**, 253902(1)-(4).
24. Jiang, W. X., Ma, H. F., Cheng, Q., and Cui, T. J. (2010). Illusion media: generating virtual objects using realizable metamaterials, *Appl. Phys. Lett.*, **96**, 121910(1)-(3).
25. Li, C., Meng, X. K., Liu, X., Li, F., Fang, G. Y., Chen, H. Y., and Chan, C. T. (2010). Experimental realization of a circuit-based broadband illusion-optics analogue, *Phys. Rev. Lett.*, **105**, 233906(1)-(4).

26. Jiang, W. X., Qiu, C.-W., Han, T., Zhang, S., and Cui, T. J. (2013). Creation of ghost illusions using wave dynamics in metamaterials, *Adv. Funct. Mater.*, **23**, 4028–4034.
27. Kuester, E. F., Mohamed, M. A., and Holloway, C. L. (2003). Averaged transition conditions for electromagnetic fields at a metafilm, *IEEE Trans. Antennas Propag.*, **51**, 2641–2652.
28. Senior, T. B. A., and Volakis, J. L. (1995). *Approximate Boundary Conditions in Electromagnetics* (IEE Press, UK).
29. Wait, J. R. (1955). Scattering of a plane wave from a circular dielectric cylinder at oblique incidence, *Can. J. Phys.*, **33**, 189–194.
30. Marqués, R., Medina, F., and Rafii-El-Idrissi, R. (2002). Role of bianisotropy in negative permeability and left-handed metamaterials, *Phys. Rev. B*, **65**, 144440(1)-(6).
31. Jiang, Z. H., Gregory, M. D., and Werner, D. H. (2011). Experimental demonstration of a broadband transformation optics lens for highly directive multibeam emission, *Phys. Rev. B*, **84**, 165111(1)-(6).
32. Jiang, Z. H., Gregory, M. D., and Werner, D. H. (2011). A broadband monopole antenna enabled by an ultrathin anisotropic metamaterial coating, *IEEE Antennas Wireless Propag. Lett.*, **10**, 1543–1546.
33. Jiang, Z. H., Wu, Q., Brocker, D. E., Sieber, P. E., and Werner, D. H. (2011). A low-profile high-gain substrate-integrated waveguide slot antenna enabled by an ultrathin anisotropic zero-index metamaterial coating, *IEEE Trans. Antennas Propag.*, **62**, 1173–1184.
34. Kettunen, H., Wallen, H., and Sihvola, A. (2008). Electrostatic resonances of a negative-permittivity hemisphere, *J. Appl. Phys.*, **103**, 094112(1)(8).
35. Alù, A., and Engheta, N. (2009). Optical nanoswitch: An engineered plasmonic nanoparticle with extreme parameters and giant anisotropy, *New J. Phys.*, **11**, 013026(1)-(14).
36. Alù, A., and Engheta, N. (2011). Optical metamaterials based on optical nanocircuits, *Proc. IEEE*, **10**, 1669–1681.
37. Wait, J. R. (1962). *Electromagnetic Waves in Stratified Media* (Pergamon).
38. Wait, J. R. (1963). Electromagnetic scattering from a radially inhomogeneous sphere, *Appl. Sci. Res. B*, **10**, 441–449.
39. Aden, L., and Kerker, M. (1951). Scattering of electromagnetic waves from two concentric spheres, *J. Appl. Phys.*, **22**, 1242–1246.
40. Johnson, B., and Christy, R. W. (1972). Optical constants of noble metals, *Phys. Rev. B*, **6**, 4370–4379.

41. Novotny L., and Hulst, V. (2011). Antennas for light, *Nat. Photonics,* **5**, 83–90.
42. Bharadwaj, P., Deutsch, B., and Novotny, L. (2009). Optical antennas, *Adv. Opt. Photonics,* **1**, 438–483.
43. Muhlschlegel, P., Eisler, H.-J., Martin, O. J. F., Hecht, B., and Pohl, D. W. (2005). Resonant optical antennas, *Science,* **308**, 1607–1608.
44. Alù, A., and Engheta, N. (2013). Theory, modeling and features of optical nanoantennas, *IEEE Trans. Antennas Propag.,* **61**, 1508–1517.
45. Agio, M., and Alu, A. (2013). *Optical Antennas* (Cambridge University Press).
46. Fischer, H., and Martin, O. J. F. (2008). Engineering the optical response of plasmonic nanoantennas, *Opt. Express,* **16**, 9144–9154.
47. Taminiau, T. H., Segerink, F. B., Moerland, R. J., Kuipers, L., and Hulst, N. F. (2007). Near-field driving of an optical monopole antenna, *J. Opt. A: Pure Appl. Opt.,* **9**, S315–S321.
48. Hofmann, H. F., Kosako, T., and Kadoya, Y. (2007). Design parameters for a nano-optical Yagi–Uda antenna, *New J. Phys.,* **9**, 217(1)(12).
49. Alù, A., and Engheta, N. (2008). Input impedance, nanocircuit loading, and radiation tuning of optical nanoantennas, *Phys. Rev. Lett.,* **101**, 043901(1)(4).
50. Alù, A., and Engheta, N. (2008). Tuning the scattering response of optical nanoantennas with nanocircuit loads, *Nat. Photonics,* **2**, 307–310.
51. Zhao, Y., Engheta, N., and Alu, A. (2011). Effects of shape and loading of optical nanoantennas on their sensitivity and radiation properties, *J. Opt. Soc. Am. B,* **28**, 1266–1274.
52. Brown, G. H. (1936). A critical study of the characteristics of broadcast antennas as affected by antenna current distribution, *Proc. IRE,* **24**, 48–81.
53. Harrison, C. Jr. (1963). Monopole with inductive loading, *IEEE Trans. Antennas Propag.,* **11**, 394–400.
54. Czerwinski, W. P. (1965). On optimizing efficiency and bandwidth of inductively loaded antennas, *IEEE Trans. Antennas Propag.,* **13**, 811–812.
55. Smith, D. (1975). The trap-loaded cylindrical antenna, *IEEE Trans. Antennas Propag.,* **23**, 20–27.
56. Fournier, M., and Pomerleau, A. (1978). Experimental study of an inductively loaded short monopole antenna, *IEEE Trans. Veh. Technol.,* **27**, 1–6.

57. Saoudy, S., and Hamid, M. (1986). Rigorous solution of a dipole antenna with lumped impedance loading, *Can. J. Phys.*, **64**, 1537–1545.
58. Boag, A., Michielssen, E., and Mittra, R. (1996). Design of electrically loaded wire antennas using genetic algorithms, *IEEE Trans. Antennas Propag.*, **44**, 687–695.
59. Locatelli, A., De Angelis, C., Modotto, D., Boscolo, S., Sacchetto, F., Midrio, M., Capobianco, A.-D., Pigozzo, F. M., and Someda, C. G. (2009). Modeling of enhanced field confinement and scattering by optical wire antennas, *Opt. Express*, **17**, 16792–16800.
60. Kristensson, G., Rikte, S., and Sihvola, A. (1998). Mixing formulas in the time domain, *J. Opt. Soc. Am. A*, **15,** 1411–1422.
61. Bergman, D. J. (1978). The dielectric constant of a composite material - a problem in classical physics, *Phys. Rep.*, **43**, 377–407.
62. Milton, G. W. (1981). Bounds on the complex permittivity of a two–component composite material, *J. Appl. Phys.*, **52**, 5286–5293.
63. Diaz, R. E., Merrill, W. M., and Alexopoulos, N. G. (1998). Analytic framework for the modeling of effective media, *J. Appl. Phys.*, **84**, 6815–6826.
64. Hansen, N., and Ostermeier, A. (2001). Completely derandomized self-adaptation in evolutionary strategies, *Evol. Comput.*, **9**, 159–195.
65. Gregory, M. D., Bayraktar, Z., and Werner, D. H. (2011). Fast optimization of electromagnetic design problems using the covariance matrix adaptation evolutionary strategy, *IEEE Trans. Antennas Propag.*, **59**, 1275–1285.

Chapter 8

Experiments on Cloaking for Surface Water Waves

Sebastien Guenneau,[a] Guillaume Dupont,[a] Stefan Enoch,[a] and Mohamed Farhat[b]

[a] *Institut Fresnel, UMR CNRS 7249, Aix-Marseille Université, Campus de St Jérôme, 13397 Marseille Cedex 20, France*

[b] *Division of Computer, Electrical, and Mathematical Sciences and Engineering, King Abdullah University of Science and Technology (KAUST), Thuwal 23955-6900, Saudi Arabia*

sebastien.guenneau@fresnel.fr

8.1 Introduction

During the last decade, research groups from all over the world have demonstrated the possibilities offered by metamaterials, thanks to their ability to control electromagnetic waves (and other types of waves, for example, acoustic and elastic). Following the advent of negative refraction and subwavelength imaging, it has proved possible to hide some objects from electromagnetic radiation and make them invisible. A first path to invisibility was investigated by Engheta and Alù [1] in 2005 and is based on plasmonic materials designed to cloak dielectric or conducting objects. This technique relies heavily on a scattering cancellation phenomenon, based on

Transformation Wave Physics: Electromagnetics, Elastodynamics, and Thermodynamics
Edited by Mohamed Farhat, Pai-Yen Chen, Sebastien Guenneau, and Stefan Enoch

ISBN 978-981-4669-95-5 (Hardcover), 978-981-4669-96-2 (eBook)
www.panstanford.com

the negative local polarizability of a cover made of low-electric-permittivity materials. It appears to be relatively robust to changes in the design parameters, geometry, and frequency of operation.

The only drawback of this process is the lack of experimental confirmation and its requirement of a priori knowledge of electromagnetic properties of the objects to hide. The use of space–time transformations as a design tool for a new class of man-made materials, coined as metamaterials, was subsequently introduced as an alternative path toward invisibility. A basic idea of this concept is that a metamaterial mimics a transformed empty space. The light rays follow straight trajectories according to Fermat's principle in this transformed (virtual electromagnetic) space and are curved in a desired way in the (real) laboratory space. This allows one to design, in an efficient way, materials with various characteristics such as invisibility cloaks, whereby light is detoured smoothly around an object.

In 2006, Pendry, Schurig, and Smith theorized that a finite size object surrounded by a coating consisting of a metamaterial might become invisible for electromagnetic waves [3]. An international team involving these authors subsequently implemented this idea using a metamaterial consisting of concentric layers of split-ring resonators (SRRs) [4], which makes a copper cylinder invisible to an incident plane wave at a specific microwave frequency (8.5 GHz). The smooth behavior of the electromagnetic field in the far-field limit could be expected in view of the numerical evidence given [3] using a geometrical optics-based software. Importantly, Leonhardt proposed, in parallel to this development of transformational optics, a conformal optics route toward invisibility [5], which is, however, constrained by the 2D geometry, as it relies heavily on complex analysis, and the ray optics limit, that is, the far field. To date, the only evidence that invisibility is preserved in the intense near-field limit is purely numerical [6].

A very different route to invisibility was proposed by McPhedran, Nicorovici, and Milton in the same year. This grouping studied a countable set of line sources using anomalous resonance when the sources lie in the close neighborhood of a cylindrical coating filled with negative permittivity material, which is nothing but a cylindrical version of the poor man's lens of Pendry [10]. These researchers attribute this cloaking phenomenon to a large field enhancement on either side of the cylindrical coating [7, 8], which is reminiscent of the field of plasmonics.

In the first, introductory, section of this chapter, we have briefly reviewed some important results in the field of acoustic cloaking. In the second section, we show that it is possible to design an invisibility cloak which is deduced from a coordinate change in a Helmholtz's equation derived from the linearized Navier–Stokes equations at the free air–liquid interface, when the liquid is incompressible. The coordinate change leads to an artificially anisotropic heterogeneous fluid. In the third section, we shall detail how the anisotropy of the fluid can be achieved with a fluid periodically perforated with a large number of small rigid inclusions. A homogenization approach is used for this purpose to design and experimentally validate a water wave cloak of a circular shape. In the fourth section, we shall explain how one can envisage water wave cloaks with arbitrary shapes and also so-called invisibility carpets. In the fifth section, we conclude that such an enhanced control of water wave trajectories, which uses techniques developed in the emerging fields of transformational and conformal optics, opens new vistas in acoustic metamaterials.

8.2 Acoustic Cloaking for Liquid Surface Waves

In this section, we describe an effective medium approach to cloaking for liquid surface waves (LSWs), which works over a finite frequency bandwidth, unlike for cloaking using resonant metamaterials, such as displaying spatially varying artificial magnetism [4]. Before we embark on our invisibility journey, we find it useful to recall under which hypotheses one can consider water waves satisfying governing equations reminiscent of those familiar to scientists working in the field of linear optics, or in other words, how can one get rid of nonlinearities in Navier–Stokes equations?

8.2.1 *From Navier–Stokes to Helmholtz*

Let us denote by Ω the open-bounded domain in the space $\mathbb{R}^3$ occupied by a fluid. If we restrict ourselves to incompressible fluids which are such that $\mathrm{div}(\mathbf{u}) = \sum_{i=1}^{3} \frac{\partial u_i}{\partial x_i} = 0$, where $\mathbf{u} = (u_1, u_2, u_3)$ is the vector velocity field, the conservation of momentum leads to the Navier–Stokes equations:

$$\frac{\partial u_i}{\partial t} + \sum_{j=1}^{3} u_j \frac{\partial u_i}{\partial x_j} + \frac{1}{\rho}\left(\frac{\partial p}{\partial x_i} - \mu \sum_{j=1}^{3} \frac{\partial^2 u_i}{\partial x_j^2}\right) = g_i \quad \text{in } \Omega\,, \tag{8.1}$$

where p is the scalar pressure field, ρ the fluid density, $\mu\Delta\mathbf{u}$ accounts for the fluid's viscosity (μ is a positive coefficient), and $\mathbf{g}$ is the vector of gravity force: $\mathbf{g} = -g\mathbf{e}_3$, where g denotes the acceleration caused by gravity and $\mathbf{e}_i$ are the basis vectors of the Euclidean space. Moreover, x_1, x_2, x_3 and t are, respectively, the space and time variables, with $t \geq 0$.

The inhomogeneous incompressible Navier–Stokes equations are supplied with an initial condition $\mathbf{u}(x_1, x_2, x_3, 0) = \mathbf{u_0}(x_1, x_2, x_3)$ which is a smooth divergence free field, but they are well known for the mathematical challenges regarding the regularity of their solutions at long time. To avoid this issue, it is usual to impose that $\mathbf{u}(x_1, x_2, x_3, t)$ does not grow large when $\sqrt{x_1^2 + x_2^2 + x_3^3} \to \infty$ and that $\mathbf{g}$ is a smooth applied field. A solution will be physically acceptable if $\int_{\mathbb{R}^3} \mid u_i \mid^2 dx_1 dx_2 dx_3 < \infty$ for all time t.

The purpose of this chapter is more modest than the physics of Navier–Stokes equations which are both theoretically and numerically difficult to handle. We shall restrict ourselves to the analysis of linear surface water waves, which are known to be solutions of a Helmholtz's equation at the free liquid–air interface. We reproduce below the classical derivation of this Helmholtz's equation reminiscent of governing equations in optics.

To start with the simplification process, we neglect the viscosity $\mu\Delta\mathbf{u}$. Such a hypothesis will bring a strong constraint on the choice of the liquid in the experimental setup. Further assuming that the velocity field $\mathbf{u}$ is curl free, we show in the sequel that it derives from a potential Φ which is related to the vertical displacement of the liquid surface ξ through a reduced potential ϕ such that $\Phi(x_1, x_2, x_3, t) = \Re\left(\phi(x_1, x_2)\cosh(\kappa x_3)e^{-i\omega t}\right)$ and $\xi(r, \theta, t) = \mathcal{R}(-\frac{i\omega}{g}\phi(r, \theta)e^{-i\omega t})$, where $r = \sqrt{x_1^2 + x_2^2}$, $\theta = \mathrm{atan}(x_2/x_1)$, ω is the angular wave frequency, and ϕ satisfies Helmholtz's equation

$$\Delta\phi + \kappa^2\phi = 0, \tag{8.2}$$

with κ the spectral parameter [2].

To derive Eq. 8.2, let $x_3 = \xi(x_1, x_2, t)$ be the equation of the free surface. The pressure is prescribed to be equal to the constant atmospheric pressure p_0 on $x_3 = \xi(x_1, x_2, t)$, and the surface tension is neglected. Hence, Eq. 8.1 leads to the well-known Bernoulli's equation

$$\frac{\partial \Phi}{\partial t} + \frac{|\nabla \Phi|^2}{2} + \frac{p_0}{\rho} + g\xi = f(t) \quad \text{on } x_3 = \xi. \tag{8.3}$$

Assuming that $f(t)$ is incorporated into ϕ and that liquid fluctuations are small, that is, $|\xi - h| \ll 1$, where h denotes the mean value of ξ, and also $|\frac{\partial \xi}{\partial x_j}| \ll 1$, $j = 1, 2$, differentiation of Eq. 8.3 with respect to t leads to the linearized equation

$$\frac{\partial^2 \Phi}{\partial t^2} + g\frac{\partial \xi}{\partial t} = 0 \quad \text{on } x_3 = h. \tag{8.4}$$

Using the small-slope ansatz

$$u_3 = \frac{dx_3}{dt} = \frac{\partial \xi}{\partial t} + \frac{\partial \xi}{\partial x_1}\frac{\partial x_1}{\partial t} + \frac{\partial \xi}{\partial x_2}\frac{\partial x_2}{\partial t} \approx \frac{\partial \xi}{\partial t}, \tag{8.5}$$

the divergence-free condition $\text{div}\mathbf{u} = 0$ (i.e., incompressible fluid) and Eq. 8.4 leads to Poisson's condition:

$$\frac{\partial^2 \phi}{\partial t^2} + g\frac{\partial \phi}{\partial x_3} = 0 \quad \text{on } x_3 = h. \tag{8.6}$$

Altogether, Φ is a solution of the following system:

$$\begin{cases} \Delta \Phi = 0 \quad \text{for } x_3 \in]0, h[, \\ \dfrac{\partial^2 \Phi}{\partial t^2} + g\dfrac{\partial \Phi}{\partial x_3} = 0 \quad \text{for } x_3 = h, \\ \mathbf{n} \cdot \nabla \Phi = 0 \quad \text{for } x_3 = 0, \end{cases} \tag{8.7}$$

where the last boundary condition accounts for a no-flow condition through the plane $x_3 = 0$ (this stands for the fixed surface at the bottom of the water tank). If one looks for a harmonic ansatz for the potential $\Phi(x_1, x_2, x_3, t) = f(x_3)e^{-i(\omega t - \kappa_1 x_1 - \kappa_2 x_2)}$, Laplace's equation in Eq. 8.7 gives $f''(x_3) - \kappa^2 f(x_3) = 0$, where $\kappa^2 = \kappa_1^2 + \kappa_2^2$ is by inspection of the Neumann boundary condition in Eq. 8.7, a parameter which forces the evanescence of the potential Φ away from the free boundary in the form $f(x_3) = \cosh(\kappa x_3)$. This means that the physics of the problem can be described by a governing equation on the free interface between liquid and air. Linear LSWs are indeed governed by Helmholtz's equation (Eq. 8.2) on the free surface where ϕ is the reduced potential related to the potential Φ via $\Phi(x_1, x_2, x_3, t) = \Re\left(\phi(x_1, x_2)\cosh(\kappa x_3)e^{-i\omega t}\right)$. We note that if the free surface is perforated by rigid cylinders immersed in the liquid, this equation is valid in the liquid region outside the rigid cylinders and is supplied with a no-flow condition $\frac{\partial\phi}{\partial n}$ on each cylinder's boundary. This will be used in our model of a structured cloak for LSWs. Furthermore, from Poisson's condition in Eq. 8.7, the spectral parameter κ in Eq. 8.2 is linked to the wave frequency via the dispersion relation [2]

$$\omega^2 = g\kappa(1 + d_c^2\kappa^2)\tanh\kappa h, \tag{8.8}$$

with $d_c = \sqrt{\sigma/(\rho g)}$ the liquid capillarity. Note that surface waves propagating at the liquid–air interface are thus always dispersive, unlike pressure waves within the liquid. This dispersive feature of surface waves is also found in the context of plasmonics, whereby electron waves propagate at a metal interface, and will be discussed in the next section.

8.2.2 *Transformed Helmholtz's Equation on the Free Surface*

Le us now consider the coordinate change $(x_1, x_2, x_3, t) \rightarrow (y_1, y_2, x_3, t)$ in the system (Eq. 8.7)

$$\begin{cases} \nabla' \cdot \left(J J^T \det(J^{-1})\nabla'\Phi\right) + \dfrac{\partial^2\Phi}{\partial x_3^2} = 0 \quad \text{for } x_3 \in]0, h[, \\ \dfrac{\partial^2\Phi}{\partial t^2} + g\dfrac{\partial\Phi}{\partial x_3} = 0 \quad \text{for } x_3 = h, \\ \mathbf{n} \cdot \left(\mathbf{J}^T\nabla'\Phi + \dfrac{\partial\Phi}{\partial x_3}\mathbf{e}_3\right) = 0 \quad \text{for } x_3 = 0, \end{cases} \tag{8.9}$$

with $\mathbf{e}_i, i = 1, \ldots, 3$ the canonical basis in $\mathbb{R}^3$, $\Phi = \Phi(y_1, y_2, x_3, t)$ and $\nabla' = \partial/\partial y_1\mathbf{e}_1 + \partial/\partial y_2\mathbf{e}_2$ and J is the 2×2 Jacobian matrix of the transformation which is such that $J_{ij} = \partial y_i/\partial x_j$, for $i, j = 1, 2$.

An elegant way to derive the transformed system is to multiply Eq. 8.7 by a smooth function ψ and to further integrate by parts. Let us consider the first equation in Eq. 8.7, we obtain following variational form

$$\int_{\Omega}\left(\sum_{i,j}\delta_{ij}(x)\frac{\partial\Phi}{\partial x_j}\right)\frac{\partial\psi}{\partial x_i}dx = 0 \tag{8.10}$$

where we assumed that ψ vanishes on the boundary of Ω.

Upon a change of variable $x = (x_1, x_2, x_3, t) \rightarrow y = (y_1, y_2, x_3, t)$ described by a Jacobian matrix $\mathbf{J}$ such that $J_{ij} = \partial y_i/\partial x_j$, this equation takes the form

$$\int_{\Omega}\sum_{i,j,k,l}\delta_{ij}(y)\left(\frac{\partial\Phi}{\partial y_k}\frac{\partial y_k}{\partial x_i}\right)\frac{\partial\psi}{\partial y_l}\frac{\partial y_l}{\partial x_j}\det\left(\frac{\partial \mathrm{x}}{\partial \mathrm{y}}\right)\mathrm{dy} = 0 \tag{8.11}$$

Subsequent integration by parts in the right-hand side gives the following:

$$\int_{\Omega}\sum_{i,j,k,l}\frac{\partial}{\partial y_l}\left(\det\left(\frac{\partial \mathrm{x}}{\partial \mathrm{y}}\right)\frac{\partial \mathrm{y_l}}{\partial \mathrm{x_j}}\delta_{\mathrm{ij}}(\mathrm{y})\frac{\partial \mathrm{y_k}}{\partial \mathrm{x_i}}\frac{\partial\Phi}{\partial \mathrm{y_k}}\right)\psi dy = 0 \tag{8.12}$$

The first equation in Eq. 8.7 takes the new form

$$\sum_{i,j}\frac{\partial}{\partial y_i}\left(\frac{1}{\det \mathrm{J_{ij}}}J_{ij}J_{ij}^{T}\frac{\partial\Phi}{\partial y_j}\right) = 0 \tag{8.13}$$

which is now referred to as the first transformed equation in Eq. 8.9. We proceed in the same way to derive the second and third equations in the transformed system Eq. 8.9. We note that the second equation remains unchanged since the coordinate change involves only x_1 and x_2.

One now considers a harmonic ansatz for the potential $\Phi(y_1, y_2, x_3, t) = f(x_3)e^{-i(\omega t-\kappa_1 y_1-\kappa_2 y_2)}$, the transformed Laplace's equation in Eq. 8.9 gives $f''(x_3) - \kappa'^2 f(x_3) = 0$, where $\kappa'^2 = J_{11}^2\kappa_1^2 + (J_{12}J_{21} + J_{21}J_{12})\kappa_1\kappa_2 + J_{22}^2\kappa_2^2$ is by inspection of the Neumann boundary condition in Eq. 8.9, a parameter which forces the evanescence of the potential Φ away from the free boundary in the form $f(x_3) = \cosh(\kappa' x_3)$. This means that the physics of the problem can be described by a governing equation on the free interface between liquid and air. Linear LSWs are indeed governed by the transformed Helmholtz's equation

$$\nabla'\cdot(JJ^T\det(J^{-1})\nabla'\phi(y_1, y_2)) = \kappa'^2\phi(y_1, y_2) \quad \text{in} \quad \Omega_f, \tag{8.14}$$

where Ω_f stands for the free surface. Upon inspection of Eq. 8.1, we decide to interpret the quantity $JJ^T\det(J^{-1})$ as an anisotropic shear viscosity, and κ' as a spectral parameter encompassing an effective density but other interpretations can be envisaged.

8.2.2.1 Coordinate change for a water wave cloak

Let us now apply Pendry's transform to the design of an invisibiliy cloak for LSWs. To do this, we consider the linear geometric transform:

$$\begin{cases} r' = \alpha r + \beta \\ \theta' = \theta \end{cases} \tag{8.15}$$

with $\alpha = (R_2 - R_1)/R_2$ and $\beta = R_1$ which is simply a radial stretch of polar coordinates. This maps a disc of radius $r \leq R_2$ onto a coronna $R_1 \leq r' \leq R_2$. We follow the algorithm detailed in the seminal work of Nicolet et al. (1994) [9] for the derivation of transformed parameters in Maxwell's equations and apply it to the special case of Helmholtz's equation in order to compute the transformed shear viscosity within Eq. 8.14.

The Jacobian matrix of the transformation is

$$\mathbf{J}_{r'r} = \frac{\partial(r', \theta')}{\partial(r, \theta)} = \begin{pmatrix} \partial r'/\partial r & \partial r'/\partial\theta \\ \partial\theta'/\partial r & \partial\theta'/\partial\theta \end{pmatrix} = \begin{pmatrix} \alpha & 0 \\ 0 & 1 \end{pmatrix} . \tag{8.16}$$

Noting that $(x, y) = (r\cos\theta, r\sin\theta)$ describes the change from polar to Cartesian coordinates $(r, \theta) \to (x, y)$, we find the Jacobian matrix

$$\mathbf{J}_{rx} = \frac{\partial(r, \theta)}{\partial(x, y)} = \begin{pmatrix} \partial r/\partial x & \partial r/\partial y \\ \partial\theta/\partial x & \partial\theta/\partial y \end{pmatrix} = \begin{pmatrix} \cos\theta & r\sin\theta \\ \sin\theta & r\cos\theta \end{pmatrix} , \tag{8.17}$$

which can be recast as $\mathbf{J}_{rx} = \mathrm{diag}(1, r)\mathbf{R}(\theta)$ where $\mathbf{R}(\theta)$ is the rotation matrix through an angle θ. We then see that $\mathbf{J}_{xr} = \mathbf{R}(-\theta)\mathrm{diag}(1, 1/r)$ since $\mathbf{J}_{xr} = \partial(x, y)/\partial(r, \theta) = (\partial(r, \theta)/\partial(x, y))^{-1} = (\mathbf{J}_{rx})^{-1}$ and the rotation matrix satifies $\mathbf{R}(\theta)^{-1} = \mathbf{R}(\theta)^T = \mathbf{R}(-\theta)$.

We start with the computation of the Jacobian matrix of the compound transformation $(x, y) \to (r, \theta) \to (r', \theta') \to (x', y')$ which can be expressed as

$$\begin{aligned} \mathbf{J}_{xx'} = \mathbf{J}_{xr}\mathbf{J}_{rr'}\mathbf{J}_{r'x'} &= \mathbf{R}(-\theta)\mathrm{diag}(1, 1/r)\mathrm{diag}(1/\alpha, 1)\mathrm{diag}(1, r')\mathbf{R}(\theta') \\ &= \mathbf{R}(-\theta)\mathrm{diag}(1/\alpha, r'/r)\mathbf{R}(\theta') , \end{aligned}$$

where $\mathbf{J}_{xr} = \partial(x, y)/\partial(r, \theta)$ and $\mathbf{J}_{r'x'} = \partial(r', \theta')/\partial(x', y')$; we note that $\det(\mathbf{J}_{xx'}) = r'/(\alpha r)$ as the rotation matrices $\mathbf{R}(\theta)$ and $\mathbf{R}(\theta')$ are unimodular.

We then compute the artificial shear viscosity

$$\begin{aligned}
&\mathbf{J}_{xx'}\mathbf{J}_{xx'}^T \det(\mathbf{J}_{xx'}^{-1}) \\
&= \mathbf{R}(-\theta)\mathrm{diag}(1/\alpha, r'/r)\mathbf{R}(\theta)\mathbf{R}(\theta)^T \mathrm{diag}(1/\alpha, r'/r)\mathbf{R}(-\theta)\alpha r/r' \\
&= \mathbf{R}(-\theta)\mathrm{diag}(\alpha^{-1}r/r', \alpha r'/r)\mathbf{R}(\theta)\,,
\end{aligned}$$

where we have used the fact that the rotation matrix satifies $\mathbf{R}(\theta)^{-1} = \mathbf{R}(\theta)^T$ and $\theta' = \theta$.

We now observe that from Eq. 8.15 $r = (r' - R_1)/\alpha$, hence the artificial shear viscosity inside the circular coating of the cloak can be expressed as

$$\underline{\underline{\mu}} = \mathbf{R}(\theta')\mathrm{diag}(\mu_{r'r'}, \mu_{\theta'\theta'})\mathbf{R}(\theta')^T\,, \tag{8.18}$$

where the eigenvalues of the diagonal matrix (principal values of conductivity) are from Eq. 8.18

$$\mu_{r'r'} = \frac{r' - R_1}{\alpha^2 r'}\,,\ \mu_{\theta'\theta'} = \frac{\alpha^2 r'}{r' - R_1}\,, \tag{8.19}$$

with R_1 and R_2 the interior and the exterior radii of the cloak. One should note that when $r' = R_1$ in the transformed coordinates, that is, $r = 0$ in the original coordinates, the shear viscosity becomes singular as its first coefficient vanishes and the second one tends to infinity.

8.2.3 *Effective Anisotropic Shear Viscosity through Homogenization*

Our aim is now to extend this principle to the LSW waves governed by Helmholtz's equation (Eq. 8.2) and the Neumann boundary conditions at the boundary of inclusions. These inclusions are the cross section of pillars perforating the free surface of a fluid. The goal is to homogenize the microstructured cloak shown in Fig. 8.1 which reveals an effective anisotropic shear viscosity underpinning the cloaking effect. Having in mind that when the shear viscosity matrix is large enough in the tangential θ direction, LSW waves will be bent around the central region of the cloak and thus make its interior almost invisible to any exterior observer.

For this, we note that when the fluid penetrates the microstructure of Fig. 8.1 (whose cross section Ω_c is evenly divided into a large

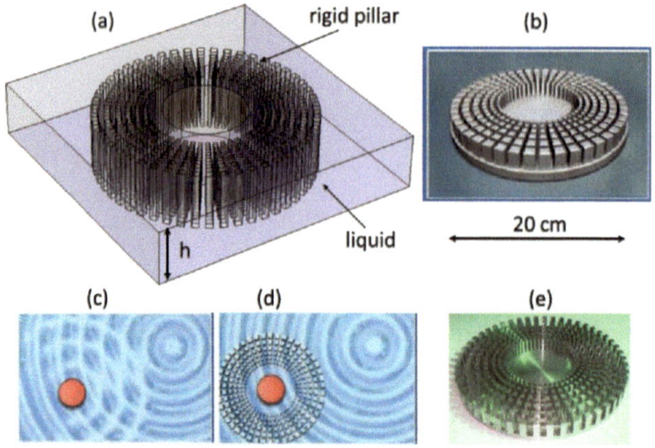

Figure 8.1 Principle of water wave cloak: (a) Geometry of a structured cloak consisting of concentric arrays of rigid pillars immersed in a vessel of liquid of depth h; (b) diagrammatic view of the cloak; (c–d) scattering of water waves on a rigid obstacle (red disc) without (c) and with (d) the water wave cloak; and (e) photo of the microstructured cloak used in experiments around 10 Hz.

number of small curved sectors ηY of radial length $\eta(R_2 - R_1)$ and azimuthal length $2\pi\eta$, where η is a small positive real parameter), it undergoes fast periodic oscillations. To filter these oscillations, we consider an asymptotic expansion of the potential field solution of the Helmholtz equation (Eq. 8.2) in terms of a macroscopic (slow) variable $\mathbf{x} = (r, \theta)$ and a microscopic (fast) variable $\mathbf{x}/\eta$ [14]:

$$\forall \mathbf{x} \in \Omega_c, \quad \phi_\eta(\mathbf{x}) = \phi_0\left(\mathbf{x}, \frac{\mathbf{x}}{\eta}\right) + \eta\phi_1\left(\mathbf{x}, \frac{\mathbf{x}}{\eta}\right) + \eta^2\phi_2\left(\mathbf{x}, \frac{\mathbf{x}}{\eta}\right) + \cdots \tag{8.20}$$

where each term $\phi^{(i)}(\mathbf{x}, \cdot)$ is Y periodic.

The differential operator is rescaled accordingly as $\nabla = \nabla_\mathbf{x} + \frac{1}{\eta}\nabla_\mathbf{y}$, and collecting terms of same powers of η, we obtain the following homogenized problem in the limit when η tends to zero (see also [14]):

$$\nabla \cdot ([\mu_{\text{hom}}]\nabla\phi_{\text{hom}}(\mathbf{x})) = \kappa^2\phi_{\text{hom}}(\mathbf{x}) \qquad \text{in} \quad \Omega_c. \tag{8.21}$$

This homogenized equation on the free surface could be interpreted as a velocity field solution of Eq. 8.1 with an anisotropic matrix of viscosity with a nontrivial part (transverse shear)

$$[\mu_{\text{hom}}] = \frac{1}{\mathcal{A}(Y^*)}\begin{pmatrix} \mathcal{A}(Y^*) - \psi_{rr} & \psi_{r\theta} \\ \psi_{\theta r} & \mathcal{A}(Y^*) - \psi_{\theta\theta} \end{pmatrix}. \quad (8.22)$$

Here, $\mathcal{A}(Y^*)$ denotes the area of the region Y^* surrounding a rigid inclusion S (subject to Neumann boundary conditions) in an elementary cell Y of the periodic array, and ψ_{ij} represent corrective terms derived from an integral over the boundary ∂S:

$$\forall i, j \in \{r, \theta\}, \ \psi_{ij} = -\int_{\partial S} \Psi_i n_j \, ds \,, \quad (8.23)$$

where $\mathbf{n}$ is the unit outward normal to ∂S, and Ψ_j , $j \in \{r, \theta\}$, are Y periodic potentials, which are unique solutions (up to an additive constant) of the following two Laplace equations $(\mathcal{L}_j)$:

$$(\mathcal{L}_j) : \ \nabla^2 \Psi_j = 0 \text{ in } Y^* \,. \quad (8.24)$$

These so-called annex problems are supplied with the effective boundary condition $\frac{\partial \Psi_j}{\partial n} = -\mathbf{n} \cdot \mathbf{e}_j$ on the boundary ∂S of the inclusion. Here, $\mathbf{e}_r$ and $\mathbf{e}_\theta$ denote the vectors of the basis in polar coordinates (r, θ).

Altogether, our interpretation is that the velocity field is solution of Eq. 8.1 with an anisotropic matrix of viscosity whose nontrivial part (transverse shear) could be calculated as [15]

$$[\mu_{\text{hom}}] = \frac{1}{\mathcal{A}(Y^*)}\begin{pmatrix} \mathcal{A}(Y^*) + 0.7 & 0 \\ 0 & \mathcal{A}(Y^*) + 7.2 \end{pmatrix}, \quad (8.25)$$

which shows that the effective fluid is strongly anisotropic along the θ direction.

We propose next to give a crash course in homogenization to support the results of this section which is based upon our earlier work [15], but readers who are already familiar with this topic could skip this section.

8.3 Homogenization of Helmholtz's Equation

Homogenization is concerned with problems of fast oscillating partial differential equations (PDEs). These PDEs depend upon a small positive parameter η characterizing the lengthscale of

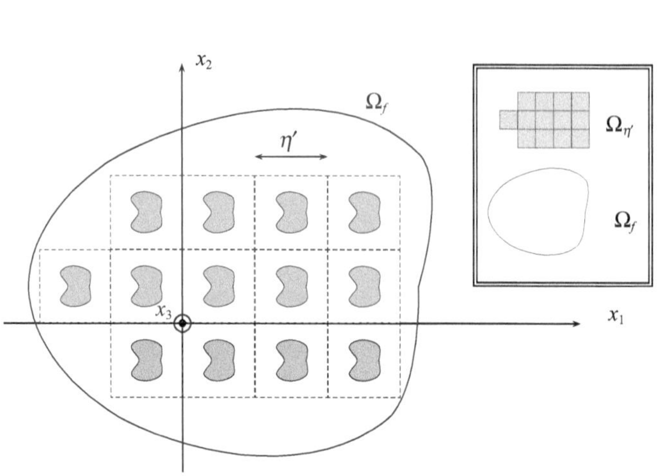

Figure 8.2 Top view of a structured fluid with 13 basic cells. The small positive parameter η' has the typical length scale of a basic cell in the array Ω'_η which is contained in the finite domain Ω_f. The smaller the value of η', the larger the number of basic cells in Ω_f, and the better the approximation of Ω_f by Ω'_η.

elementary (basic) cells constituting the structured fluid under study. One wishes to identify the homogenized PDE in the limit when η goes to zero. In what follows, we apply so-called two-scale expansions to Eq. 8.2 for a perforated domain. For this, we look for solutions of Eq. 8.2 supplied with a Neumann boundary condition in a domain Ω_η like in Figs. 8.2 and 8.3:

$$(\mathcal{P}_\eta):\begin{cases} -\displaystyle\sum_{i=1}^{2}\frac{\partial}{\partial x_i}\left(\chi(\frac{x}{\eta})\frac{\partial\phi_\eta(x)}{\partial x_i}\right) = \kappa^2\phi_\eta(x), \text{ in } \Omega_\eta, \\ \displaystyle\sum_{i=1}^{2}\chi(\frac{x}{\eta})\frac{\partial\phi_\eta}{\partial x_i}n_i = 0, \qquad \text{at } \partial\Omega_\eta, \end{cases}$$

where we assume that there is no source inside Ω_η, $x = (x_1, x_2)$ is the macrosopic (or slow) variable and $\partial\Omega_\eta$ denotes the boundary of Ω_η with an outward unit normal (n_1, n_2). Moreover, $\chi(y)$ is a (spatially varying) coefficient that is equal to a small value inside the perforations and 1 outside (it allows us to mark the location of perforations). This is a piecewise constant function of the microscopic (or fast) variable $y = (y_1, y_2)$, of period 1 (in y_1 and y_2)

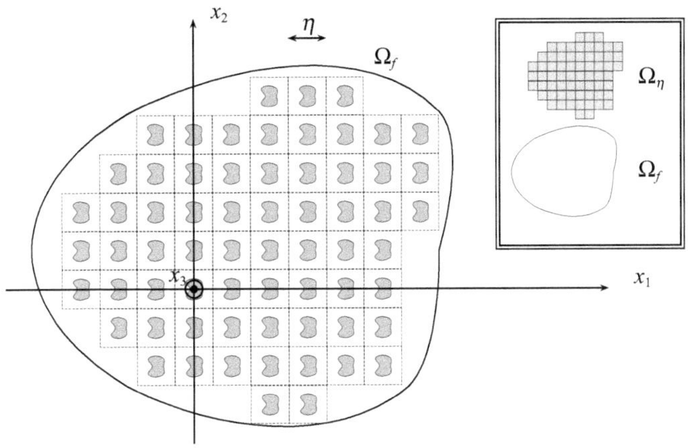

Figure 8.3 Top view of a structured fluid with 5 times more basic cells as in Fig. 8.1. Their typical length scale is $\eta \approx \eta'/2 \ll 1$ and one notes that Ω_η is a better approximation of Ω_f than $\Omega_{\eta'}$. In the limit when η tends to zero, one can replace the structured medium in Ω_f by a homogenized one.

such that:

$$\chi(y) \geq c > 0\,, \ \forall y \in Y = [0,1]^2\,.$$

It is useful to rewrite $(\mathcal{P}_\eta)$ as follows:

$$-\sum_{i=1}^{2} \frac{\partial}{\partial x_i}\sigma_i(\phi_\eta) = \kappa^2 \phi_\eta(x)$$

where σ is the flux vector defined by

$$\sigma_i(\phi_\eta) = \chi\left(\frac{x}{\eta}\right)\frac{\partial \phi_\eta(x)}{\partial x_i}\,.$$

Let us consider the following two-scale asymptotic expansion:

$$\phi_\eta = \phi_0(x) + \eta\phi_1(x, y) + \eta^2\phi_2(x, y) + \ldots$$

where $\phi_i(x, y)$ is a periodic function of period 1 in the microscopic variable y.

Moreover, we rescale the differential operator in the following way:

$$\frac{\partial \phi_\eta}{\partial x_i} = \left(\frac{\partial \phi_0}{\partial z_i}_{|z_i=x_i} + \frac{\partial \phi_1}{\partial y_i}\right) + \eta\left(\frac{\partial \phi_1}{\partial z_i}_{|z_i=x_i} + \frac{\partial \phi_2}{\partial y_i}\right) + \ldots$$

by noting that $1/\eta(\partial\phi_0(x)/\partial y_i) = 0$. Moreover, we consider the ansatz:

$$\sigma_i(\phi_\eta) = \sigma_i^0 + \eta\sigma_i^1 + \eta^2\sigma_i^2 + ...$$

We are led to:

$$\sigma_i^0(\phi_\eta) = \chi(y)\left(\frac{\partial\phi_0}{\partial z_i}_{|z_i=x_i} + \frac{\partial\phi_1}{\partial y_i}\right)$$

$$\sigma_i^1(\phi_\eta) = \chi(y)\left(\frac{\partial\phi_1}{\partial z_i}_{|z_i=x_i} + \frac{\partial\phi_2}{\partial y_i}\right)$$

and so forth, all terms being periodic of period 1. For the sake of simplicity, we assume in the sequel that whenever we differentiate with respect to z_i, the expression is estimated at $z_i = x_i$.

The problem (P_η) takes the form

$$-\left(\frac{1}{\eta}\frac{\partial}{\partial y_i} + \frac{\partial}{\partial z_i}\right)(\sigma_i^0 + \eta\sigma_i^1 + ...) = \kappa^2(\phi_0(x) + ...)$$

so that collecting the same factors in front of the same powers of η, we obtain at order η^{-1}

$$(A) : -\frac{\partial}{\partial y_i}\sigma_i^0 = 0$$

and at order η^0

$$(H) : -\frac{\partial}{\partial z_i}\sigma_i^0 - \frac{\partial}{\partial y_i}\sigma_i^1 = \kappa^2\phi_0(x)$$

(equations at next orders won't be needed).

In the sequel, we shall see that Eq. H leads to the homogenized Helmholtz's equation, with coefficients of the homogenized tensor of shear viscosity given by the annex problem Eq. A, which is an equation on the microscopic scale. In this way we can compute ϕ_0 and ϕ_1 which are the first two terms of the asymptotic expansion of ϕ_η.

To do this, we define the mean operator on Y, denoted by $< . >$, that acts on a function g of the microscopic variable y:

$$< g >= \frac{1}{\mid Y \mid}\int\int_Y g(y_1, y_2)dy_1dy_2 ,$$

where $\mid Y \mid$ is the measure (area) of Y.

Applying the mean operator to both sides of Eq. H, we get

$$(H): -\frac{\partial}{\partial z_i} < \sigma_i^0 > - < \frac{\partial}{\partial y_i}\sigma_i^1 >= \kappa^2\phi_0(x) < 1 > ,$$

where we used the fact that the order upon which we apply the operators $< . >$ and $\partial/\partial z_i$ does not matter.

We now note that from the divergence (Gauss's) theorem:

$$< \frac{\partial}{\partial y_i}\sigma_i^1 >= \frac{1}{| Y |}\int\int_Y \frac{\partial}{\partial y_i}\sigma_i^1(y)dy = \frac{1}{| Y |}\int_{\partial Y} \sigma_i^1(y)n_i ds ,$$

where $\mathbf{n} = (n_1, n_2)$ is the outward unit normal to the boundary ∂Y of Y. This normal points in opposite directions on opposite sides of Y, so the integral over the boundary vanishes.

We finally get

$$(H): -\frac{\partial}{\partial z_i} < \sigma_i^0 >= \kappa^2\phi_0(x) ,$$

which is a homogenized equation of the macroscopic variable x with partial derivatives $\partial/\partial z_i$ with respect to the macroscopic variable only.

We would like to find a relationship between $< \sigma^0 >$ and the gradient in the variable x of ϕ_0. Indeed, we have seen that

$$\sigma_i^0(\phi_\eta) = \chi(y)\left(\frac{\partial\phi_0}{\partial z_i} + \frac{\partial\phi_1}{\partial y_i}\right)$$

which using Eq. A leads to

$$(A1): -\frac{\partial}{\partial y_i}\left(\chi(y)\frac{\partial\phi_1}{\partial y_i}\right) = \left(\frac{\partial\phi_0}{\partial z_i}\right)\left(\frac{\partial}{\partial y_i}\chi(y)\right) .$$

We can interpret Eq. A1 like an equation for the unknown $\phi_1(x, y)$, periodic of period 1 in y and parameterized by x. This equation admits a unique solution up to an additive constant. Bearing in mind that the parameters x and t only appear through the factor $\partial\phi_0/\partial z_j$, by linearity of the PDE, we can write the solution $\phi_1(x, y)$ as follows:

$$\phi_1(x, y) = w^j(y)\partial\phi_0(x)/\partial z_j ,$$

where the two functions $w^j(y)$, $j = 1, 2$ are the solutions of Eq. A1 corresponding to the partial derivative $\partial\phi_0(x, t)/\partial z_j$, $j = 1, 2$ equal to 1 (the other partial derivative being zero), that is to say solutions of

$$(A2): -\frac{\partial}{\partial y_i}\left(\chi(y)\frac{\partial w^k}{\partial y_i}\right) = \delta_{ik}\left(\frac{\partial}{\partial y_i}\chi(y)\right) ,$$

with $w^k(y)$, $k = 1, 2$ periodic functions of the y variable with period 1.

We note that Eq. A2 is two equations that depend only on coefficients $\kappa(y)$, that is, on microscopic properties of the medium. The two functions w^k (defined up to an additive constant) can be computed once for all, independently of Ω_f.

Once the functions $w^k(y)$ are known, we note that

$$\sigma_j^0(x, y, t) = \chi(y)\left(\frac{\partial \phi_0}{\partial z_j} + \frac{\partial \phi_1}{\partial y_j}\right) = \chi(y)\left(\frac{\partial \phi_0}{\partial z_j} + \frac{\partial \phi_0}{\partial z_k}\frac{\partial w^k(y)}{\partial y_j}\right),$$

which can be recast as

$$\sigma_i^0(x, y) = \chi(y)\left(1 + \frac{\partial w^k(y)}{\partial y_i}\right)\frac{\partial \phi_0(x, t)}{\partial z_k}.$$

Applying the mean operator to this equation, we obtain

$$< \sigma_i^0 > (x) = -\chi_{ik,\mathrm{hom}}\frac{\partial \phi_0(x)}{\partial z_k}.$$

We have thus derived that the homogenized problem is

$$(\mathcal{P}_0): \begin{cases} -\displaystyle\sum_{i,k=1}^{2} \frac{\partial}{\partial z_i}\left(\chi_{ik,\mathrm{hom}}\frac{\partial \phi_0(x)}{\partial z_k}\right) = \kappa^2 \phi_0(x), \text{ in } \Omega_f, \\ \displaystyle\sum_{i,k=1}^{2} \chi_{ik,\mathrm{hom}}\frac{\partial \phi_0}{\partial z_k} n_i = 0, \qquad \text{at } \partial\Omega_f, \end{cases}$$

where $\chi_{ik,\mathrm{hom}}$ are the entries of the homogenized shear viscosity tensor defined by

$$\chi_{ik,\mathrm{hom}} = \frac{1}{\mid Y \mid}\int\int_Y \chi(y)\left(1 + \frac{\partial w^k(y)}{\partial y_i}\right) dy_1 dy_2 .$$

One can then identify $(\mathcal{P}_0)$ with Eq. 8.21 and $\chi_{ik,\mathrm{hom}}$ with $[\mu_{\mathrm{hom}}]$ in Eq. 8.22 by assuming that $\chi(y) = 1$ in $Y^* = Y \setminus S$ and $\chi(y) = \varepsilon$, with ε a small positive parameter $\varepsilon \ll 1$ in S. A first-order ansatz provides the expected result.

8.3.1 *Numerical Analysis of LSW Cloaking*

In this section, we give some numerical illustrative examples of cloaking of an acoustic vibration (radiated by a cylindrical point source) located close to an obstacle through the phononic crystal

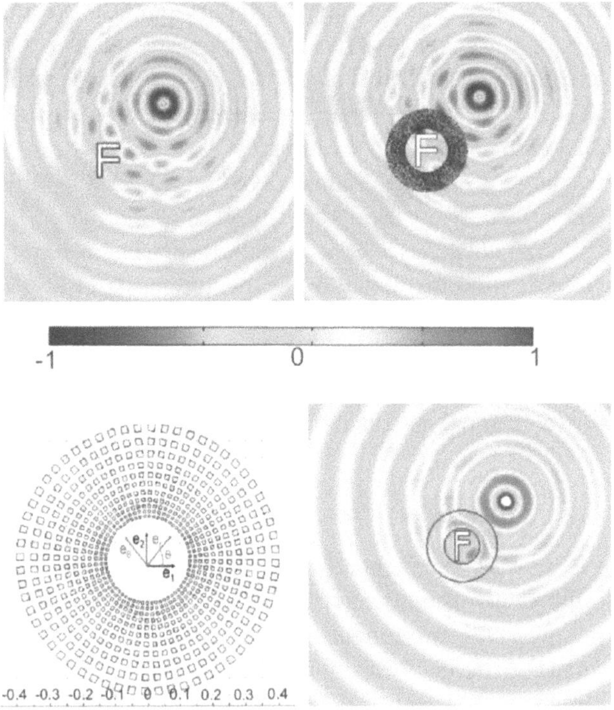

Figure 8.4 Upper panel: 2D plot of the real part $\mathcal{R}e(\phi)$ of the potential of velocity **u** radiated by an acoustic point source of normalized wavelength, $\lambda = c/\nu = c/2.5$, in the presence of an F-shaped obstacle (left) and a microstructured cloak surrounding the obstacle (right). Lower panel: Magnification of the geometry of the microstructured cloak (left) and diffraction by a cloak (with inner and outer boundaries $r = R_1 = 0.164$ and $r = R_2 = 0.4$) whose parameters are given by Eq. 8.19. The similarities between the upper right and lower right panels are noted. Importantly, the radii are normalized here and are related to those of Fig. 8.5 through a factor 2.5 (for meters).

fiber making the cloak (see Figs. 8.1 and 8.4). As detailed in the previous section, we first replace the microstructured cloak immersed in the liquid by an effective transversely anisotropic fluid whose homogenized shear viscosity is deduced from the numerical solution (Eq. 8.25) of the two annex problems (Eq. 8.24). This provides us with a qualitative pictorial representation of the cloaking mechanism shown in the lower right panel of Fig. 8.4.

We then compare this asymptotic theory against the numerical solutions of the same scattering problems when we model the complete microstructured cloak, as shown in the upper right panel of Fig. 8.4. We notice the strong similarities between the asymptotic solution and the numerical solution.

According to transformation acoustics, the effective fluid should be characterized by a varying density ρ, as well as a varying radial and azimuthal shear viscosities $\mu'_{r'r'}$ and $\mu'_{\theta'\theta'}$. These requirements seem to be beyond the actual experimental possibilities. Nevertheless, we can introduce some variation in the radial length of the sectors for which it seems reasonable to assume that the improved cloak is characterized by an effective anisotropic fluid whose shear viscosity (a diagonal matrix in polar basis) is

$$\mu'_{r'r'} = \left(\frac{R_2(r' - R_1)}{(R_2 - R_1)r'}\right)^2, \qquad \mu'_{\theta'\theta'} = \left(\frac{(R_2 - R_1)r'}{R_2(r' - R_1)}\right)^2, \tag{8.26}$$

where R_1 and R_2 are, respectively, the inner and outer radii of the ring. Importantly, the effective fluid's density $\rho' = \rho$; that is, it does not play any prominent role. Figure 8.5 shows simulations performed for the ideal case when the obstacle is cloaked and for a vacuum for comparison.

The upper right panel of Fig. 8.5 clearly shows that the structured cloak consisting of a very large number (100) of identical curved sectors small compared to the working wavelength and regularly arranged along the radial and azimutal directions will enable one to gain control over surface waves. We have also verified the effect of varying the size of sectors along the radial direction and numerically checked that the cloaking is thus further improved: It is enough to design a cloak with identical sectors to gain a good control of the velocity field, but the azimuthal shear wave speed of liquid particles will increase linearly with their distance to the center of the cloak only in the case of sectors with increasing size. We numerically checked that in the case of a cloak with identical sectors, this shear wave speed does not vary; hence, a shadow region revealing the presence of the hidden object is observed behind the cloak (through phase shift). We checked also that such a structured cloak with 256 curved sectors (Fig. 8.4, upper right panel) is more appropriate for

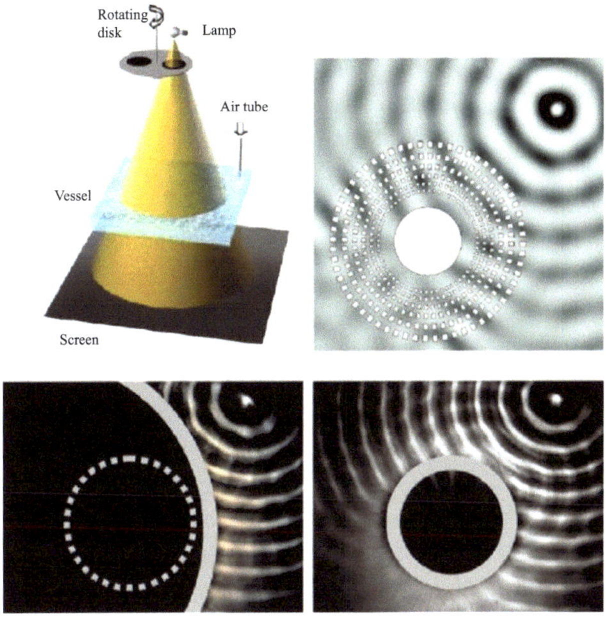

Figure 8.5 Upper left (experimental setup): A halogen lamp modulated by a perforated rotating disc illuminates a transparent vessel containing the liquid (methoxynonafluorobutane). The surface waves are excited by a localized pressure, due to air pulsed in a small tube at the same frequency as the modulation of the light (stroboscopic effect). Upper right (simulation): Diffraction pattern of the surface waves generated by a harmonic acoustic source at frequency 10 Hz for a cloak with 100 rigid sectors. The depth of liquid in the vessel is $h = 9$ mm and its capillarity is $d_c = 0.95$ mm. Lower left: Snapshot of diffraction pattern by a rigid cylinder of radius 38 mm surrounded by the structured cloak (outlined as the gray coating) of Fig. 8.4 with inner and outer radii $R_1 = 41$ mm and $R_2 = 100$ mm. Lower right: Snapshot of diffraction pattern by the rigid cylinder on its own (outlined by a dashed gray circle in the left panel for comparison). Reprinted (figure) with permission from Ref. [15]. Copyright (2008) by the American Physical Society.

the cloaking than with 100 curved sectors. But we had to find a compromise between the realizable structures using conventional machining, the limits imposed by the viscosity of the liquids, and the constraints imposed by homogenization.

8.3.2 *Experimental Measurements of LSW Cloaking*

Regarding the experimental setup, we refer the reader to Fig. 8.5, which clearly demonstrates a reduced backscattering for an acoustic source generating LSW at 10 Hertz a couple of wavelengths away from a structured cloak consisting of 100 rigid pillars arranged, as in the lower left panel of Fig. 8.4.

The liquid used for the experiments was methoxynonafluorobutane, which is chosen for its physical properties and especially for having a low kinematic viscosity [2] $\nu = \mu/\rho = 0.61\ \text{mm}^2/\text{s}$) so that $\mu\nabla^2\mathbf{u}$ can be neglected outside the cloak in Eq. 8.1, a small surface tension $\sigma = 13.6\text{N/cm}$ and a large density ($\rho = 1.529$ g/mL), ensuring a small capillarity length $d_c = \sqrt{\sigma/(\rho g)} = 0.95$ mm. The vessel is filled with a depth of liquid $h = 9$ mm.

The basic principle behind the experiments is very simple: the light of a halogen lamp modulated by a perforated rotating disc illuminates a transparent vessel containing the liquid. The surface waves are excited by a localized pressure, thanks to air pulsed in a small tube at the same frequency as the modulation of the light (to take advantage of the stroboscopic effect for the observation). The surface waves create local curvatures of the liquid and the light is refracted when crossing the surface. Thus, on the screen the dark and light zones allow for visualizing the LSWs. Note that the low viscosity of the liquid is important for such experiments and we were unable to produce similar results with water: due to its larger viscosity, the water profile flattens within the microstructured cloak, much like in thin channels [2], and water cannot flow.

8.4 Water Wave Cloaks and Invisibility Carpets of an Arbitrary Shape

Let us now apply Pendry's transform to the design of an invisibility cloak of an arbitrary cross section for LSW. To do this, we consider the linear geometric transform [16]:

$$\begin{cases} r' = \alpha(\theta)r + \beta(\theta) \\ \theta' = \theta \end{cases} \tag{8.27}$$

where $\beta(\theta) = R_1(\theta)$ and $\alpha(\theta) = (R_2(\theta) - R_1(\theta))/R_2(\theta)$. This transform is simply a radial stretch of polar coordinates but also depends upon the azimuthal angle in order to model domains of a (smooth) arbitrary shape. The cloak's boundary is defined by two embedded smooth closed curves $R_1(\theta) < R_2(\theta)$, for $0 \leq \theta < 2\pi$, as shown in Fig. 8.2. The transform maps the field within the domain $r \leq R_2(\theta)$ onto the annular domain $R_1(\theta) \leq r' \leq R_2(\theta)$. The region $r' \leq R_1(\theta)$ defines the invisibility zone, while the annulus in the cloak consists of a material with heterogeneous anisotropic conductivity. We follow the algorithm detailed in the seminal work of Nicolet et al. (1994) ([9]) for the derivation of transformed parameters in Maxwell's equations and apply it to the special case of Helmholtz's equation in order to compute the transformed shear viscosity within Eq. 8.14.

One should note that when $r' = \beta$ in the transformed coordinates, that is, $r = 0$ in the original coordinates, the transformation matrix becomes singular as its first coefficient vanishes and the other three tend to infinity.

The Jacobian matrix of the transformation is

$$\mathbf{J}_{rr'} = \begin{pmatrix} 1 & 0 & 0 \\ -\alpha & c_{22} & 0 \\ 0 & 0 & 1 \end{pmatrix}, \tag{8.28}$$

After some elementary algebra, we obtain the following shear viscosity:

$$\Longrightarrow \quad J J^T \det(J^{-1}) = \mathbf{R}(\theta) \begin{pmatrix} \frac{(r'-\beta)^2 + c_{12}^2.\alpha^2}{(r'-\beta)r'} & -\frac{c_{12}.\alpha}{r'-\beta} \\ -\frac{c_{12}\alpha}{r'-\beta} & \frac{r'}{r'-\beta} \end{pmatrix} \mathbf{R}(\theta)^T \,, \tag{8.29}$$

where $c_{12} = \partial r/\partial\theta'$ and $\mathbf{R}(\theta)$ denotes the rotation matrix through an angle θ, which amounts to expressing the metric tensor in a Cartesian coordinate basis. The details of the derivation can be found in Nicolet et al. (1994) ([16]), where a similar analysis was performed for control of transverse time-harmonic electromagnetic waves propagating in cylindrical invisibility cloaks.

The above coefficient takes the developed expression

$$c_{12} = \frac{-1}{(R_2(\theta) - R_1(\theta))^2} \left\{ \frac{dR_1(\theta)}{d\theta}(R_2(\theta) - r')(R_2(\theta) - R_1(\theta))^2 R_2(\theta) + \frac{dR_2(\theta)}{d\theta} R_1(\theta)(R_1(\theta) - r') \right\} \tag{8.30}$$

Let us consider three embedded smooth closed curves defined by $R_1(\theta) < R_2(\theta) < R_3(\theta)$, for $0 \leq \theta < 2\pi$, as shown in Fig. 8.2 and the transformation (Eq. 8.15) with $\alpha = \frac{R_1(\theta)}{R_2(\theta)}$ and $\beta = 0$ for $0 \leq r \leq R_2(\theta)$ and with $\alpha = \frac{R_3(\theta)-R_1(\theta)}{R_3(\theta)-R_2(\theta)}$ and $\beta = R_3(\theta)\frac{R_1(\theta)-R_2(\theta)}{R_3(\theta)-R_2(\theta)}$ for $R_2(\theta) \leq r \leq R_3(\theta)$. This transformation maps the field in the region $0 \leq r \leq R_2(\theta)$ onto $0 \leq r' \leq R_1(\theta')$ (i.e., compression of fluid) and the field in the region $R_2(\theta) \leq r \leq R_3(\theta)$ onto $R_1(\theta') \leq r' \leq R_3(\theta')$ (i.e., extension of fluid). Importantly, the compression and extension compensate each other for $0 < r' \leq R_3(\theta')$ and the transformation should be the identity from $R_3(\theta) < r$ to $R_3(\theta') < r'$.

The resulting LSW concentrator (in transformed variables) then consists of two parts:

- $0 \leq r' \leq R_1(\theta')$ (inner core):

$$c_{12} = \frac{\partial r}{\partial \theta'} = -\frac{r}{R_1(\theta)^2}\left(R_2(\theta)\frac{\partial R_1(\theta)}{\partial \theta} - R_1(\theta)\frac{\partial R_2(\theta)}{\partial \theta}\right), \quad (8.31)$$

- $R_1(\theta') \leq r' \leq R_3(\theta')$ (coating):

$$c_{12} = \frac{-1}{(R_2(\theta) - R_1(\theta))^2}\left\{\frac{\partial R_1(\theta)}{\partial \theta}(R_3(\theta) - r)(R_3(\theta) - R_2(\theta)) + \frac{\partial R_2(\theta)}{\partial \theta}(R_1(\theta) - R_3(\theta))(R_3(\theta) - r) - \frac{\partial R_3(\theta)}{\partial \theta}(R_2(\theta) - R_1(\theta))(R_1(\theta) - r)\right\}. \quad (8.32)$$

Moreover, the effective density is as follows:

$$\begin{cases} \left(\dfrac{R_2}{R_1}\right)^2 & , \text{if } 0 \leq r' \leq R_1 , \\ \dfrac{r' + R_3\frac{R_2-R_1}{R_3-R_2}}{r'}\left(\dfrac{R_3 - R_2}{R_3 - R_1}\right)^2 & , \text{if } R_1 \leq r' \leq R_3 , \end{cases}$$

which is a product that cannot vanish. This makes an LSW concentrator difficult to realize in practice. However, the numerical illustration can be performed fairly easily. We consider an arbitrary shaped rotator described as follows:

$$R_1(\theta) = 0.4\ R\ (1 + 0.2\sin(3\theta))$$
$$R_2(\theta) = R\ (1 + 0.2\sin(3\theta) + 0.2\cos(4\theta)); \quad R = 0.4\,. \quad (8.33)$$

The numerical results are shown in Fig. 8.6.

Let us propose now an LSW rotator. More precisely, we consider the geometric transform

$$\begin{cases} r' = r \\ \theta' = \alpha.r + \beta \quad \text{with } \alpha = \dfrac{\theta_o}{R_1(\theta) - R_2(\theta)}; \quad \beta = \theta + \dfrac{R_2(\theta).\theta_o}{R_2(\theta) - R_1(\theta)} \\ z' = z \end{cases} \tag{8.34}$$

We obtain the following shear viscosity:

$$R(\theta) \begin{pmatrix} c_{22} & \alpha.r \\ \alpha.r & \dfrac{1+\alpha^2.r^2}{c_{22}} \end{pmatrix} R(\theta)^T \tag{8.35}$$

where

$$c_{22} = \frac{\partial\theta}{\partial\theta'} = 1 - \frac{\theta_o.r}{(R_2(\theta) - R_1(\theta))^2}.\left(\frac{\partial R_2(\theta)}{\partial\theta'} - \frac{\partial R_1(\theta)}{\partial\theta'}\right) - \frac{\theta_o}{(R_2(\theta) - R_1(\theta))^2}\left(R_2(\theta)\frac{\partial R_1(\theta)}{\partial\theta'} - R_1(\theta)\frac{\partial R_2(\theta)}{\partial\theta'}\right) \tag{8.36}$$

We note that the determinant of the shear viscosity is 1. The numerical results are shown in Fig. 8.6.

We finally propose an invisibility carpet for water waves. Let us consider the linear geometric transform

$$\begin{cases} x' = x\,,\ a < x < b\,, \\ y' = \frac{y_2(x) - y_1(x)}{y_2(x)} y + y_1(x)\,,\ 0 < y < y_2(x)\,, \end{cases} \tag{8.37}$$

where y' is a stretched vertical coordinate. It is easily seen that this linear geometric transform maps the segment (a, b) of the

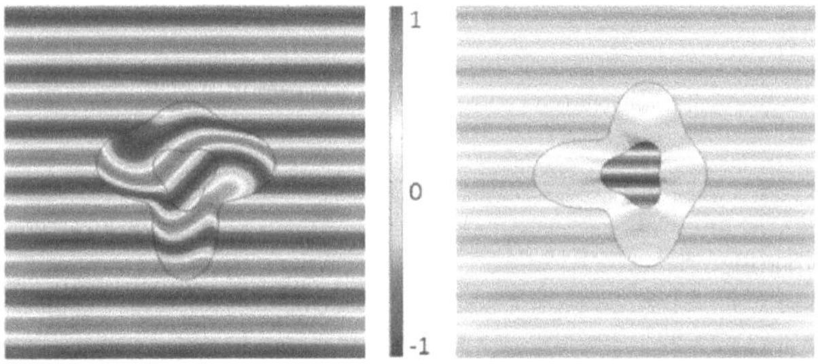

Figure 8.6 2D plot of the real part $\mathcal{R}e(\phi)$ of the potential of velocity **u** for an acoustic plane wave of normalized wavelength , $\lambda = c/\nu = c/2.5$ incident from top on a rotator (left) and a concentrator (right).

horizontal axis $y = 0$ onto the curve $y' = y_1(x)$, and it leaves the curve $y = y_2(x)$ unchanged. Importantly, there is a one-to-one correspondence between the segment and y_1. The curves y_1 and y_2 are assumed to be differentiable, and this ensures that the carpet won't display any singularity on its inner boundary, as we shall now derive.

The linear transform (Eq. 8.37) is expressed in a Cartesian basis as $\mathbf{J}_{xx'} = \begin{pmatrix} 1 & 0 \\ \frac{\partial y}{\partial x'} & \frac{1}{\alpha} \end{pmatrix}$ where $\alpha = (y_2 - y_1)/y_1$ and from the chain rule

$$\frac{\partial y}{\partial x'} = y_2 \frac{y' - y_2}{(y_2 - y_1)^2} \frac{\partial y_1}{\partial x} - y_1 \frac{y' - y_1}{(y_2 - y_1)^2} \frac{\partial y_2}{\partial x} . \tag{8.38}$$

This leads to the shear viscosity tensor which is fully described by three nonvanishing entries in a Cartesian basis:

$$\begin{aligned} \mu_{11} &= \frac{1}{\alpha} , \; \mu_{12} = \mu_{21} = -\frac{\partial y}{\partial x'} \\ \mu_{22} &= \left(1 + \left(\frac{\partial y}{\partial x'}\right)^2\right) \alpha , \end{aligned} \tag{8.39}$$

For the sake of illustration, let us now consider a 2D carpet that has inner and outer boundaries given by

$$y_i(x) = h_i \left(e^{-\frac{1}{2}(\frac{x}{\sigma})^2} - \frac{1}{8}\right) + c_i \sin\left(d_i . h_i \left(e^{-\frac{1}{2}(\frac{x}{\sigma})^2} - \frac{1}{8}\right)\right) , \tag{8.40}$$

$i = 1, 2$, with $h_1 = 0.2$, $h_2 = 0.4$, $c_1 = c_2 = 0.01$, $d_1 = 60$, $d_2 = 50$ and $\sigma = 0.3$.

Numerical results are shown in Fig. 8.7.

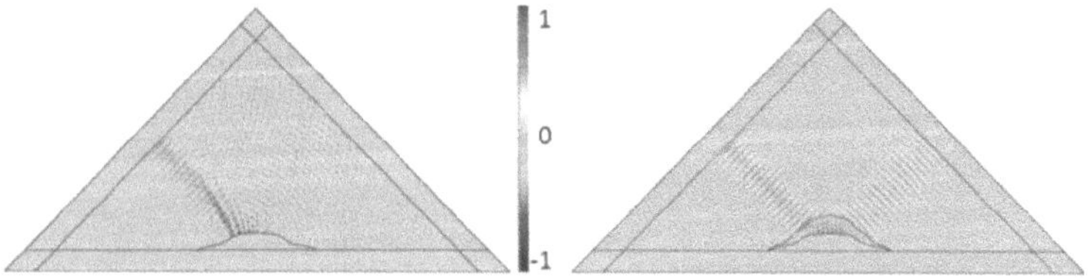

Figure 8.7 Diffraction by an LSW Gaussian beam under oblique incidence at wavelength $\lambda = c/\nu = 0.06$: we set $\phi^i = \exp(-1/2(1/\sqrt{2}(x+y))^2/0.1^2)$ on the left side of the inner triangle. 2D plot of the real part of ϕ. (Left) Deformed mirror; (right) deformed mirror with a carpet. Perfectly matched layers were implemented by applying a rotation of $\pi/4$ radians in the coordinate axes.

8.5 Conclusion

In this chapter, we have reviewed some results of our group in the control of LSWs. Water wave cloaks, concentrators, rotators, and carpets can be designed via geometric transforms, with recent experimental proofs of concepts in Refs. [18, 19]. However, this leads to metafluids with shear anisotropic heterogeneous viscosity. There is also the issue of the effective density which is heterogeneous with very low values in some cases, except for rotators. One thus needs to make some compromise between control of an LSW and feasibility of the hydrodynamic metamaterial. Actually, a practical design of a broadband water wave cloak has been discussed based upon homogenization theory. This cloak turns out to also work for acoustic pressure waves and microwaves [17]. This is to the best of our knowledge the first prototype of a multiwave cloak.

References

1. Alù, A., and Engheta, N. (2005). Achieving transparency with plasmonic and metamaterial coatings, *Phys. Rev. E*, **72**, 016623.
2. Acheson, D.J. (1990). *Elementary Fluid Dynamics* (Oxford University Press, Oxford).
3. Pendry, J. B., Schurig, D., and Smith, D. R. (2006). controlling electromagnetic fields, *Science*, **312**, 1780–1782.
4. Schurig, D., Mock, J.J., Justice, J.B., Cummer, S.A., Pendry J.B., Starr, A.F., and Smith, D.R. (2006). Metamaterial electromagnetic cloak at microwave frequencies, *Science*, **314**, 977–980.
5. Leonhardt, U. (2006). Optical conformal mapping, *Science*, **312**, 1777–1780.
6. Zolla, Z., Guenneau, S., Nicolet, A., and Pendry, J.B. (2007). Electromagnetic analysis of cylindrical invisibility cloaks and the mirage effect, *Opt. Lett.*, **32**, 1069.
7. Milton, G.W., and Nicorovici, N.A.P. (2006). On the cloaking effects associated with anomalous localised resonance, *Proc. R. Soc. Lond. A*, **462**, 3027.
8. Nicorovici, N.A.P., Milton, G.W., McPhedran, R.C., and Botten, L.C. (2007). Quasistatic cloaking of two-dimensional polarizable discrete systems by anomalous resonance, *Opt. Express*, **15**, 6314–6323.

9. Nicolet, A., Remacle, J.F., Meys, B., Genon, A., and Legros, W. (1994). Transformation methods in computational electromagnetics, *J. Appl. Phys.*, **75**, 6036–6038.
10. Pendry, J.B. (2000). Negative refraction makes a perfect lens, *Phys. Rev. Lett.*, **86**, 3966–3969.
11. Li, J., and Pendry, J.B. (2008). Hiding under the carpet: a new strategy for cloaking, *Phys. Rev. Lett.*, **101**, 203901.
12. Greenleaf, A., Kurylev, Y., Lassas, M., and Uhlmann, G. (2007). Full-wave invisibility of active devices at all frequencies, *Comm. Math. Phys.*, **275**, 3, 749–789.
13. Rahm, M., Schurig, D., Roberts, D.A., Cummer, S.A., Smith, D.R., and Pendry, J.B. (2008). Design of electromagnetic cloaks and concentrators using form-invariant coordinate transformations of Maxwell's equations, *Photon. Nanostruct. Fundam. Appl.*, **6**, 87–95.
14. Jikov, V.V., Kozlov, S.M., and Oleinik, O.A. (1994). *Homogenization of Differential Operators and Integral Functionals* (Springer-Verlag, New York).
15. Farhat, M., Enoch, S., Guenneau, S., and Movchan, A.B. (2008). Broadband cylindrical acoustic cloak for linear surface waves in a fluid, *Phys. Rev. Lett.*, **101**, 1345011.
16. Nicolet, A., Zolla, Z., and Guenneau, S. (2008). Electromagnetic analysis of cylindrical cloaks of an arbitrary cross section, *Opt. Lett.*, **33**, 1584–1586.
17. Xu, J., Jiang, X., Fang, N., Georget, E., Abdeddaim, R., Geffrin, J.-M., Farhat, M., Sabouroux, P., Enoch, S., and Guenneau, S. (2015). Molding acoustic, electromagnetic and water waves with a single cloak, *Sci. Rep.*, **5**.
18. Berraquero, C.P., Maurel, A., Petitjeans, P., and Pagneux, V. (2013). Experimental realization of a water-wave metamaterial shifter, *Phys. Rev. E*, **88**, 051002(R).
19. Dupont, G., Kimmoun, O., Molin, B., Guenneau, S., and Enoch, E. (2015). Numerical and experimental study of an invisibility carpet in a water channel, *Phys. Rev. E*, **91**(2), 023010.

Chapter 9

Cloaking for Heat and Mass Diffusion

Sebastien Guenneau,[a] David Petiteau,[a] Myriam Zerrad,[a] Claude Amra,[a] and Tania M. Puvirajesinghe[b]

[a]*Institut Fresnel, UMR CNRS 7249, Aix-Marseille Université, Campus de St Jérôme, 13397 Marseille Cedex 20, France*
[b]*Institut Paoli-Calmettes, UMR INSERM 1068, UMR CNRS 7258, Aix-Marseille Université, Marseille, France*
sebastien.guenneau@fresnel.fr

9.1 Introduction

In the present chapter, we discuss miscellaneous functionalities of cloaks for heat and mass diffusion processes first proposed in the transient regime by Guenneau et al. (2012) [1] and Guenneau and Puvirajesinghe (2013) [2]. Such designs are based upon the extension of transformational optics to the academic disciplines of thermodynamics and biophysics. More precisely, in a similar manner to the English physicist Sir John Pendry and his American colleagues David Schurig and David Smith who proposed in Ref. [3] to design an invisibility cloak by mapping Maxwell's equations on a curvilinear space with a hole in it (where an object can be hidden), one can make coordinate changes in equations of thermodynamics

Transformation Wave Physics: Electromagnetics, Elastodynamics, and Thermodynamics
Edited by Mohamed Farhat, Pai-Yen Chen, Sebastien Guenneau, and Stefan Enoch

ISBN 978-981-4669-95-5 (Hardcover), 978-981-4669-96-2 (eBook)
www.panstanford.com

and biochemistry [1, 2] governing convection-diffusion phenomena. Interestingly, the transformed Fick's equation described by Guenneau and Puvirajesinghe (2013) [2] can also be applied to models of invisibility cloaks for light in diffusive media [4]. A multitude of research papers have been published in the past three years on the control of diffusive processes via artificial anisotropy, therefore justifying the fact that thermal and biochemical metamaterials are no longer a niche but rather a fast-growing subject [5].

We shall see in the sequel that quite unexpectedly cloaking for diffusion processes is more subtle than cloaking for waves: Unlike for diffusion where heat or mass concentration inside the inner region of an invisibility cloak depends crucially on the cloak's position versus the source and the time elapsed (we stress here that transient regime is for short times, before the steady-state regime is reached, which corresponds to a different physics akin to thermostatics [6]), in the latter case it is widely accepted that the field vanishes in the invisibility region irrespective of the cloak's material constituent, time parameter, and its distance to the source. This fact generated a small controversy as to whether or not one can speak of some form of thermal protection with heat cloaks. On the one hand, asymptotic arguments can lead to the assumption that heat/mass concentration need not be continuous across the cloak's inner boundary, only the flux, in which case one can show that temperature field (resp. mass concentration) keeps its initial value (i.e., that at time step $t = 0$) with time [13].

On the other hand, the form of protection displayed by diffusion cloaks can be assumed to be related to heat/mass concentration fluxes if one assumes that temperature (resp. mass concentration) need be continuous across the cloak's inner boundary (let us recall that usually one has the continuity of the field and its flux at interfaces). In other words, protection is related to gradients of temperature (resp. mass concentration) and no longer to temperature (resp. mass concentration) themselves. With such kind of assumption, made by Guenneau et al. (2012) [1], fluxes vanish inside the invisibility region, as confirmed by Guenneau et al. (2015) [8]. From a physical standpoint, one of two things can happen, either scenario 1 with a perfect insulating-like boundary condition at the inner cloak's boundary that prevents any temperature

Figure 9.1 Artistic representation of the potential use of a thermal cylindrical cloak for protection of a component on a computer motherboard.

(resp. mass concentration) increase in the invisibility region or scenario 2 that allows some field increase but in such a way that it is constant at every time step. Both scenarios seem plausible when compared to the experiment led by Wegener's group [7]. It is therefore legitimate to envisage some thermal protection for electronic components through either minimization of temperature increase or the minimization of heat gradients, as depicted by the artistic view in Fig. 9.1.

What is the secret of this magical trick? The trick lies in the extension of the design of transformation-based metamaterials to the area of thermodynamics and life sciences with convection-diffusion phenomena. In this chapter, we explain the underlying cloaking mechanism for the diffusion equation. We first show that the diffusion equation retains its form under geometric transform and we derive the expression of its heterogeneous anisotropic diffusivity in a general coordinate system. We then apply specific transforms in order to design a cloak, a concentrator and a rotator for heat flux or for concentration flux. We use full pseudo-wave computations to back up our claims of an unprecedented control of heat or concentration flux through a change of metric in the thermic or chemical space. We finally propose a multilayered device working either as a thermic or chemical protection in the homogenization regime.

The plan of the chapter is as follows: In Section 9.2, we apply a change of coordinates to the diffusion-convection equation and show that the diffusivity becomes in general heterogeneous (i.e., spatially varying) and anisotropic (i.e., matrix valued). In Section 9.3, we apply specific mappings in order to make a hole in the transformed coordinates so as to design a cloak (with a blow-up of a point), a concentrator (with a compression of a region), and a rotator (with a rotation of axes) for heat flux or for mass concentration flux. We use finite element computations to validate our hypothesis of the control of heat or concentration flux in the transformed thermal or chemical spaces. We also notably analyze the potential protection offered by diffusion cloaks, which could have potential applications in microelectronics (see Fig. 9.1 for an artistic view). In Section 9.4, we propose a multilayered spherical device which approximates the functionality of a diffusion cloak in the homogenization regime [13, 18]. In Section 9.5, we look at so-called diffusion carpet cloaks and we conclude the chapter in Section 9.6.

9.2 Coordinates Changes as a Magic Potion to Control Convection-Diffusion Phenomena

We consider the convection-diffusion equation which is a parabolic partial differential equation combining the diffusion equation and the advection equation. This equation describes physical phenomena where particles or energy (or other physical quantities) are transferred inside a physical system due to two processes: diffusion and convection. In its simplest form (when the diffusion coefficient and the convection velocity are constant and there are no sources or sinks), the convection-diffusion equation in a bounded cylindrical domain Ω (with a source outside) can be written as

$$\rho(\mathbf{x})c(\mathbf{x})\frac{\partial u}{\partial t} = \nabla \cdot (\kappa(\mathbf{x})\nabla u - \mathbf{v}(\mathbf{x})u) + p(\mathbf{x}, t) , \qquad (9.1)$$

where u represents the distribution of temperature (in thermodynamics) at each point $\mathbf{x} = (x, y, z)$ in Ω evolving with time $t > 0$. Moreover, κ is the thermal conductivity ($W{\cdot}m^{-1}{\cdot}K^{-1}$, that is, watt per meter kelvin in SI units); ρ is the density ($kg{\cdot}m^{-3}$, that is, kilogram per cubic meter in SI units); and c the specific heat (or thermal) capacity ($J{\cdot}K^{-1}{\cdot}kg^{-1}$, that is, joule per kilogram kelvin in SI units). The bulk velocity $\mathbf{v}$ has the unit of **length by time**. It is customary to let κ go in front of the spatial derivatives when the medium is homogeneous. The diffusion flux $-\kappa\nabla u$ measures the amount of substance that will flow through a small volume during a small time interval ($mol{\cdot}m^{-3}{\cdot}s^{-1}$). However, here we consider a heterogeneous medium; hence the spatial derivatives of κ might suffer some discontinuity (derivatives are taken in distributional sense; hence transmission conditions ensuring continuity of the heat flux $\kappa\nabla u$ are encompassed in Eq. 9.1). In this chapter, we consider a source with a time step (Heaviside) variation and a singular (Dirac) spatial variation, that is, $p(\mathbf{x}, t) = p_0 H(t)\delta(\mathbf{x} - \mathbf{x}0)$, with H the Heaviside function and Delta the Dirac distribution. This means that the source term is constant throughout time $t > 0$, while it is spatially localized on the line $\mathbf{x} = \mathbf{x}_0$.

Upon a change of variable $(x, y, z) \to (x', y', z')$ described by a Jacobian matrix $\mathbf{J} = \partial(x', y', z')/\partial(x, y, z)$, this equation takes the form

$$\rho(\mathbf{x}')c(\mathbf{x}')\det(\mathbf{J})\frac{\partial u}{\partial t} = \nabla \cdot \left(\mathbf{J}^{-T}\kappa(\mathbf{x}')\mathbf{J}^{-1}\det(\mathbf{J})\nabla u \right.$$
$$\left. -\mathbf{J}^{-T}\det(\mathbf{J})\mathbf{v}(\mathbf{x}')u\right) + \det(\mathbf{J})\,p(\mathbf{x}', t)\,. \quad (9.2)$$

We note that Eqs. 9.1 and 9.2 have the same structure, except that the transformed diffusivity takes the form

$$\underline{\underline{\kappa}}' = \mathbf{J}^{-T}\kappa\mathbf{J}^{-1}\det(\mathbf{J}) = \kappa\mathbf{J}^{-T}\mathbf{J}^{-1}\det(\mathbf{J}) = \kappa\mathbf{T}^{-1}\,, \quad (9.3)$$

which is matrix-valued conductivity and the transformed velocity has the form

$$\mathbf{v}' = \mathbf{J}^{-T}\det(\mathbf{J})\mathbf{v}\,. \quad (9.4)$$

which is vector valued. In Eq. 9.3, $\mathbf{T}$ is the metric tensor, and the time derivative on the left-hand side of Eq. 9.2 is multiplied by the determinant of the Jacobian matrix $\mathbf{J}$ of the transformation.

An elegant way to derive Eq. 9.2 is to multiply Eq. 9.1 by a smooth function $\phi \in C_0^\infty(\Omega)$ (i.e., an infinitely differentiable function with a compact support on Ω) and to further integrate by parts, which leads to the following variational form:

$$\int_{\Omega} \rho c \left(\frac{\partial u}{\partial t} \phi \right) dxdydz + \int_{\Omega} (\kappa \nabla u \cdot \nabla \phi)\, dxdydz$$
$$- \int_{\partial\Omega} (\kappa \nabla u \cdot \mathbf{n}\phi - \kappa \nabla \phi \cdot \mathbf{n} u)\, ds$$
$$- \int_{\Omega} \mathbf{v} u \cdot \nabla \phi dxdydz + \int_{\partial\Omega} (\mathbf{v} u \cdot \mathbf{n}\phi)\, ds + < p, \phi >= 0\,,$$

where $\mathbf{n}$ is the unit outward normal to the boundary $\partial\Omega$ of the integration domain Ω. Moreover, $<, >$ denotes the duality product between the space of distributions ($\mathcal{D}'(\Omega)$) and the space of smooth functions ($\mathcal{D}(\Omega) = C_0^{\infty}(\Omega)$), that is, a pairing in which one integrates a distribution against a test function, so denoted because it corresponds to a bilinear map from vector spaces to scalars (see Ref. [18] and references therein).

We now apply to the above equation the coordinate change $(x, y, z) \to (x', y', z')$, and noting that $\nabla = \mathbf{J}^{-1}\nabla'$, where ∇' is the gradient in the new coordinates, we end up with

$$\begin{aligned}
&\int_{\Omega} \left(\rho c \det(\mathbf{J}) \frac{\partial u}{\partial t} \phi \right) dx'dy'dz' \\
&+ \int_{\Omega} \left\{ \left(\mathbf{J}^{-1} \nabla' \phi \cdot \kappa \mathbf{J}^{-1} \nabla' u \right) \det(\mathbf{J}) \right\} dx'dy'dz' \\
&- \int_{\partial\Omega} \left(\kappa \mathbf{J}^{-1} \nabla' u \cdot \mathbf{n}\phi - \kappa \mathbf{J}^{-1} \nabla' \phi \cdot \mathbf{n} u \right) ds' \\
&- \int_{\Omega} \left\{ \left(\mathbf{J}^{-1} \nabla' \phi \cdot \mathbf{v} \nabla' u \right) \det(\mathbf{J}) \right\} dx'dy'dz' \\
&+ \int_{\partial\Omega} \{ (\mathbf{J}^{-T} \mathbf{v} u \cdot \mathbf{n}\phi) \det(\mathbf{J}) \} ds' + < \det(\mathbf{J}) p, \phi >= 0\,.
\end{aligned} \tag{9.5}$$

Upon integration by parts and noting that $\mathbf{J}^{-1}\nabla'\phi \cdot \kappa \mathbf{J}^{-1}\nabla' u = (\nabla'\phi)^T \mathbf{J}^{-T} \kappa \mathbf{J}^{-1} \nabla' u$ we obtain the variational form of Eq. 9.2 which lays the foundation of transformation thermodynamics with a convection term. We further note that Fick's equation which is used in the context of mass diffusion is encompassed by the convection-diffusion equation (to see this one need only take the product ρc to be equal to 1 and assume that the velocity field $\mathbf{v}$ vanishes). Thus, what we discuss in the sequel applies straightforwardly to mass diffusion.

Let us now apply the transformed Eq. 9.2 to the design of three metamaterials of potential practical interest: an invisibility cloak, a concentrator, and a rotator for heat flux (or mass diffusion).

9.3 Invisibility Cloak, Concentrator, and Rotator of an Arbitrary Shape for Diffusion Processes

To describe boundaries of general shapes, a finite Fourier expansion, following [17]

$$\rho(\theta) = a_0 + \sum_{k=1}^{n} (a_k \cos(k\theta) + b_k \sin(k\theta)) \,, \tag{9.6}$$

may be used. Throughout this section, we consider a linear transform

$$\rho'(\rho, \theta) = \alpha\rho + \beta, \tag{9.7}$$

where α and β are θ dependent coefficients, with $0 < \theta \leq 2\pi$. In this way, we can design some cloak, concentrator, and rotator of a complex shape.

We shall consider three boundaries described as follows:

$$\begin{cases} R_1(\theta) = 0.4R(1 + 0.2\sin(3\theta)); \quad R_2(\theta) = 0.6R(1 + 0.2\sin(3\theta)) \\ R = 0.4; \quad R_3(\theta) = R(1 + 0.2(\sin(3\theta) + \cos(4\theta))). \end{cases} \tag{9.8}$$

9.3.1 *Diffusion Cloaks*

As depicted in Fig. 9.2, heat flux is smoothly detoured within the thermal cloak. Moreover, isothermal lines are unperturbed outside this metamaterial, that is, they are straight, parallel lines.

To achieve such a magic trick, we consider the geometric transform Eq. 9.7 with

$$\alpha = \frac{R_3(\theta) - R_1(\theta)}{R_3(\theta)} \,, \; \beta = R_1(\theta) \,, \; R_1(\theta) \leq r \leq R_3(\theta) \,. \tag{9.9}$$

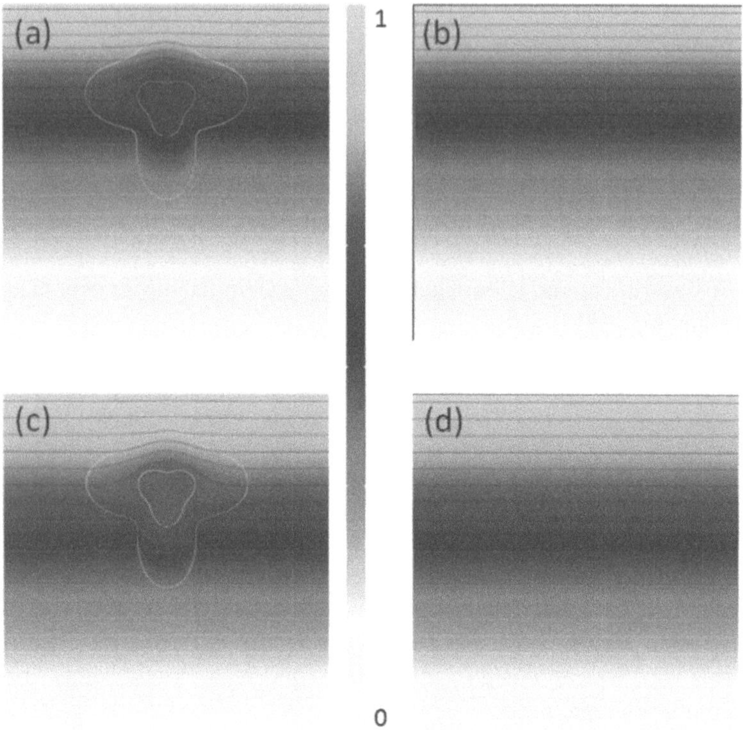

Figure 9.2 Diffusion cloak: Diffusion of heat (or mass) from the top on a cloak of complex shape of inner radius $R_1(\theta) = 0.4R(1 + 0.2\sin(3\theta))$ and outer radius $R_3(\theta) = R(1 + 0.2(\sin(3\theta) + \cos(4\theta)))$, with $R = 0.4\ 10^{-4}$ m. The cloak is placed close to the source (left) and removed (right). Snapshots of heat (resp. concentration) distribution at $t = 0.07$ s (a, b) and $t = 0.21$ s (c, d) show that isovalues of temperature (resp. concentration) are nearly unperturbed outside the cloak. One notes that in the core of the cloak the maximum temperature (resp. concentration) increases with time and is almost uniform at a given time step.

We obtain the following Jacobian and transformation matrices (where $\mathbf{R}(\theta)$ denotes the matrix of rotation through an angle θ):

$$\mathbf{J_{rr'}} = \begin{pmatrix} 1 & 0 \\ -\alpha & c_{22} \end{pmatrix} \quad \Longrightarrow \quad \mathbf{T^{-1}} = \mathbf{R}(\theta) \begin{pmatrix} \frac{(r-\beta)^2 + c_{22}^2 . \alpha^2}{(r-\beta).r} & -\frac{c_{22}.\alpha}{r-\beta} \\ -\frac{c_{22}.\alpha}{r-\beta} & \frac{r}{r-\beta} \end{pmatrix} \mathbf{R}(\theta)^{\mathbf{T}}. \tag{9.10}$$

with $\det(\mathbf{J}) = (r-\beta)/(\alpha^2 r)$ and

$$c_{22} = \frac{-1}{(R_2(\theta)-R_1(\theta))^2}\left\{\frac{\partial R_1(\theta)}{\partial\theta}(R_3(\theta)-r)(R_3(\theta))\right\}. \qquad (9.11)$$

It is interesting to inspect the diffusion cloak parameter values at its nonreflecting outer boundary $R_2(\theta)$. This can be done through the analysis of the entries of the inverse of the metric tensor $\mathbf{T}$ in the polar basis. We first notice that the off-diagonal terms $(T^{-1})_{r\theta} = (T^{-1})_{\theta r}$ are generally nonzero unlike for the circular case when $\mathbf{T}^{-1}$ is diagonal. In general we observe that $-1 < (T^{-1})_{r\theta} < 1$, which reflects the rotation of the tensor $\mathbf{T}^{-1}$ with respect to its eigenbasis. We further note that $(T^{-1})_{rr}$ also varies with θ, unlike for circular thermal cloaks. Last, $0 < (T^{-1})_{rr} < 1$ and $(T^{-1})_{\theta\theta} > 1.5$, in agreement with the fact that the cloak ought to exhibit a strong azimuthal anisotropy for heat to flow around the inner core.

It is interesting to compare these numerical results carried out in the intense near-field limit when the cloak is located nearby the source to those obtained from an analytical approach in order to better understand how this thermal cloak works. The temperature (or mass concentration) field u is given in the transformed coordinates by

$$u'(\rho', \theta') := u(\rho(\rho', \theta'), \theta(\theta')), \qquad (9.12)$$

with $\rho(\rho', \theta')$ and $\theta(\theta')$ given by the inverse map of the map defined by Eq. 9.9. We used the software Mathematica to produce the map of isothermal values (see Fig. 9.3). Comparing this ideal diffusion cloaking with that of Fig. 9.2, we notice that the temperature (or mass concentration) inside the invisibility region (inner core) of the cloak has the same value at any time point as the temperature (mass concentration) at the point that we blow up in the original coordinate system. Importantly, all these computations are done in the case of scenario 2, where the assumption is that the temperature (resp. mass concentration) field is continuous at the cloak's inner boundary. In scenario 1, temperature does not evolve inside the invisibility region. As previously mentioned in the introduction, the fact that two scenarios coexist makes the analysis of diffusion cloaks all the more challenging from a mathematical viewpoint.

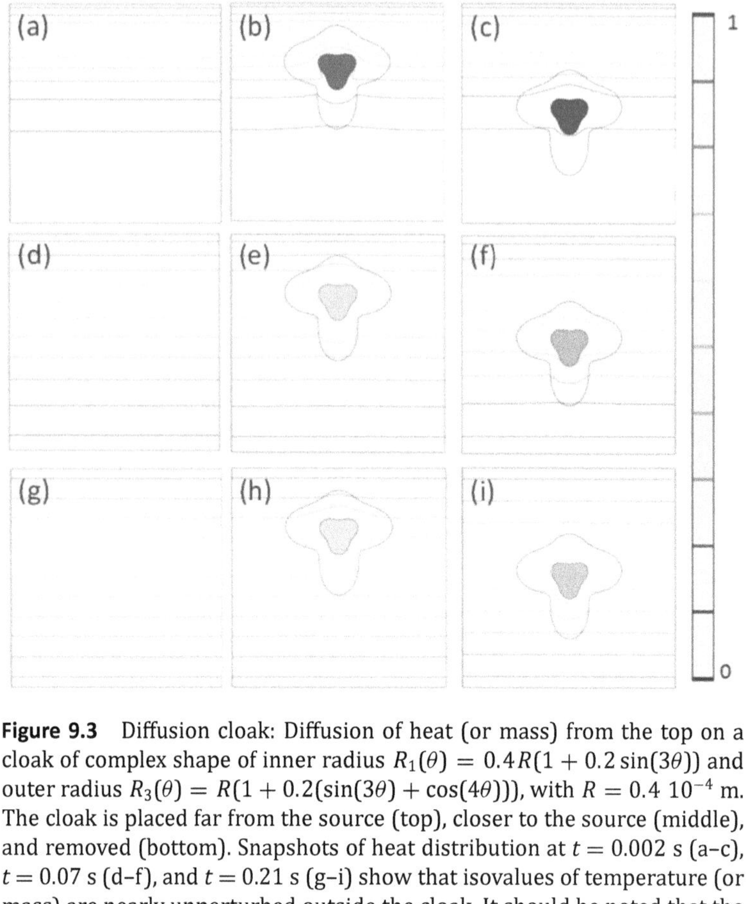

Figure 9.3 Diffusion cloak: Diffusion of heat (or mass) from the top on a cloak of complex shape of inner radius $R_1(\theta) = 0.4R(1 + 0.2\sin(3\theta))$ and outer radius $R_3(\theta) = R(1 + 0.2(\sin(3\theta) + \cos(4\theta)))$, with $R = 0.4\ 10^{-4}$ m. The cloak is placed far from the source (top), closer to the source (middle), and removed (bottom). Snapshots of heat distribution at $t = 0.002$ s (a–c), $t = 0.07$ s (d–f), and $t = 0.21$ s (g–i) show that isovalues of temperature (or mass) are nearly unperturbed outside the cloak. It should be noted that the maximum temperature in the core of the cloak is achieved for $t \geq 21$ s, is almost uniform in the core (vanishing flux), and depends upon the distance between the center of the cloak to the source.

9.3.2 *Diffusion Concentrators and Rotators*

In the same spirit as the previous section, we propose to extend the design of circular concentrators for heat [1] to more complex shapes. For the sake of simplicity, we consider a concentrator with the same boundaries as in Eq. 9.8. One can see in Fig. 9.4 that heat (resp. mass concentration) isovalues are smoothly squeezed (hence the flux is enhanced) within the concentrator. Moreover, the concentrator is itself invisible for heat (resp. mass) as the isothermal lines are unperturbed outside this metamaterial.

More precisely, we consider the geometric transform with

$$\begin{cases} \alpha = \frac{R_1(\theta)}{R_2(\theta)} & \beta = 0 \qquad (0 \leq r \leq R_1(\theta)) \\ \alpha = \frac{R_3(\theta)-R_1(\theta)}{R_3(\theta)-R_2(\theta)} & \beta = R_3(\theta)\frac{R_1(\theta)-R_2(\theta)}{R_3(\theta)-R_2(\theta)} \quad (R_1(\theta) \leq r \leq R_3(\theta)) \end{cases} \tag{9.13}$$

We obtain the following Jacobian and transformation matrices (where $\mathbf{R}(\theta)$ denotes the matrix of rotation through an angle θ):

$$\mathbf{J}_{\mathbf{rr}'} = \begin{pmatrix} 1 & 0 \\ -\alpha & c_{22} \end{pmatrix} \implies \mathbf{T}^{-1} = \mathbf{R}(\theta) \begin{pmatrix} \frac{(r-\beta)^2 + c_{22}^2.\alpha^2}{(r-\beta).r} & -\frac{c_{22}.\alpha}{r-\beta} \\ -\frac{c_{22}.\alpha}{r-\beta} & \frac{r}{r-\beta} \end{pmatrix} \mathbf{R}(\theta)^{\mathrm{T}} \tag{9.14}$$

with $\det(\mathbf{J}) = \frac{r-\beta}{\alpha^2 r}$.

Also, we have

- $0 \leq r \leq R_1(\theta)$:

$$c_{22} = \frac{\partial r}{\partial \theta'} = -\frac{r}{R_1(\theta)^2}\left(R_2(\theta)\frac{\partial R_1(\theta)}{\partial \theta} - R_1(\theta)\frac{\partial R_2(\theta)}{\partial \theta}\right) \tag{9.15}$$

- $R_1(\theta) \leq r \leq R_3(\theta)$:

$$\begin{aligned} c_{22} = \frac{-1}{(R_2(\theta) - R_1(\theta))^2} \Bigg\{ & \frac{\partial R_1(\theta)}{\partial \theta}(R_3(\theta) - r)(R_3(\theta) - R_2(\theta)) \\ & + \frac{\partial R_2(\theta)}{\partial \theta}(R_1(\theta) - R_3(\theta))(R_3(\theta) - r) - \frac{\partial R_3(\theta)}{\partial \theta}(R_2(\theta) \\ & - R_1(\theta))(R_1(\theta) - r)\Bigg\} \end{aligned} \tag{9.16}$$

We now extend the design of circular cylindrical rotators for heat [8] to more complex geometries. To simplify the mathematical setup of the problem, we consider a rotator with the same boundaries as the concentrator (see Eq. 9.8). One can see in Fig. 9.4 that heat (resp. mass) flux is smoothly rotated within the rotator (through an angle $\pi/4$). Moreover, the rotator is itself invisible for heat (resp. mass) as the isothermal values are unperturbed outside this metamaterial. It could be used to enhance thermal (or mass) exchanges.

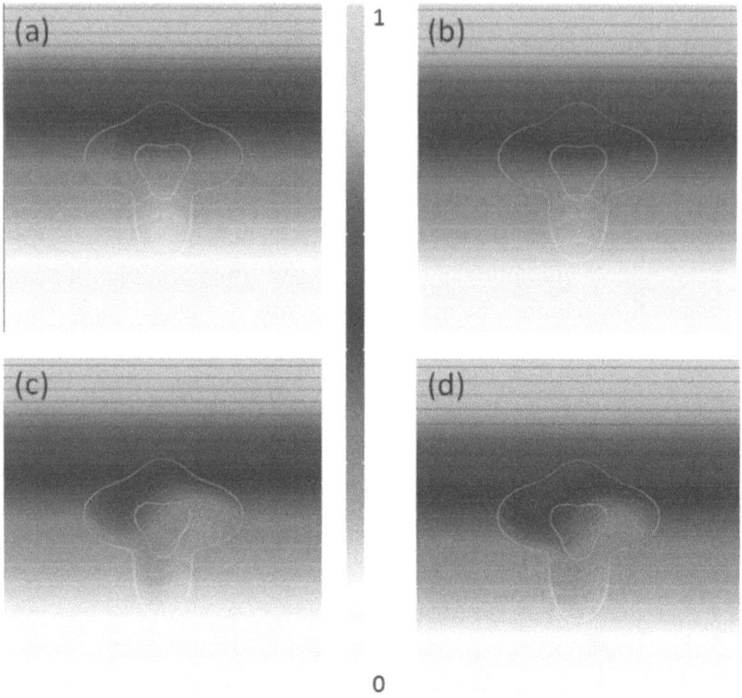

Figure 9.4 Diffusion concentrator and rotator: Diffusion of heat (or mass) from the top onto a concentrator (a, b) and rotator (c, d) of complex shape of inner radius $R_1(\theta) = 0.4R(1 + 0.2\sin(3\theta))$ and outer radius $R_3(\theta) = R(1 + 0.2(\sin(3\theta) + \cos(4\theta)))$, with $R = 0.4\ 10^{-4}$m. Snapshots of heat distribution at $t = 0.004$ s (a, c) and $t = 0.01$ s (b, d) show that isovalues of temperature (resp. mass concentration) are nearly unperturbed outside the concentrator and the rotator, whereas these are smoothly squeezed and rotated in the concentrator and the rotator, respectively.

To design the rotator, we consider the geometric transform

$$\begin{cases} r' = r \\ \theta' = \alpha.r + \beta \end{cases} \text{ with } \alpha = \frac{\theta_o}{R_1(\theta) - R_2(\theta)}; \ \beta = \theta + \frac{R_2(\theta).\theta_o}{R_2(\theta) - R_1(\theta)}, \tag{9.17}$$

where $\theta_0 = \pi/4$ denotes the tilt. We obtain the following Jacobian and transformation matrices:

$$\mathbf{J}_{\mathbf{rr}'} = \begin{pmatrix} 1 & 0 \\ -\alpha & c_{22} \end{pmatrix} \implies \mathbf{T}^{-1} = \mathbf{R}(\theta) \begin{pmatrix} c_{22} & \alpha r \\ \alpha.r & \frac{1+\alpha^2 r^2}{c_{22}} \end{pmatrix} \mathbf{R}(\theta)^{\mathbf{T}}, \tag{9.18}$$

with $\det(\mathbf{J}) = \frac{r-\beta}{\alpha^2 r}$. Also, we have

$$c_{22} = \frac{\partial\theta}{\partial\theta'} = 1 - \frac{\theta_o.r}{(R_2(\theta) - R_1(\theta))^2}.\left(\frac{\partial R_2(\theta)}{\partial\theta'} - \frac{\partial R_1(\theta)}{\partial\theta'}\right)$$
$$-\frac{\theta_o}{(R_2(\theta) - R_1(\theta))^2}\left(R_2(\theta)\frac{\partial R_1(\theta)}{\partial\theta'} - R_1(\theta)\frac{\partial R_2(\theta)}{\partial\theta'}\right). \tag{9.19}$$

One notes that the determinant of T equals 1, which mathematically means that we use an area-preserving transform. This should facilitate potential devices compared to cloaks and concentrators of an arbitrary shape.

9.3.3 *Three-Dimensional Cloak of a Complex Shape for Diffusion Processes*

We have explored diffusion cloaks, concentrators, and rotators of an arbitrary shape in two dimensions. A nontrivial question to ask is whether one can design thermal or mass diffusion cloaks, concentrators, and rotators of nonspherical shapes. The extension of the previous section to the general 3D case requires us to describe the inner and outer boundaries of the cloak with varying radii in the spirit of [17]

$$\rho(\theta,\phi) = a_{0,0} + \left\{\sum_{(m,n)\in N^2\setminus\{(0,0)\}} a_{m,n}\cos(m\theta + n\phi)\right.$$
$$\left. + b_{m,n}\sin(m\theta + n\phi)\right\}. \tag{9.20}$$

We further assume there is some revolution for the design with an arbitrary cross section described by two functions $R_1(\phi)$ and $R_2(\phi)$ giving an angle-dependent distance from the origin. These functions correspond respectively to the interior and exterior boundaries of the cloak. The only assumption is that these two boundaries can be represented by a differentiable function. Their finite Fourier expansions are thought in the form $R_j(\phi) = a^j_{0,0} + \sum_{n=1}^{p} a^j_{0,n}\cos(n\phi)$, $j = 1, 2$, where p can be a small integer and $a^j_{0,n} = 0$ for $n \neq p$ for computational ease.

The geometric transformation which maps the field within the full domain $\rho \leq R_2(\phi)$ onto the annular domain $R_1(\phi) \leq \rho' \leq R_2(\phi)$ can be expressed as

$$
\begin{aligned}
\rho'(\rho, \phi) &= R_1(\phi) + \rho \frac{R_2(\phi) - R_1(\phi)}{R_2(\phi)} , \\
\theta' &= \theta \; , \; 0 < \theta \leq 2\pi \; , \\
\phi' &= \phi \; , \; -\pi/2 < \phi \leq \pi,
\end{aligned}
\tag{9.21}
$$

where $0 \leq \rho \leq R_2(\phi)$. Note that the transformation maps the field for $\rho > R_2(\phi)$ onto itself through the identity transformation.

This change of coordinates is characterized by the transformation of the differentials through the Jacobian

$$
\mathbf{J}(\rho', \phi') = \frac{\partial(\rho(\rho', \phi'), \theta, \phi)}{\partial(\rho', \theta', \phi')} . \tag{9.22}
$$

After some elementary algebra, we find that

$$
\mathbf{T}^{-1} = \begin{pmatrix} \dfrac{c_{13}^2 + \rho^2}{c_{11}\rho'^2} & 0 & -\dfrac{c_{13}}{\rho'} \\ 0 & c_{11} & 0 \\ -\dfrac{c_{13}}{\rho'} & 0 & c_{11} \end{pmatrix} , \tag{9.23}
$$

with $\det(\mathbf{J}) = \frac{r-\beta}{\alpha^2 r}$. Also, we have
$c_{11}(\phi') = R_2(\phi')/R_2(\phi') - R_1(\phi')$ and

$$
\begin{aligned}
c_{13} = {} & R_2(\phi') \frac{\rho' - R_2(\phi')}{(R_2(\phi') - R_1(\phi'))^2} \frac{dR_1(\phi')}{d\phi'} \\
& + R_1(\phi') \frac{R_1(\phi') - \rho'}{(R_2(\phi') - R_1(\phi'))^2} \frac{dR_2(\phi')}{d\phi'} ,
\end{aligned}
\tag{9.24}
$$

for $R_1(\phi') \leq \rho' \leq R_2(\phi')$. Elsewhere, $\mathbf{T}^{-1}$ reduces to the identity matrix ($c_{11} = 1$ and $c_{13} = 0$ for $\rho' > R_2(\phi')$).

In a similar way, one can design 3D concentrators and rotators for diffusion processes.

9.4 Multilayered Cloak with Simplified Isotropic Parameters

We would like to engineer more realistic diffusion cloaks. This has been achieved using a systematic approach by Petiteau et al. (2014)

[12] and Petiteau (2015) [13]. In his PhD thesis, David Petiteau establishes that one needs about 10,000 layers to achieve layered diffusion cloaks with an efficiency close to that of ideal diffusion cloaks. The interested reader can find all the mathematical and numerical details therein. In the sequel, we give some first examples of such approximate cloaks in two and three dimensions but with a relatively low cloaking efficiency in order to simplify their design (hundreds to thousands of concentric layers would be needed to achieve a very good efficiency).

9.4.1 *Two-Dimensional Multilayered Thermal Cloaks*

Figure 9.5 shows snapshots of heat (or concentration in the context of mass diffusion) distribution for a six-layer cloak and a two-layer cloak. At $t = 0.02$ s (Fig. 9.5a,b,e,f) and $t = 0.12$ s (Fig. 9.5c,d,g,h) isovalues of temperature (or concentration in the context of mass diffusion) are nearly unperturbed for the six-layer cloak but not quite so for the two-layer cloak. However, normalized temperature (or concentration) is nonvanishing inside the inner disc of both types of cloaks (and slightly higher for the six-layer model) and depends upon the distance between the center of the cloak to the source. The left-hand side has a constant temperature (or concentration) of 1 and the right-hand side of 0. Results are in line with the works of Han et al. (2013), Dede et al. (2013), and Xu et al. (2014) [9, 10, 11], that is, with a low cloaking efficiency, unlike for Petiteau et al. (2014) [12].

9.4.2 *Three-Dimensional Multilayered Thermal Cloaks*

We now propose a design of a layered spherical cloak for diffusion of heat (resp. mass) based on the homogenization approach described by Guenneau and Puvirajesinghe (2013) [2]. We show in Fig. 9.6 snapshots of heat (resp. mass concentration) for a 20-layer diffusion cloak. Heat (resp. mass concentration) is clearly shown to be constant inside the invisibility region at every time step, which is consistent with the fact that here we have the continuity of the field and its flow at every interface (including the cloak's inner boundary). In practice, scenario 2 seems to be the preferred one for layered cloaks with realistic parameters.

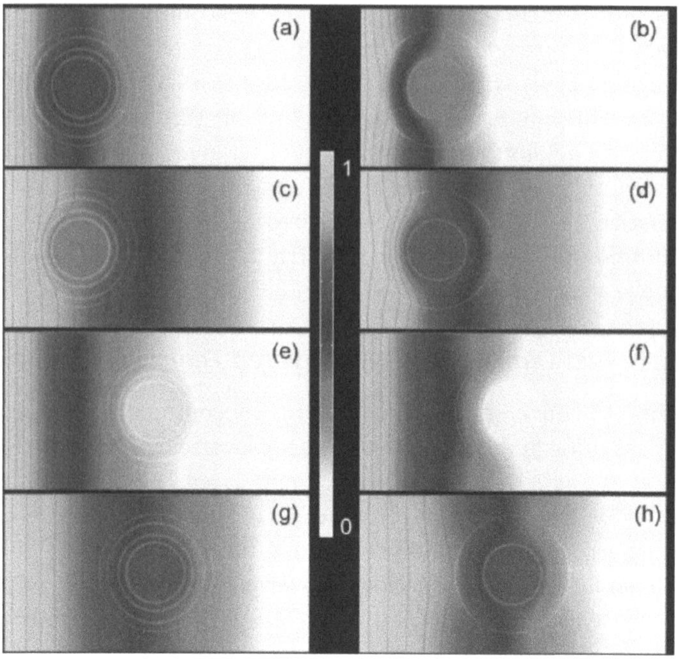

Figure 9.5 Layered diffusion cloaks. (Left) Diffusion of heat (or mass) from the left side on a six-layer cloak of inner radius $R_1 = 1.5\ 10^{-4}$m and outer radius $R_2 = 3\ 10^{-4}$m, consisting of 6 homogeneous layers alternating with interfaces at radii 1.575, 1.9125, 1.9875, 2.5125, and 2.5875 (in units of $10^{-4}m$) and of respective diffusivities 80, 0.2, 40, 0.4, 10, 0.8 in a bulk material of diffusivity 1 (in units of $10^{-5}\mathrm{m}^2\cdot\mathrm{s}^{-1}$). (Right) Diffusion of heat (or mass) from the left on a two-layer cloak of inner radius $R_1 = 1.5\ 10^{-4}$m and outer radius $R_2 = 3\ 10^{-4}$m, consisting of 2 homogeneous with interfaces at radii 1.575 and 2.5875 (in units of 10^{-4}m) and of respective diffusivities 80, 0.8 in a bulk material of diffusivity 1 (in units of $10^{-5}\mathrm{m}^2\cdot\mathrm{s}^{-1}$).

9.5 Invisibility Carpet for Diffusion Processes: Mapping a Curved Surface on a Flat Surface

Let us finally consider the design of an invisibility carpet for diffusion processes with the linear geometric transform

$$\begin{cases} x' = x\,, \quad a < x < b\,, \\ y' = \frac{y_2(x)-y_1(x)}{y_2(x)} y + y_1(x)\,, \quad 0 < y < y_2(x)\,, \\ z' = z\,, \quad -\infty < z < +\infty\,, \end{cases} \tag{9.25}$$

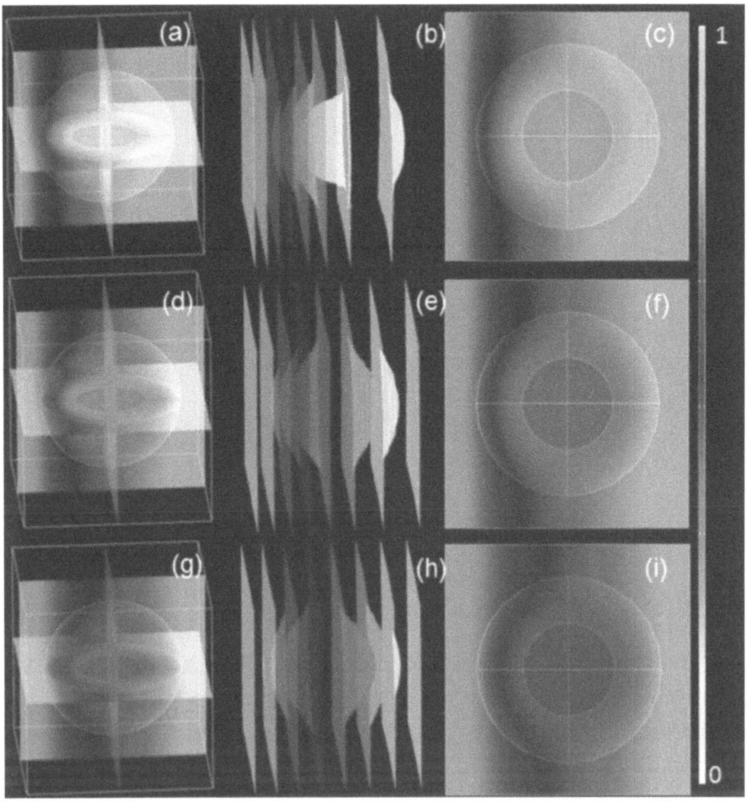

Figure 9.6 Layered spherical cloak: Diffusion of heat (or mass) from the left on a cloak of inner radius $R_1 = 1.5\ 10^{-6}$ m and outer radius $R_3 = 3.0\ 10^{-6}$ m consisting of 20 homogeneous layers of equal thickness and of respective diffusivities (in units of $10^{-5} \cdot \mathrm{m}^2 \cdot \mathrm{s}^{-1}$), which are 1700, 0.25, 81, 0.25, 30, 0.25, 16.5, 0.25, 11, 0.25, 8, 0.25, 6.5, 0.25, 5.5, 0.25, 4.5, 0.25, 4, 0.25 (from inner to outer layer). Snapshots of heat distribution at $t = 0.01$ s (a–c), $t = 0.015$ s (d–f), and t — 0.02 s (g–i) show that isovalues of heat (resp. mass) are nearly unperturbed outside the cloak and the core has uniform heat (resp. mass concentration).

where y' is a stretched vertical coordinate. It is easily seen that this linear geometric transform maps the segment $[a, b]$ within the horizontal axis $y = 0$ onto the curve $y' = y_1(x)$, and leaves the curve $y = y_2(x)$ unchanged. Importantly, there is a one-to-one correspondence between the segment and y_1. The curves y_1 and y_2

are assumed to be differentiable, and this ensures that the carpet won't display any singularity on its inner boundary.

9.5.1 *Two-Dimensional Carpets*

The linear transform (Eq. 9.25) is expressed in a Cartesian basis as $\mathbf{J}_{xx'} = \begin{pmatrix} 1 & 0 \\ \frac{\partial y}{\partial x'} & \frac{1}{\alpha} \end{pmatrix}$ where $\alpha = (y_2 - y_1)/y_1$ and from the chain rule

$$\frac{\partial y}{\partial x'} = y_2 \frac{y' - y_2}{(y_2 - y_1)^2} \frac{\partial y_1}{\partial x} - y_1 \frac{y' - y_1}{(y_2 - y_1)^2} \frac{\partial y_2}{\partial x} \,. \tag{9.26}$$

This leads to the inverse symmetric tensor $\mathbf{T}^{-1}$

$$(T^{-1})_{11} = \frac{1}{\alpha} \,, \ (T^{-1})_{12} = (T^{-1})_{21} = -\frac{\partial y}{\partial x'} \,,$$
$$(T^{-1})_{22} = \left(1 + \left(\frac{\partial y}{\partial x'}\right)^2\right) \alpha \,. \tag{9.27}$$

We note that the transformed product of density and specific heat has the factor $\det(\mathbf{J}) = \alpha^{-1} = y_1/(y_2 - y_1)$, which is spatially varying but nonvanishing.

Such invisibility carpets for diffusion processes seem to be easier to achieve than invisibility cloaks. However, one has to keep in mind that the temperature (resp. mass concentration) needs to be uniform on the lower boundary of the computational domain, which is hard to enforce in a real experimental setup. Some advances toward a manufactured carpet have been achieved by Petiteau during his PhD thesis [13] in collaboration with the group of Nathalie Rolland at the University of Lille.

9.5.2 *Three-Dimensional Carpets*

Let us consider the linear geometric transform parameterized by $t \in (0, 2\pi)$:

$$\begin{cases} x'(t) = x(t) \,, \ a(t) < x(t) < b(t) \,, \\ y'(t) = y(t) \,, \ c(t) < y(t) < d(t) \,, \\ z'(t) = \dfrac{z_2(x(t), y(t)) - z_1(x(t), y(t))}{z_2(x(t), y(t))} z(t) \\ \qquad + z_1(x(t), y(t)) \,, \ 0 < z(t) < f_2(x(t), y(t)) \,, \end{cases} \tag{9.28}$$

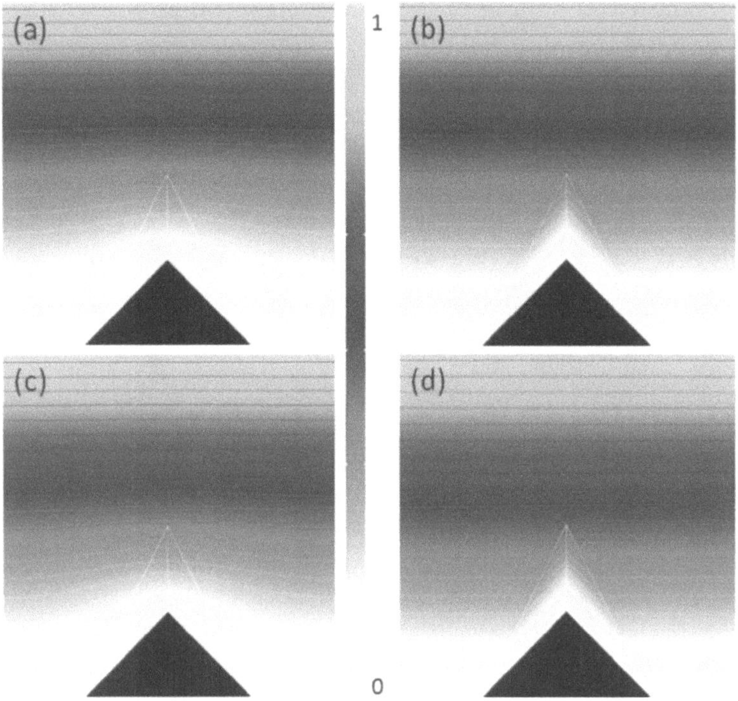

Figure 9.7 Thermal (resp. mass diffusion) carpet cloak: (a) carpet (20 ms), (b) no carpet (20 ms), (c) carpet (50 ms), and (d) no carpet (50 ms). One notes that Dirichlet boundary conditions are set on upper and lower boundaries, with normalized temperature (resp. mass) $u = 1$ on the top and $u = 0$ on the bottom.

where y' is a stretched vertical coordinate. It is easily seen that this linear geometric transform maps the arbitrary domain $D = \bigcup_{t\subset(0,2\pi)}\{[a(t), b(t)] \times [c(t), d(t)]\}$ within the plane xy onto the surface $z' = z_1(x, y)$ and leaves the surface $z = z_2(x, y)$ unchanged. Importantly, there is a one-to-one correspondence between the domain D and the surface $z' = z_1$. The surfaces z_1 and z_2 are assumed to be differentiable, and this ensures that the carpet won't display any singularity on its inner boundary.

The linear transform (Eq. 9.28) is expressed in a Cartesian basis as $\mathbf{J}_{xx'} = \begin{pmatrix} 1 & 0 & 0 \\ 0 & 1 & 0 \\ \frac{\partial z}{\partial x'} & \frac{\partial z}{\partial y'} & \frac{1}{\alpha} \end{pmatrix}$ where $\alpha = [z_2(x(t), y(t)) - z_1(x(t), y(t)]/z_2(x(t), y(t))$ and from the chain rule

$$\frac{\partial z}{\partial x'} = z_2 \frac{z' - z_2}{(z_2 - z_1)^2} \frac{\partial z_1}{\partial x} - z_1 \frac{z' - z_1}{(z_2 - z_1)^2} \frac{\partial z_2}{\partial x},$$

$$\frac{\partial z}{\partial y'} = z_2 \frac{z' - z_2}{(z_2 - z_1)^2} \frac{\partial z_1}{\partial y} - z_1 \frac{z' - z_1}{(z_2 - z_1)^2} \frac{\partial z_2}{\partial y}. \tag{9.29}$$

This leads to the inverse symmetric tensor $\mathbf{T}^{-1}$, which is fully described by seven nonvanishing entries in a Cartesian basis:

$$(T^{-1})_{11} = (T^{-1})_{22} \frac{1}{\alpha}, (T^{(-1)})_{13} = (T^{-1})_{31} = -\frac{\partial z}{\partial x'},$$

$$(T^{-1})_{23} = (T^{-1})_{32} = -\frac{\partial z}{\partial y'},$$

$$(T^{-1})_{33} = \left(1 + \left(\frac{\partial z}{\partial x'}\right)^2 + \left(\frac{\partial z}{\partial y'}\right)^2\right) \alpha.$$

We note that the transformed product of density and specific heat has the factor $\det(\mathbf{J}) = \alpha^{-1} = z_1/(z_2 - z_1)$, which is spatially varying but nonvanishing.

One might wonder what kind of applications could be found in invisibility carpets for heat diffusion. One could envision some uniform heat distribution over a rough boundary that diffuses heat light over a flat boundary.

9.6 Concluding Remarks

In this chapter, we have shown that one can control diffusion processes in the context of heat or mass processes, thanks to geometric transformations in the convection-diffusion equations. The required heterogeneous anisotropic diffusivity inside the cloaks can be replaced by concentric layers with isotropic homogeneous diffusivity which achieve some approximate control of diffusion in the case of circular, or spherical, geometries. In the case of cloaks, concentrators, and rotators of an arbitrary shape, the path toward an approximate design remains a challenge. We have also shown that one can design invisibility carpets for diffusion processes. In this latter case, one only needs to achieve some control on backscattered diffusion, which makes the metamaterial parameters easier to achieve: there is no singularity in the required tensor of diffusivity due to the fact that we use a one-to-one mapping. Significant preliminary advances have been made by Petiteau during his PhD thesis toward designs of highly efficient diffusion cloaks and carpets, the interested reader will find all mathematical and numerical details in Ref. [13]. Thermal and biochemical metamaterials are young research axes in metamaterials, and we look forward to seeing future studies, notably implicating graphene composites.

References

1. Guenneau, S., Amra, C., and Veynante, D. (2012). Transformation thermodynamics: cloaking and concentrating heat flux, *Opt. Express*, **20**(7), 8207–8218.
2. Guenneau, S., and Puvirajesinghe, T. M. (2013). Fick's second law transformed: one path to cloaking in mass diffusion, *J. Roy. Soc. Interface*, **10**, 20130106.
3. Pendry, J. B., Schurig, D., and Smith, D. R. (2006). Controlling electromagnetic fields, *Science*, **312**, 1780–1782.
4. Schittny, R., Kadic, M., Buckmann, T., and Wegener, M. (2014). Invisibility cloaking in a diffusive light scattering medium, *Science*, **345**(6195), 427–429.
5. Maldovan, M. (2013). Sound and heat revolutions in phononics, *Nature*, **503**, 209–217.
6. Narayana, S., and Sato, Y. (2012). Heat flux manipulation with engineered thermal materials, *Phys. Rev. Lett.*, **108**, 214303.
7. Schittny, R., Kadic, M., Guenneau, S., and Wegener, M. (2014). Experiments on transformation thermodynamics: molding the flow of heat, *Phys. Rev. Lett.*, **110**(19), 195901.
8. Guenneau, S., Petiteau, D., Zerrad, M., Amra, C., and Puvirajesinghe, T. (2015). Transformed Fourier and Fick equations for the control of heat and mass diffusion, *AIP Adv.*, **5**, 053404.
9. Han, T., Yuan, T., Li, B., and Qiu, C.-W. (2013). Homogeneous thermal cloaks with constant conductivity and tunable heat localization, *Sci. Rep.*, **3**, 1593.

10. Dede, E. M., Nomura, T., Schmalenberg, P., and Lee, J. S. (2013). Heat flux cloaking, focusing, and reversal in ultra-thin composites considering conduction-convection effects, *Appl. Phys. Lett.*, **103**, 063501.
11. Xu, H., Shi, X., Gao, F., Sun, H., and Zhang, B. (2014). Ultrathin three-dimensional thermal cloak, *Phys. Rev. Lett.*, **112**, 054301.
12. Petiteau, D., Guenneau, S., Bellieud, M., Zerrad, M., and Amra, C. (2014). Spectral effectiveness of engineered thermal cloaks in the frequency regime, *Sci. Rep.*, **4**, 7386.
13. Petiteau, D. (2015). *Ingénierie des métamatériaux thermiques: transformation d'espace et techniques d'homogénéisation appliquées à l'équation de la chaleur*, PhD thesis, Aix-Marseille Université.
14. Leonhardt, U. (2006). Optical conformal mapping, *Science*, **312**, 1777–1780.
15. Greenleaf, A., Kurylev, Y., Lassas, M., and Uhlmann, G. (2007). Full-wave invisibility of active devices at all frequencies, *Comm. Math. Phys.*, **275**(3), 749–789.
16. Rahm, M., Schurig, D., Roberts, D. A., Cummer, S. A., Smith, D. R., and Pendry, J. B. (2008). Design of electromagnetic cloaks and concentrators using form-invariant coordinate transformations of Maxwell's equations, *Photon. Nanostruct. Fundam. Appl.*, **6**, 87–95.
17. Nicolet, A., Zolla, F., Agha, Y.O., and Guenneau, S. (2008). Geometrical transformations and equivalent materials in computational electromagnetism, *COMPEL*, **27**, 806–819.
18. Jikov, V. V., Kozlov, S. M., and Oleinik, O. A. (1994). *Homogenization of Differential Operators and Integral Functionals* (Springer-Verlag, New York).

Chapter 10

Experiments on Cloaking in Electromagnetism, Mechanics, and Thermodynamics

Muamer Kadic,[a,b] Robert Schittny,[a] Tiemo Bückmann,[a] and Martin Wegener,[a,b]

[a]*Institute of Applied Physics, Karlsruhe Institute of Technology (KIT), Wolfgang-Gaede Straße 1, 76131 Karlsruhe, Germany*

[b]*Institute of Nanotechnology, Karlsruhe Institute of Technology (KIT), Hermann von Helmholtz Platz 1, 76344 Eggenstein Leopoldshafen, Germany*

muamer.kadic@kit.edu

10.1 Introduction

Cloaking can be seen as a look-alike contest: The goal is to make some object $\mathcal{A}$ appear like another object $\mathcal{B}$ with respect to some physical observable. The early literature has indeed spoken of the nonuniqueness of the tomography problem [9, 27, 41]. The experiments to be discussed in this chapter concern the observables light, sound, elastic waves, static elasticity, electric conduction, heat conduction, and particle diffusion.

Transformation Wave Physics: Electromagnetics, Elastodynamics, and Thermodynamics
Edited by Mohamed Farhat, Pai-Yen Chen, Sebastien Guenneau, and Stefan Enoch

ISBN 978-981-4669-95-5 (Hardcover), 978-981-4669-96-2 (eBook)
www.panstanford.com

10.1.1 *True Cloaks*

In optics, you can, for example, aim at making an object appear like vacuum [47, 48, 63]. This means that it becomes invisible with respect to vacuum—in all aspects. First of all, for all incident directions, light must emerge in the same direction as in vacuum: The cloak should be omnidirectional rather than unidirectional. Vacuum does not reflect light; neither must the cloak. Vacuum has no polarization dependence; hence the cloak must not exhibit any polarization dependence either. Many self-acclaimed cloaks do not pass this hurdle. Vacuum leads to a particular propagation time of the wave or of the light ray; thus the cloak must recover the same time. This means that simple arrangements that one finds on the Internet based on mirrors guiding light around an object, leading to an increased propagation time with respect to vacuum due to the detour, are not really invisibility cloaks. The same deficiency holds true for several self-acclaimed cloaks in the literature.

More precisely, we distinguish between a cloak and an invisible object. In optics, both may appear like vacuum or water. For example, polyacrylate spheres soaked up with water can have a refractive index practically identical to that of bulk water. They are thus invisible in water. However, you cannot hide an arbitrary object, for example, a piece of metal, by positioning it into a polyacrylate sphere. A cloak must allow for inserting arbitrary objects $\mathcal{C}$ into the arrangement $\mathcal{A}$. Regardless of what one inserts, the result ideally appears like $\mathcal{B}$. In contrast, for an invisible object, the structure of $\mathcal{A}$ needs to be changed if $\mathcal{C}$ is altered. In this chapter, we focus on the more demanding case of true cloaks. For example, if one succeeds in making a rigid hollow sphere $\mathcal{A}$ appear like a homogeneous compliant material $\mathcal{B}$ in elasticity, the arrangement is a cloak because one can insert arbitrary objects $\mathcal{C}$ into the hollow sphere.

Such arrangement obviously also isolates an object $\mathcal{C}$ from the surrounding. This isolation may be useful to protect the object from, for example, heat, light, or vibrations. However, we emphasize that a true cloak is more than just isolation. In optics, you can simply isolate a person from the rest of the world by locating him/her inside a hollow metal sphere. You will still see the metal shell though by its reflections and image distortions. In heat conduction, you can isolate a person by a hollow Styrofoam shell, but the shell is going to modify the flux of heat. A cloak not only isolates but also makes the hollow sphere $\mathcal{A}$ appear like something else $\mathcal{B}$, that is, like vacuum or like a homogeneous heat-conducting medium, respectively.

10.1.2 *Role of the Environment*

It is important to realize that no cloak is a cloak *per se*: Suppose a fictitious optical cloak operating perfectly in vacuum enters a foggy or cloudy region in which diffusive light scattering dominates rather than ballistic wave propagation according to Maxwell's equations. In the fog, the perfect cloak still appears like vacuum. Vacuum does not scatter light. Thus, the cloak appears like a nonscattering medium inside a scattering medium, that is, it becomes visible by reduced scattering with respect to the foggy environment. Likewise, an object that appears like standard air with respect to acoustic sound propagation will not be detectable in air. It will be detectable though for sound waves once you put it into water with a much higher speed of sound. Yet another object may be unfeelable in an elastic sense with respect to some homogeneous compliant medium but not with respect to a stiffer one.

More generally, the object $\mathcal{B}$ in the look-alike contest does not need to be vacuum or some homogeneous dielectric, elastic, conductive, or diffusive medium. Broadly speaking, object $\mathcal{A}$ may be an apple and $\mathcal{B}$ an orange. For example, object $\mathcal{B}$ could be a metal mirror. Indeed, the so-called carpet cloak makes an object underneath a curved metal surface appear like a flat metal mirror. The acoustic counterpart to a metal mirror in optics would be a rigid wall. Object $\mathcal{B}$ might also be an optical lens. For example, the goal could be to make a bad optical lens appear as though it is a very good optical lens. The last example emphasizes that cloaking can also be seen as a particularly demanding case of aberration correction. You may even consider a person in an optical or acoustic path as an aberration. Likewise, a hole in an otherwise perfect membrane of a drum can be seen as an aberration as well.

10.1.3 *Design Approaches*

We would also like to distinguish between passive and active cloaks. Passive cloaks do not require an external power supply. In contrast, active cloaks do need some sort of power supply to control emitters, actuators, heaters, or other active elements within. Obviously, one has additional design opportunities for active cloaks (examples are [60, 79, 83]). However, passive cloaks are more stunning in that they even do the job without a power supply. We will focus on passive cloaks in this chapter.

Many passive cloaks are based on coordinate transformations to design them [7, 14, 20, 26, 28, 29, 47, 57, 59, 63, 74]. In the following section, we give a brief intuitive explanation for experimentalists—without using any formulas. Mathematical treatments are outlined in other chapters of this book. Other cloaks exploit simpler single-shell architectures based on the idea of scattering cancellation [3, 12, 35, 39, 40, 56, 87]. We will see that the two approaches are connected in that the core-shell structures are the most drastic simplification of the anisotropic multiple-shell structures emerging from transformation optics. Finally, cloaks can also be designed by numerical approaches like topology optimization [5, 43, 67, 77], which can be seen as trial and error in a certain subspace of all the possibilities for microstructures to be wrapped around an object which shall be cloaked. In this chapter, we will focus on analytically designed cloaks.

10.2 From Transformations to Materials

Figure 10.1a shows an image of a rectangular grid of streets in a city, for example, like displayed on your computer monitor. Cars can drive on these streets with constant velocity, for example, with one block per second. The flux of cars shall serve to illustrate the flux of energy in a physical system. For example, the cars may stand for photons propagating in vacuum. A car entering from the bottom middle going upwards will need 15 seconds for the 15 blocks. Now suppose we distort this image on our computer screen by using software like Photoshop to get the image in (b). Objects can be hidden in the

Figure 10.1 The concept underlying coordinate-transformation-based cloaking can intuitively be explained by a square grid of streets in a city. Distorting the original city in (a) on a computer by software, for example, Adobe Photoshop, leads to an apparently inhomogeneous and anisotropic distribution of car velocities in (b). The trick of transformation optics is to turn this fictitious velocity distribution into a real one by introducing a corresponding material distribution. Reproduced with permission from Ref. [70], Robert Schittny, Karlsruhe Institute of Technology (KIT).

white circle in the middle. The same car still needs 15 seconds for the 15 blocks and it also comes out in the same direction as before, but on the computer screen it appears as though the velocity of the car changes inside in the central part. In regions of compressed streets, the car appears to drive slower (i.e., measured in cm/s on the computer monitor), whereas it appears to drive faster in regions of stretched streets. The latter is the case for streets going around the inner circle, that is, the azimuthal component of the velocity appears faster than in (a). Likewise, the radial component of the velocity appears slower than in (a).

So far, everything has been fictitious in that no car in the city is actually driving faster than in (a). Here comes the trick: We turn the apparent velocity of cars on the computer screen into an actual velocity distribution in the city. To do so, we have to ignore the fixed grid of streets and rather consider an open area. We assume that car drivers act rationally in the sense that they take that route which corresponds to the shortest travel time. Photons in a graded index structure do that anyway according to Fermat's principle. The cars will then follow exactly the same tracks as the cars in Fig. 10.1b. The mathematics of transformation optics allows calculating explicitly the corresponding anisotropic velocity distribution [48]. Usually, however, the result is rather cast into a distribution of the electric permittivity and magnetic permeability tensors [48]. This is much less intuitive because the phase velocity is inversely proportional to the square root of electric permittivity times magnetic permeability and because the indices "azimuthal" and "radial" (they commonly refer to the polarization in electromagnetism rather than to the propagation direction) interchange with respect to that of the velocity.

In any case, it is clear that the azimuthal component of the phase velocity close to the circle in Fig. 10.1b needs to be much larger than the constant velocity in (a). Translated to photons, this means that the local azimuthal phase velocity of light in (b) needs to be much larger than the vacuum speed of light. According to Maxwell's equations for dispersive media, both phase and group velocity of light can even be infinitely large at a given frequency [13, 15, 25, 65]. However, it is fundamentally impossible to get such superluminal behavior over a broad frequency range: For a strictly frequency independent response, the electromagnetic energy velocity is strictly identical to the phase velocity. According to Albert Einstein's theory of relativity, energy and mass are equivalent and mass cannot propagate faster than the vacuum speed of light. Thus, broadband electromagnetic free-space cloaking is fundamentally inconsistent with the laws of physics. One can get cloaking for a specific frequency or for a narrow frequency range. In this case, causality further spoils things because a frequency-dependent phase velocity (or refractive index) is inherently connected with finite losses via the Kramers–Kronig relations. Simple estimates [34] show that these losses will render the "cloak" essentially opaque for structures that are macroscopic compared to the wavelength of light.

These considerations are not meant to frustrate the reader but they rather point towards situations were these fundamental restrictions do not apply. For example, the energy velocity of a mechanical wave cannot exceed the vacuum speed of light either. This limitation is nearly irrelevant though because practical speeds of elastic or acoustic waves are orders of magnitude below the vacuum speed of light anyway. It is thus possible to increase the azimuthal component of the energy velocity with respect to some reference medium ($\mathcal{A}$ above) by large factors, albeit not by infinity.

10.2.1 *Laminate Metamaterials*

Many realizations of anisotropic material distributions are based on laminates, that is, on alternating shells of low/high velocity for waves or on alternating low/high conductivity for static electric conduction. Let us use the latter example for illustration.

Figure 10.2a shows a square array of identical resistors [71]. The array becomes distorted after performing a coordinate transformation in (b)—just like the streets in Fig. 10.1. Nevertheless, an outside observer would obviously still measure the same resistances between any two points on the four borders for (a) and (b). The Calderón tomography problem [9] is not unique [27, 41]. After all, the same set of resistors is just arranged differently on the table. Next, this discrete model can be mapped onto a continuous material distribution by noting that, for current flow in the azimuthal direction, one has more resistors per area in parallel in (b) than in (a). Likewise, one has more resistors in serial for current flow in the radial direction in (b) compared to (a). This means that the radial component of the effective conductivity is reduced in (b), whereas the azimuthal component of the effective conductivity is increased

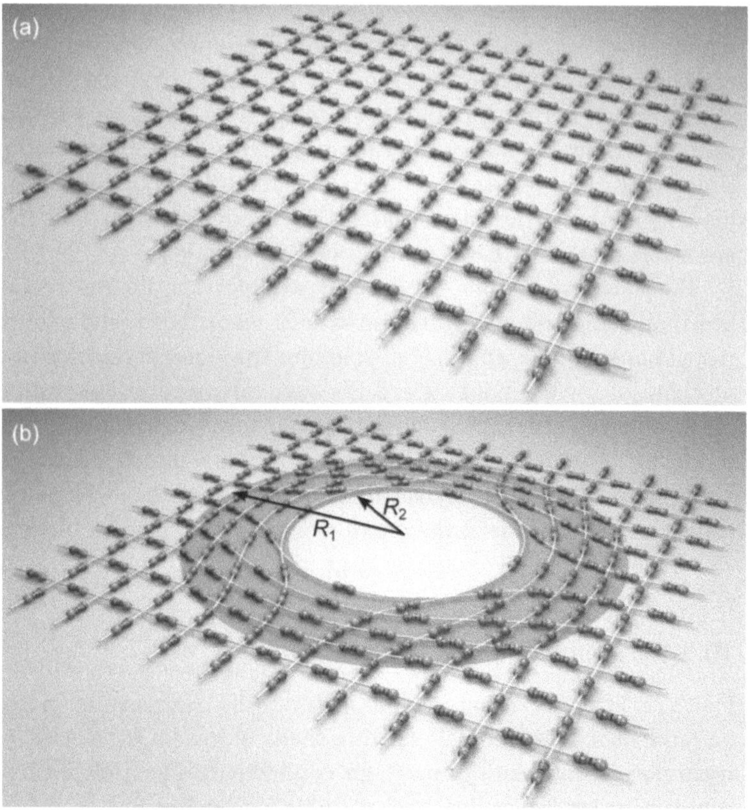

Figure 10.2 A square grid of resistors in (a) is subject to the same coordinate transformation as used in Fig. 10.1. This leads to the resistor network in (b), which can immediately be translated into an effective inhomogeneous and anisotropic conductivity distribution. Reproduced (figure) with permission from Ref. [72]. Copyright (2014) by the American Physical Society.

in (b) with respect to the reference situation (a). Interestingly, the effective conductivity behaves just like the wave velocity in Fig. 10.1.

The shell structure in Fig. 10.3a mimics this anisotropy. If, for example, one layer does not conduct at all whereas the other one does, no current can flow normal to the layers, whereas the composite is conductive for current flowing parallel to the surfaces. Less pronounced anisotropies can be obtained if the conductivities

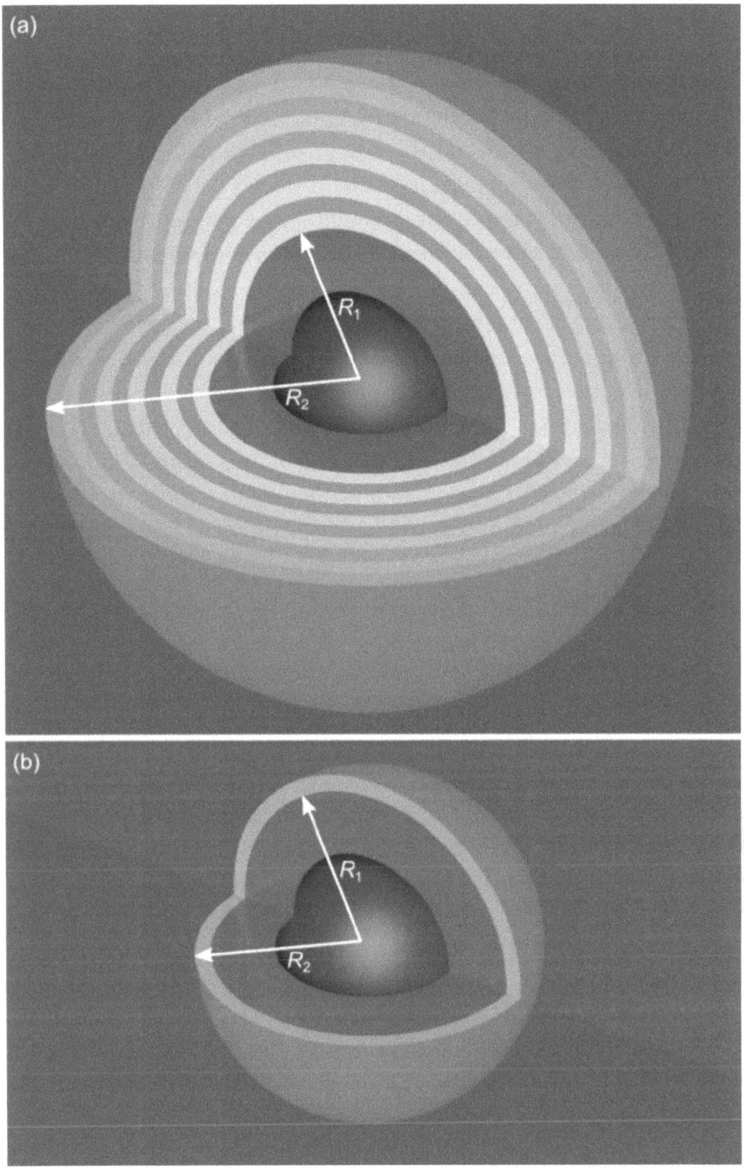

Figure 10.3 (a) The anisotropic conductivity distribution in Fig. 10.2 can be implemented by a laminate metamaterial, that is, by a structure composed of alternating layers of high/low conductivity. (b) Reducing the number of layers leads to a simple core-shell cloaking structure.

of the two constituent materials are more similar. In the limit of many thin shells, one obtains a perfect approximation of a continuous anisotropic material variation [56]. In the opposite limit of only one nonconducting and one high-conduction layer, one obtains a simple core-shell structure [56] shown in Fig. 10.3b. Note that this particular structure is a true cloak (see discussion above) in the sense that the interior of the hollow core can contain objects with arbitrary conductivities. It has been shown mathematically [3, 12, 35, 39, 56, 87] that perfect cloaking is still possible with such simple arrangements provided the gradient of the electric potential is constant across the cloak. In contrast, a structure designed by coordinate transformations principally works perfectly for any arbitrary potential distribution. However, numerical calculations show that even the simple core-shell structure exhibits good (albeit not perfect) cloaking performance for many different potential distributions. The tremendous beauty of the core-shell approach lies in the fact that simple closed analytical expressions are available in the literature [56] that connect the radii of core and shell with the conductivities of core, shell, and surrounding (likewise for other physical quantities). This holds true for cylindrical as well as for spherical geometries [56]. For reasonable radii, the required conductivity contrasts are also rather moderate (see below). Another advantage of the single-shell cloak is that the size of the cloak compared to the object to be cloaked can be much smaller than for the case of multiple shells.

10.3 Electromagnetism

Early work on cloaking by Smith and coworkers based on coordinate transformations has been performed at microwave frequencies, in two-dimensional waveguide geometries, and for one polarization of the electromagnetic wave only [74]. As to be expected from our introductory remarks regarding fundamental limitations due to Albert Einstein's theory of relativity, the cloaking operation bandwidth has been only 2% of the center frequency. At wavelengths outside of this interval, the cloak actually scattered much more strongly than the object alone.

10.3.1 *Optical Carpet Cloaks*

How can one circumvent the narrow-bandwidth problem? The proposal of the carpet cloak [50] avoids superluminal light propagation altogether. This becomes possible by considering an edge of space—a mirror in optics. If the mirror is curved to a bump, objects can be hidden underneath. To eliminate the resulting optical aberrations due to the bump, a tailored structure is put on top which makes the bump appear like a flat mirror (see introduction). If vacuum is the reference medium, superluminal light propagation in some regions is still fundamentally unavoidable. However, provided the bump is gentle in the sense that its height is not too large compared to its width, the relative variation in the speed of light (and equivalently in the refractive index) can be limited to some 10%. Thus, by using water or a dielectric as the reference medium, speeds beyond the vacuum speed of light can be avoided altogether. Given such small refractive index variations, the Kramers–Kronig relations also allow for large bandwidths and for fairly small losses. The original theoretical work [50] used a quasi-conformal coordinate transformation to obtain approximately locally isotropic material properties. This means that only a graded index structure is necessary in a finite-size cloak. Later, we showed [73] that one can even use strictly conformal coordinate transformations [47]. The price to be paid though is that a perfect cloak becomes infinitely large. Nevertheless, very good cloaking can still be obtained with finite cloaks—albeit still more than ten times larger than the object to be cloaked.

Early experimental work [51] on the carpet cloak in 2009 was again at microwave frequencies in two-dimensional waveguides and for one polarization only. Later in 2009, silicon-based two-dimensional waveguide experiments at telecom frequencies were published as well [24, 78]. In 2010, we realized the first three-dimensional and polarization independent version of the carpet cloak [18]. Further experiments followed [46, 66, 68, 85].

The polymer microstructure made by three-dimensional direct-laser-writing optical lithography [80] is schematically illustrated in Fig. 10.4. Herein, the graded index obtained from the coordinate transformation is mimicked by a three-dimensional woodpile

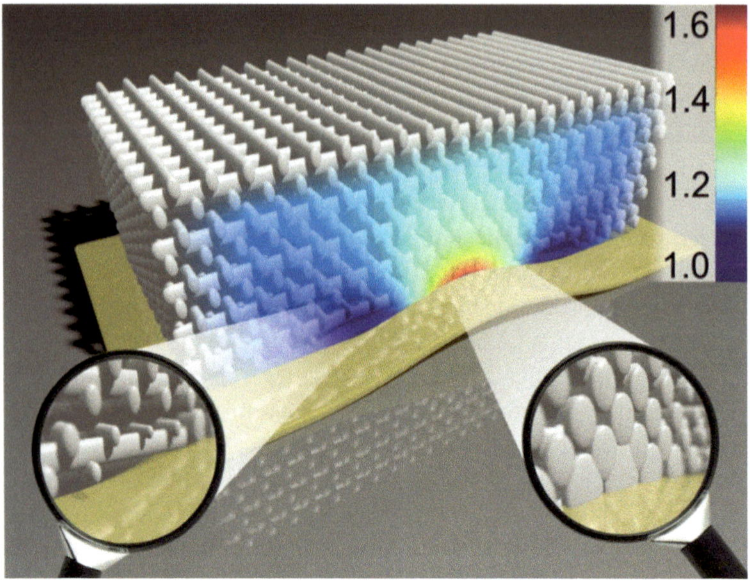

Figure 10.4 Illustration of the three-dimensional optical carpet cloak. The locally isotropic refractive index n is shown by the false-color scale. Adapted from [18] with permission from AAAS.

photonic crystal used at frequencies below the fundamental stop band. In this regime, the dispersion relation of the woodpile photonic band structure is nearly a straight line and nearly isotropic [18]. This means that one gets an effectively isotropic metamaterial refractive index that is wavelength independent for wavelengths between infinity down to about twice the spacing of the polymer rods, that is, the design is inherently broadband. By changing the local polymer woodpile volume filling fraction between 0% and 100%, effective refractive indices between that of air ($n = 1$) and that of the polymer ($n = 1.52$) can be adjusted.

In our optical characterization experiments [18], we have simply imaged the structure in an optical microscope. Excellent cloaking in the range between 1.4 μm and 2.8 μm wavelength—one octave bandwidth—has been found for many viewing directions and for all polarizations of light. Later [22], we have realized similar structures miniaturized by a bit more than a factor of two. This miniaturization to beyond the Abbe diffraction limit has become possible by using stimulated-emission-depletion-inspired direct-laser-writing optical lithography [23]. Due to the scalability of Maxwell's equations, the cloak operation wavelengths scale down by the same factor, bringing invisibility cloaking to the visible part of the electromagnetic spectrum for the first time [16, 22].

Figure 10.5 shows corresponding measured (a) amplitude images [22] as well as (b) phase images [16] at 700 nm wavelength (red light). The latter data have been obtained by imaging interferometry [16]. In (a), the bump shows up by two vertical dark stripes in the middle of the sample. These aberrations disappear for the cloak. The other dark lines are just due to the edges of the sample. Likewise, the physical bump translates into a bump in phase or propagation time of light in Fig. 10.5b for the reference structure. In the presence of the cloak, this distortion is strongly suppressed and one nearly sees a constant phase as expected for a flat metal mirror. All experimental observations are in excellent agreement with numerical calculations based on the ray optics limit [17, 73].

Other three-dimensional versions of the carpet cloak [11, 52, 84, 88] have used piecewise homogeneous birefringent constituent materials like calcite. This allows for truly macroscopic cloaking structures, however, at the price of obtaining invisibility for one linear polarization of light only. This means that an observer wearing polarization goggles would indeed not see the object. If the observer rotates her/his head with respect to the cloak or if he/she simply takes off the goggles, the object immediately becomes half visible. The result is very unsatisfactory [30]. Furthermore, some of these designs work for a very limited range of viewing directions only [44]. The same holds true for two carpet cloaks put back to back, which allows for unidirectional free-space cloaking [30].

10.4 Mechanics

Mechanics is more involved than electromagnetism or optics in the sense that optics usually only supports two transverse polarizations. Perfect vacuum-borne cloaking in optics based on anisotropic materials thus necessarily requires magnetodielectric distributions

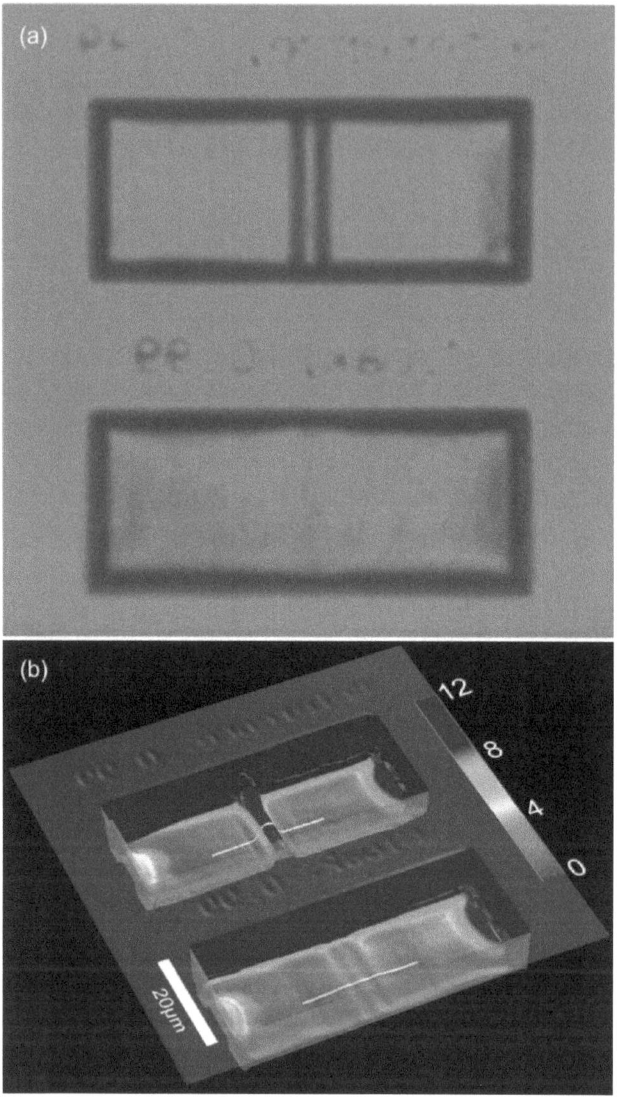

Figure 10.5 Measurements on an optical carpet cloak at 700 nm wavelength. (a) Intensity images as obtained by optical microscopy and (b) phase images as obtained by interferometry. In both cases, the bare bump is shown on the top and the cloaked bump on the bottom. (a) Reproduced (figure) with permission from Ref. [16]. Copyright (2011) by the American Physical Society. (b) Reproduced with permission from Ref. [22] Copyright The Optical Society.

in which the electric permittivity tensor is equal to the magnetic permeability tensor at each point in space. Mechanics generally supports three modes, namely two transverse polarizations and one longitudinal polarization. Perfect cloaking requires dealing with all of them at the same time.

Reference [57] showed that the underlying equations of elastic wave propagation are not form-invariant under general coordinate transformations—at least not for normal elastic materials and scalar mass densities. The same holds true in the static limit, where the mass density drops out. As a result, ideal cloaking is not possible in general.

However, it is possible exactly in special cases and approximately in others. For example, in acoustic airborne or liquid-borne wave propagation, only a single longitudinal polarization is supported. This simplification is connected to the fact that the shear modulus of gases and liquids is dead zero. In contrast, they exhibit a finite bulk modulus, which is equal to the inverse of their compressibility. Compressibility and mass density together determine the speed of sound. As a result, one gets a scalar wave equation, which is again form-invariant under general coordinate transformations [14, 57, 59]. Experiments on acoustic narrow-bandwidth two-dimensional free-space cloaks [86] and broadband three-dimensional carpet cloaks [64] designed by coordinate transformations have been reported. Narrow-bandwidth unidirectional three-dimensional free-space cloaks [67] designed by topology optimization have been realized as well. Fibrous materials have been used for approximates unidirectional cloaking in solid mechanics at ultrasound frequencies [4].

The situation is much more involved in elastic solid mechanics. Theory has suggested metamaterials with anisotropic dynamic mass density tensors [59] (also see [6, 36]) or Cosserat metamaterials [7, 61] as solutions in three dimensions. However, no corresponding experiments have been reported so far.

10.4.1 *Flexural-Wave Cloaks*

Special cases are mechanical waves in thin plates or membranes and mechanical waves at the surface of a thick mechanical solid. For both, in-plane waves have only one polarization that is normal to the plane of propagation. For thin plates or membranes these are called flexural waves [21, 76], for the surface of a bulk material they are called Rayleigh waves [1]. For dispersion-less constituent materials, both waves are in the regime of anomalous dispersion (the group velocity is larger than the phase velocity for all frequencies). The phase velocity can be influenced by the Young's modulus and scales proportional to its square root [76], that is, the stiffer the material, the larger its phase velocity.

In 2012, we have reported corresponding model experiments [76]. Holes are milled into a 1 mm thin plate made of polyvinyl chloride (PVC). A photograph of part of the structure at this stage is shown in Fig. 10.6. Next, the holes are filled with polydimethylsiloxane (PDMS). The Young's modulus of PVC is more

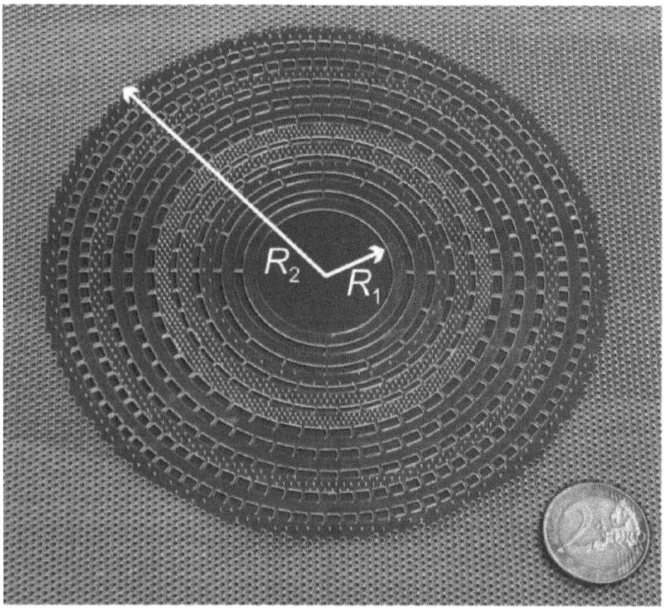

Figure 10.6 Photograph of part of the cloak for flexural waves in thin plates at an intermediate stage of fabrication. After backfilling of the holes with another more compliant material, one recovers a plate with constant thickness and little optical contrast. Reproduced (figure) with permission from Ref. [76]. Copyright (2012) by the American Physical Society.

than three orders of magnitude larger than that of PDMS. As illustrated in Fig. 10.3, the alternating rings mimic the anisotropy demanded by the coordinate transformation [21]. The perforation of the rings leads to the desired gradual change of the velocity. In particular, the radial component tends to zero at the inner radius. Here, a pair of soft/stiff rings can be seen as two Hooke springs in serial. This leads to an effectively soft spring, hence to slow wave propagation. In contrast, for wave propagation in the azimuthal direction, one has a soft and a stiff spring in parallel, which effectively leads to a stiff spring constant, hence to fast wave propagation. Importantly, the azimuthal component of the velocity must exceed that of the isotropic surrounding (see Section 10.1.2). Otherwise, the wave cannot make up for the detour around the object and the phase cannot properly be recovered. However, by mere perforation of a plate with constant thickness, the wave speed can only be reduced with respect to that of a homogeneous plate. Thus, we have also perforated the surrounding to lower its effective wave speed such that the azimuthal component of the velocity inside the cloak can exceed that of the surrounding. We will use the same trick in the following section for the cloaking of heat flow.

In the experiments, we have excited plane waves from the left-hand side of the structure by a loudspeaker. Reflections at the two sides and on the right-hand side are suppressed by tailored impedance matching [76] analogous to perfectly matched layers used in many numerical calculations. Wave propagation is monitored by stroboscopically illuminating the structures with diffuse light and recording movies of the light scattered off of the thin plates from the top. Excellent cloaking has been measured for frequencies in the range between 200 Hz and 400 Hz (one octave bandwidth). Theoretically, the cloak should work for frequencies down to the static limit (where we could not perform experiments), leading to infinitely many octaves operation bandwidth. On the high-frequency side, either the thickness of the plate or the characteristic scale of the in-plane perforation set the limit. In our experiments, the former has been dominating. In principle, the high-frequency cutoff can be shifted to infinity by using very thin plates and smaller-scale perforation.

Quantitative analysis (which measures deviations from plane waves) of the performance of the fabricated cloak is presented in the Supplemental Material of [76].

While originally designed merely as a simple demonstration experiment, this quasi two-dimensional flexural-wave cloak [76] is the only free-space and truly broadband cloak for waves realized in any physical system to date to the best of our knowledge.

10.4.2 *Three-Dimensional Elastostatic Cloaks*

Things are much more difficult in three dimensions because of the mentioned three polarizations. In so-called pentamode metamaterials, the two transverse polarizations can be well separated from the longitudinal one. Corresponding metamaterial microstructures have independently been suggested by Sigmund in 1995 using numerical topology optimization [75] and by Milton and Cherkaev using analytical considerations [58]. The structure shown in Fig. 10.7 is composed of double-cone elements, the tips of which touch each

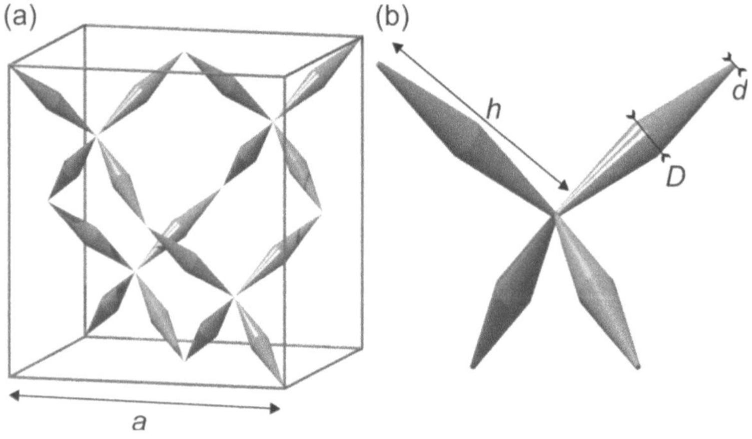

Figure 10.7 (a) One extended face-centered cubic unit cell of a three-dimensional pentamode mechanical metamaterial. (b) Provided the connection diameter d is small compared to the lattice constant a, this structure mimics a liquid in that its effective shear modulus can be orders of magnitude smaller than its effective bulk modulus. Reproduced with permission from Ref. [37]. Copyright 2012, AIP Publishing LLC.

other on a set of points forming a diamond lattice. The effective metamaterial shear modulus becomes much smaller than its bulk modulus if the connection diameter is small compared to the face-centered cubic lattice constant [37]. For example, geometrical ratios of 1% lead to bulk-to-shear modulus ratios in excess of 1000 [37, 69]. Correspondingly, the velocity of transverse shear waves is much smaller than that of the longitudinal compression waves [54]. Large intentional anisotropies are possible, too [38]. Two-dimensional versions of these three-dimensional structures are bimode metamaterials [45, 55] (but misleadingly sometimes called pentamode as well).

For ordinary liquids, cloaking based on coordinate transformations is possible (see above). One might be tempted to think that three-dimensional pentamode metamaterials are metaliquids behaving approximately like liquids. This reasoning is valid in the sense that the pentamode-metamaterial shear modulus can asymptotically approach zero indeed. However, a stable mechanical structure necessarily requires that the shear modulus is nonzero. This makes a pentamode metamaterial conceptually distinct from a liquid. In the Euler description of gases and liquids, one does not track individual molecules but rather investigates velocity fields, pressure fields, and mass density fields. In fact, in practice, one cannot easily track individual molecules in a gas or liquid. As a result, a displacement field like commonly used in solid mechanics has no meaning for true gases/liquids. In sharp contrast, one can track the displacement of each and every unit cell of a stable pentamode metamaterial. Even if one does not exploit this opportunity and only detects pressure fields, the difference is that ordinary gases/liquids can literally flow away, for example, along an interface to a solid, whereas the unit cells of a stable pentamode metamaterial stay where they are forever. Thus, one cannot simply apply acoustic equations to pentamode metamaterials.

This situation has led us to investigate a supposedly simple static core-shell elastic cloak based on three-dimensional pentamode metamaterials [8]. The geometry is illustrated in Fig. 10.8. A rigid hollow cylinder mechanically isolates its interior from the outside. A cylindrical compliant pentamode-metamaterial shell aims at cloaking the rigid cylinder with respect to a stiffer pentamode-

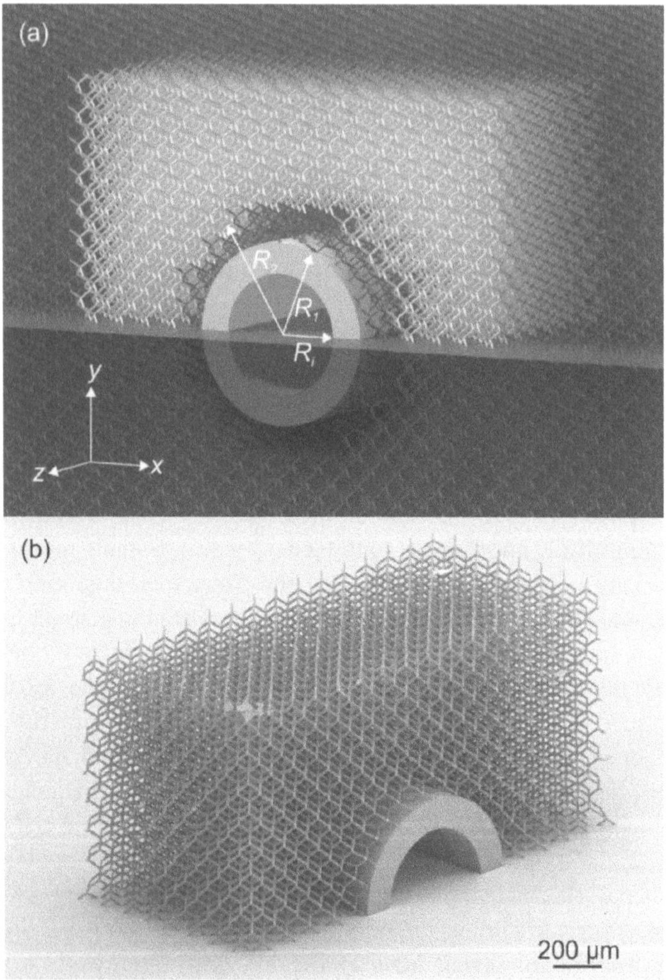

Figure 10.8 (a) Illustration of a core-shell elastic cloak based on pentamode metamaterials. (b) Electron micrograph of a corresponding polymer microstructure fabricated by three-dimensional dip-in direct-laser-writing optical lithography. Here, a more compliant pentamode-metamaterial shell is wrapped around a rigid hollow cylinder (under which objects can be hidden) to make the cylindrical obstacle appear like the pentamode-metamaterial surrounding—an "unfeelability cloak." Reprinted by permission from Macmillan Publishers Ltd: [*Nature Communications*] (Ref. [8]), copyright (2014).

metamaterial surrounding. This means that it must make the cylinder appear as the surrounding with respect to compression as well as with respect to shear. It is this combination that makes cloaking in solid mechanics harder than in other areas of physics. The theory of neutral inclusions based on core-shell geometries has been worked out [35]. Subsequent theoretical work is comprehensively reviewed in the textbook [56]. Hashin and Shtrikman showed that perfect neutrality requires that the three shear moduli of cylinder, shell, and surrounding are all equal. This condition would immediately be fulfilled for three different liquids with strictly zero shear modulus, but liquids would flow away or intermix. Zero shear modulus is also incompatible with a rigid hollow cylinder (with finite and even large shear modulus), which is necessary to isolate its interior from the outside. This has led us to conclude that perfect cloaking is not possible in three-dimensional solid mechanics [8].

Nevertheless, approximate cloaking is possible. For relatively small shear moduli of shell and surrounding, one even gets a simple relation connecting the bulk moduli of shell and surrounding with the radii of the geometry [8]. For example, for a shell radius (R_2) that is 4/3 times larger than the outer radius (R_1) of the rigid cylinder, (a much smaller cloak-to-object ratio than in the above optical and flexural wave experiments), the shell bulk modulus must be smaller than that of the surrounding by a factor of 7/16. The softer shell compensates the incompressible interior. In our experiments, we have pushed with a flat rigid stamp from the top onto corresponding polymer structures [8] shown in Fig. 10.8b. The resulting displacement vector fields have been measured by analyzing recorded movies of the sample surface using an autocorrelation approach, delivering a spatial resolution better than the camera pixels. We find excellent cloaking in that the derived displacement vectors of the cloaking structure closely resemble that of a homogeneous pentamode-metamaterial reference structure [8]. This is true even at locations just outside of the cloaking shell, where the displacements within obstacle and reference are different by as much as a factor of three [8].

10.5 Thermodynamics

10.5.1 *Heat Conduction Cloaks*

In 2012, Guenneau and coworkers [28] showed that the thermodynamic (scalar) equations of heat conduction and particle diffusion are form-invariant with respect to curvilinear spatial coordinate transformations (earlier theoretical work addressed the static case [10, 19, 33, 49]). Simply speaking, this is due to the fact that the spatial parts of the heat and diffusion differential equations are just the same as the spatial part of the wave equation. The equations only differ in that the wave equation is second order in time, whereas the heat and diffusion equation are first order in time. As a result, thermodynamics supports no waves and phenomena like interference, scattering, or reflection are completely absent. Thermodynamics is all about the increase of entropy, or, loosely speaking, about smearing out of inhomogeneities. The static case of heat conduction (and of particle diffusion) is strictly equivalent to that of electrical conduction. Experiments on static electric cloaking using simple resistor networks have been performed in 2012 [82]. We have already discussed electric cloaking intuitively in Fig. 10.2 in the second section of this chapter. There, we have seen that the conductivity plays the role of the phase velocity for wave problems. Consequently, the mathematics is very closely related [38].

The design of the experimental cloaking structures can be closely similar as well. This becomes apparent by comparing Fig. 10.6 and Fig. 10.9.

In our experiments, we expose the structure (previously brought to a homogeneous temperature) to a hot and cold water tank on the left and right-hand side, respectively. The resulting heat flow in two dimensions is monitored by an infrared camera looking onto the structure from the top. This provides us with movies of the temperature distribution. We find that the cloak not only transiently protects its interior from heating but also leaves the heat flow behind the cloak as though one had just a homogeneous medium. Note that the homogeneous surrounding is again periodically perforated to allow the azimuthal component of the heat conductivity inside the cloak to become larger than the heat

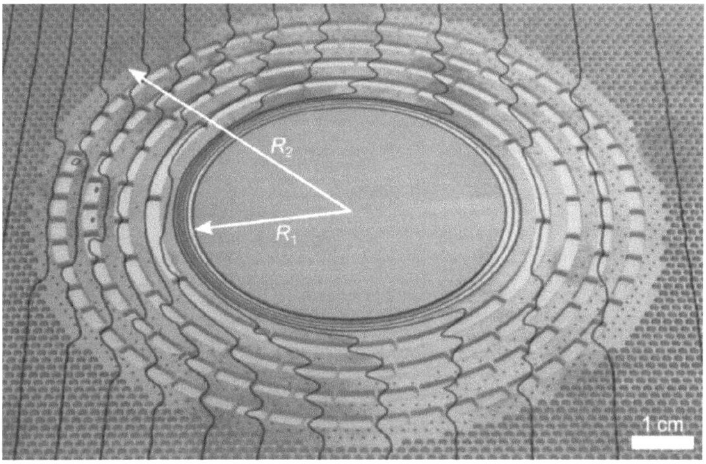

Figure 10.9 Photograph of a fabricated two-dimensional thermal cloak. Holes are milled into a 2 mm thin copper plate that are subsequently filled with PDMS, which has a heat conductivity that is 2600 times smaller than that of Cu. Note that the cloaking structure is closely similar to that for mechanical flexural waves shown in Fig. 10.6. The black lines are measured iso-temperature contours taken two minutes after exposure to a hot temperature bath on the left and a cold temperature bath on the right. Adapted (figure) with permission from Ref. [71]. Copyright (2013) by the American Physical Society.

conductivity of the surrounding (also see Fig. 10.2 and our above discussion for flexural waves in mechanics).

More recently, experiments have been published in which the multiple-ring structure shown in Fig. 1.9. has been simplified to a more compact two-ring or core-shell structure [2, 32, 81] (also compare Fig. 1.3.). However, for thin shells, such arrangement does not really work under transient conditions—in sharp contrast to designs based on coordinate transformations [29].

10.5.2 *Light Diffusion Cloaks*

The mathematics for particle diffusion is the same as that for heat conduction [29]. Three-dimensional multiple-shell structures have been proposed [29]. Corresponding experiments have not yet been reported though.

We conclude this chapter with a simpler core-shell type of cloak within the diffusive limit of optics, which brings us back to the discussion in our introduction. It is known that multiple scattering of light in clouds, fog, or in milk can effectively be described by a diffusion equation for light [53]. Due to the large speed of light, it is natural and usually sufficient to consider the static limit, because dynamic experiments would require picosecond time resolution. The effective light diffusivity can, for example, be controlled via the concentration of microparticles acting as scattering centers in an otherwise homogeneous optical material. Larger concentration leads to lower diffusivity. To again turn a core-shell invisible object into a true cloak (see introduction), we need to isolate the inside from the outside. This can be accomplished by a zero-diffusivity hollow core, where zero diffusivity is analogous to zero heat conductivity (see above) or zero electric conductivity (see Fig. 10.2). We mimic zero light diffusivity by a reflecting metal wall coated with a thin strongly scattering layer, which prevents any flow of photons from the outside to the inside. Again, simple analytical formulas are available in the literature [56] to connect the required shell diffusivity to that of the surrounding and to the radii of the structure. As pointed out above, these are only strictly valid in the limit of constant particle gradient across the object to be cloaked. In diffusive optics, this corresponds to spatially homogenous illumination. However, our numerical calculations and our experiments show that the same core-shell structure yields very good approximate cloaking for many other illumination conditions as well.

Figure 10.10 shows a photograph of a fabricated cylindrical core-shell structure made off PDMS doped with dielectric microparticles [70]. The imaging experiments using water with some added white paint as surrounding show good cloaking performance. Notably, this diffusive free-space cloak works for unpolarized light and throughout the entire visible part of the spectrum. Our analysis and estimates [70] show that the longest time constants in the problem are a few tens of nanoseconds. This means that the validity of the static regime reaches down to a microsecond scale or even below. This corresponds to image frame rates beyond one million per second.

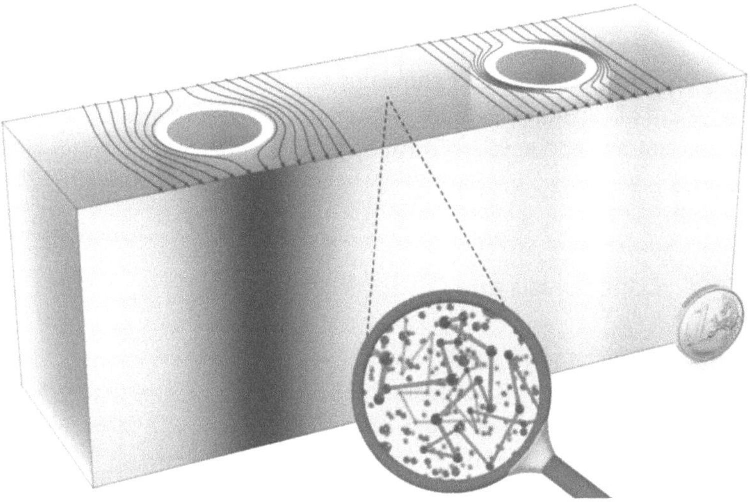

Figure 10.10 Experiments on cloaking in diffusive light-scattering media. The front shows measured true-color photographs taken under white-light illumination of a tank containing water with white paint added as the surrounding. The left image is for the bare core, the obstacle, which is a hollow metal cylinder coated with white paint. It casts a diffusive shadow. The right image is for the core-shell cloak. The shell is made of PDMS doped with microparticles. Here, the diffusive shadow is largely reduced. The scheme superimposed on the top of the tank exhibits the streamlines of the photon current (neglecting any absorption effects), the counterpart to the Poynting vector in electromagnetism. The euro coin on the lower right-hand side corner gives an intuitive impression for the dimensions. The tank has a thickness of 6.0 cm, the outer diameter of the core is 3.2 cm, and that of the shell 4.0 cm. The magnifying glass illustrates the multiple scattering events of light. Adapted from Ref. [70], Robert Schittny, Karlsruhe Institute of Technology (KIT).

Similar experiments have also been performed for other illumination conditions as well as for spherical rather than cylindrical geometry [70]. Theoretical calculations based on the diffusion equation for the photon density of light and for the experimental parameters are in excellent agreement with all of these experiments [70].

The fundamental restrictions imposed by Albert Einstein's theory of relativity onto free-space cloaks discussed in Section 10.2 are not relevant here because the effective speed of information transport is far below the vacuum speed of light (due to the numerous scattering events, which can be seen as a large detour for photons). This trick of choosing an adequate reference medium (instead of vacuum) is closely related to that used for the broadband polarization-independent carpet cloak in the ballistic regime of optics (see above).

10.6 Conclusions and Outlook

Taking advantage of spatial coordinate transformations has a long tradition in physics. Conformal maps have successfully been used in hydromechanics for more than a century [42] to transform boundaries and hence find analytical solutions for boundary value problems that are otherwise difficult to solve. The same general idea has recently also been applied in nano-optics [62].

The spirit of transformation optics/physics as outlined in this chapter is distinct from that. Here, coordinate transformations are used to design inhomogeneous and generally anisotropic material distributions. What could be more striking than demonstrating the design of something that seemed impossible for centuries? Likely because of that, invisibility cloaking and its counterparts have attracted so much attention. In this chapter, we have shown that the spirit of transformation physics can be applied in optics, thermodynamics, and mechanics and that it has led to a variety of experimentally functioning cloaking devices in all of these areas.

What is next? The theory of transformation optics on the basis of macroscopic Maxwell's equations for continua is fairly well developed. Fundamental restrictions due to relativity forbid three-dimensional macroscopic omnidirectional broadband invisibility cloaking for ballistic light propagation–which some may have dreamed of for military purposes. However, it does, for example, seem in reach to hide electrical contacts on solar cells, which typically block a few percent of the usable area, in the sense that the light is guided around the metal contacts onto the active semiconductor area. Thereby, the overall conversion efficiency increases. This down-to-earth function can be accomplished by two carpet cloaks arranged back to back [31] on top of the wire, even by graded index structures with low material contrast. Corresponding experiments have not been published yet, but will likely appear in the near future.

For light transport in scattering (turbid) media, which can be described by Fick's diffusion equation of thermodynamics, three-dimensional macroscopic omnidirectional broadband invisibility cloaking is possible and has been demonstrated experimentally for the entire visible spectrum. These experiments have been inspired by other experiments based on coordinate transformations of the thermodynamic heat conduction equation. Applications might arise in diffuse backlighting or in homogenizing diffusive light emission from large-area organic light-emitting diodes (OLEDs), where front-side electric contact wires are needed electrically but unwanted regarding their optical appearance. Ideally, the living room would be equipped with diffusively reflecting homogeneous walls that appear like white wall paint and that can serve as large-area light emitters, if needed.

Conceptually, mechanics still poses severe challenges regarding using coordinate transformations as a design tool, because the general continuum mechanics equations are not form-invariant under arbitrary coordinate transformations, at least not for ordinary elastic solids in three dimensions. Striking experiments have become possible though for two-dimensional flexural waves in thin plates as well as for airborn sound waves. Cosserat metamaterials or metamaterials with anisotropic mass density tensors do provide general mathematical solutions in three dimensions, but too little is known at this point to realize functional devices on this basis. In contrast, tailored pentamode metamaterials have emerged experimentally in recent years, but they are restricted to the limit of vanishingly small shear forces (by definition), corresponding to longitudinal waves only. Applications of coordinate transformations in mechanics might lie in protecting sensitive infrastructure like a power plant from seismic waves by guiding the surface earthquake wave around the power plant, or by making a void in a material (e.g., a tunnel underneath a river) appear to the outside as though the void was not there in a static elastic sense.

Acknowledgments

We thank our group members Tolga Ergin, Joachim Fischer, Nicolas Stenger, and Michael Thiel, as well as our collaborators Sebastien Guenneau (Marseille) and Manfred Wilhelm (Karlsruhe), for their contributions to the work reviewed here. We acknowledge support by the Deutsche Forschungsgemeinschaft (DFG) and the State of Baden-Württemberg through the DFG-Center for Functional Nanostructures (CFN) via subprojects A1.4 and A1.5, as well as through the Karlsruhe School of Optics and Photonics (KSOP). We also thank the Hector Fellow Academy for support.

References

1. Aki, K., and Richards, P. (2002). *Quantitative Seismology, 2nd ed. (University Science Books).*
2. Alù, A. (2014). Thermal cloaks get hot, *Physics*, **7**, 12.
3. Alù, A., and Engheta, N. (2005). Achieving transparency with plasmonic and metamaterial coatings, *Phys. Rev. E*, **72**, 016623.
4. Amirkhizi, A., Tehranian, A., and Nemat-Nasser, S. (2010). Stress-wave energy management through material anisotropy, *Adv. Funct. Mater.* **47**, 519–536.
5. Andkjaer, J., and Sigmund, O. (2011). Topology optimized low-contrast all-dielectric optical cloak, *Appl. Phys. Lett.*, **98**, 021112.
6. Banerjee, B. (2011). *An Introduction to Metamaterials and Waves in Composites* (Taylor and Francis).
7. Brun, M., Guenneau, S., and Movchan, A. (2009). Achieving control of in-plane elastic waves, *Appl. Phys. Lett.*, **94**, 061903.
8. Bückmann, T., Thiel, T., Kadic, M., Schittny, R., and Wegener, M. (2014). An elasto-mechanical unfeelability cloak made of pentamode metamaterials, *Nat. Commun.*, **5**, 4130.
9. Calderon, A.-P. (1980). On an inverse boundary value problem, in *Seminar on Numerical Analysis and its Applications to Continuum Physics*, (Rio de Janeiro, 1980), Soc. Brasil. Matter, pp. 65–73.
10. Chen, T., Weng, C., and Chen, J. (2008). Cloak for curvilinearly anisotropic media in conduction, *Appl. Phys. Lett.*, **93**, pp. 114103–114103-3.

11. Chen, X., Luo, Y., Zhang, J., Jiang, K., Pendry, J., and Zhang, S. (2011). Macroscopic invisibility cloaking of visible light, *Nat. Commun.*, **2**, 176.
12. Chew, H., and Kerker, M. (1976). Abnormally low electromagnetic scattering cross sections, *J. Opt. Soc. Am.*, **66**, 445.
13. Chu, S., and Wong, S. (1982). Linear pulse propagation in an absorbing medium, *Phys. Rev. Lett.*, **48**, pp. 738–741.
14. Cummer, S., Popa, B.-I., Schurig, D., Smith, D., Pendry, J., Rahm, M., and Starr, A. (2008). Scattering theory derivation of a 3d acoustic cloaking shell, *Phys. Rev. Lett.*, **100**, 024301.
15. Dolling, G., Enkrich, C., Wegener, M., Soukoulis, C., and Linden, S. (2006). Simultaneous negative phase and group velocity of light in a metamaterial, *Science*, **312**, 892–894.
16. Ergin, T., Fischer, J., and Wegener, M. (2011). Optical phase cloaking of 700 nm light waves in the far field by a three-dimensional carpet cloak, *Phys. Rev. Lett.*, **107**, 173901.
17. Ergin, T., Fischer, J., and Wegener, M. (2012). Detailed optical characterization of three-dimensional visible-frequency polarization-independent carpet invisibility cloak, *Physica B: Condens. Matter*, **407**, 4075–4077.
18. Ergin, T., Stenger, N., Brenner, P., Pendry, J., and Wegener, M. (2010). Three-dimensional invisibility cloak at optical wavelengths, *Science*, **328**, 337–339.
19. Fan, C., Gao, Y., and Huang, J. (2008). Shaped graded materials with an apparent negative thermal conductivity, *Appl. Phys. Lett.* **92**, 251907–251907-3.
20. Farhat, M., Enoch, S., Guenneau, S., and Movchan, A. (2008). Broadband cylindrical acoustic cloak for linear surface waves in a fluid, *Phys. Rev. Lett.*, **101**, 134501.
21. Farhat, M., Guenneau, S., and Enoch, S. (2009). Ultrabroadband elastic cloaking in thin plates, *Phys. Rev. Lett.*, **103**, 024301.
22. Fischer, J., Ergin, T., and Wegener, M. (2011). Three-dimensional polarization-independent visible-frequency carpet invisibility cloak, *Opt. Lett.*, **36**, 2059–2061.
23. Fischer, J., and Wegener, M. (2013). Three-dimensional optical laser lithography beyond the diffraction limit, *Laser Photon. Rev.* **7**, 22–44.
24. Gabrielli, L., Cardenas, J., Poitras, C., and Lipson, M. (2009). Silicon nanostructure cloak operating at optical frequencies, *Nat. Photon.*, **3**, 461.

25. Garrett, C. G. B., and McCumber, D. E. (1970). Propagation of a gaussian light pulse through an anomalous dispersion medium, *Phys. Rev. A*, **1**, 305–313.

26. Greenleaf, A., Kurylev, Y., Lassas, M., and Uhlmann, G. (2008). Isotropic transformation optics: approximate acoustic and quantum cloaking, *New J. Phys.*, **10**, 11, 115024.

27. Greenleaf, A., Lassas, M., and Uhlmann, G. (2003). On nonuniqueness for calderon's inverse problem, *Math. Res. Lett.*, **10**, 685–693.

28. Guenneau, S., Amra, C., and Veynante, D. (2012). Transformation thermodynamics: cloaking and concentrating heat flux, *Opt. Express*, **20**, 8207–8218.

29. Guenneau, S., and Puvirajesinghe, T. (2013). Fick's second law transformed: One path to cloaking in mass diffusion, *J. R. Soc. Interface*, **10**, 20130106.

30. Halimeh, J., and Wegener, M. (2012). Photorealistic ray tracing of free-space invisibility cloaks made of uniaxial dielectrics, *Opt. Express*, **20**, 27, 28330–28340.

31. Halimeh, J., and Wegener, M. (2013). Photorealistic rendering of unidirectional free-space invisibility cloaks, *Opt. Express*, **21**, 9457–9472.

32. Han, T., Bai, X., Gao, D., Thong, J. T. L., Li, B., and Qiu, C.-W. (2014). Experimental demonstration of a bilayer thermal cloak, *Phys. Rev. Lett.*, **112**, 054302.

33. Han, T., Yuan, T., Li, B., and Qiu, C. (2013). Homogeneous thermal cloak with constant conductivity and tunable heat localization, *Sci. Rep.*, **3**, 1593.

34. Hashemi, H., Zhang, B., Joannopoulos, J. D., and Johnson, S. G. (2010). Delay-bandwidth and delay-loss limitations for cloaking of large objects, *Phys. Rev. Lett.*, **104**, 253903.

35. Hashin, Z., and Shtrikman, S. (1962). A variational approach to the theory of the effective magnetic permeability of multiphase materials, *J. Appl. Phys.*, **33**, 3125.

36. Kadic, M., Bückmann, T., Schittny, R., and Wegener, M. (2013). Metamaterials beyond electromagnetism, *Rep. Prog. Phys.*, **76**, 126501.

37. Kadic, M., Bückmann, T., Stenger, N., Thiel, M., and Wegener, M. (2012). On the practicability of pentamode mechanical metamaterials, *Appl. Phys. Lett.*, **100**, 191901.

38. Kadic, M., Schittny, R., Bückmann, T., and Wegener, M. (2013). On anisotropic versions of three-dimensional pentamode metamaterials, *New J. Phys.*, **15**, 023029.
39. Kerker, M. (1975). Invisible bodies, *J. Opt. Soc. Am.*, **65**, 376.
40. Kerner, E. H. (1956). The electrical conductivity of composite media, *Proc. Phys. Soc. B*, **69**, 802.
41. Kohn, R., and Vogelius, M. (1984). Determining conductivity by boundary measurements, *Commun. Pure Appl. Math.*, **37**, 289–298.
42. Lamb, H. (1895). *Hydrodynamics* (At the University Press).
43. Lan, L., Sun, F., Liu, Y., Ong, C., and Ma, Y. (2013). Experimentally demonstrated a unidirectional electromagnetic cloak designed by topology optimization, *Appl. Phys. Lett.*, **103**, 121113.
44. Landy, N., and Smith, D. (2013). A full-parameter unidirectional metamaterial cloak for microwaves, *Phys. Rev. Lett.*, **12**, 25–28.
45. Layman, C., Naify, C., Martin, T., Calvo, D., and Orris, G. (2013). Highly anisotropic elements for acoustic pentamode applications, *Phys. Rev. Lett.*, **111**, 024302.
46. Lee, J. H., Blair, J., Tamma, V. A., Wu, Q., Rhee, S. J., Summers, C. J., and Park, W. (2009). Direct visualization of optical frequency invisibility cloak based on silicon nanorod array, *Opt. Express*, **17**, 12922–12928.
47. Leonhardt, U. (2006). Optical conformal mapping, *Science*, **312**, 1777–1780.
48. Leonhardt, U., and Philbin, T. (2010). *Geometry and Light: The Science of Invisibility* (Dover Books on Physics).
49. Li, J., Gao, Y., and Huang, J. (2010). A bifunctional cloak using transformation media, *J. Appl. Phys.*, **108**, 074504–074504–5.
50. Li, J., and Pendry, J. (2008). Hiding under the carpet: a new strategy for cloaking, *Phys. Rev. Lett.*, **101**, 203901.
51. Liu, R., Ji, C., Mock, J. J., Chin, J. Y., Cui, T. J., and Smith, D. R. (2009). Broadband ground-plane cloak, *Science*, **323**, 366–369.
52. Luo, Y., Zhang, J., Chen, H., Ran, L., Wu, B.-I., and Kong, J. (2009). A rigorous analysis of plane-transformed invisibility cloaks, *IEEE T. Antenna Propag.*, **57**, 3926–3933.
53. Martelli, F., Bianco, S., Ismaelli, A., and Zaccanti, G. (2010). *Light Propagation through Biological Tissue and Other Diffusive Media: Theory, Solutions, and Software* (SPIE Digital Library).
54. Martin, A., Kadic, M., Schittny, R., Bückmann, T., and Wegener, M. (2012). Phonon band structures of three-dimensional pentamode metamaterials, *Phys. Rev. B*, **86**, 155116.

55. Martinsson, P., and Movchan, A. (2003). Vibrations of lattice structures and phononic band gaps, *Q. J. Mech. Appl. Math.*, **56**, 4, 45–64.
56. Milton, G. (2002). *The Theory of Composites* (Cambridge University Press).
57. Milton, G., Briane, M., and Willis, J. (2006). On cloaking for elasticity and physical equations with a transformation invariant form, *New J. Phys.*, **8**, 248.
58. Milton, G., and Cherkaev, A. (1995). Which elasticity tensors are realizable? *J. Eng. Mater.-T.*, **117**, 483–493.
59. Norris, A. (2008). Acoustic cloaking theory, *P. R. Soc. A-Math. Phys.*, **464**, 2411–2434.
60. Norris, A., Amirkulova, F., and Parnell, W. (2014). Active elastodynamic cloaking, *Math. Mech. Solids*, **19**, 603–625.
61. Norris, A., and Shuvalov, A. (2011). Elastic cloaking theory, *Wave Motion*, **48**, 525 – 538.
62. Pendry, J. B., Aubry, A., Smith, D., and Maier, S. A. (2012). Transformation optics and subwavelength control of light, *Science*, **337**, 549–552.
63. Pendry, J. B., Schurig, D., and Smith, D. (2006). Controlling electromagnetic fields, *Science*, **312**, 1780–1782.
64. Popa, B.-I., Zigoneanu, L., and Cummer, S. (2011). Experimental acoustic ground cloak in air, *Phys. Rev. Lett.*, **106**, 253901.
65. Puri, A., and Birman, J. (1983). Pulse propagation in spatially dispersive media, *Phys. Rev. A*, **27**, 1044–1052.
66. Renger, J., Kadic, M., Dupont, G., Aćimović, S., Guenneau, S., Quidant, R., and Enoch, S. (2010). Hidden progress: broadband plasmonic invisibility, *Opt. Express*, **18**, 15757–15768.
67. Sanchis, L., García-Chocano, V., Llopis-Pontiveros, R., Climente, A., Martínez-Pastor, J., Cervera, F., and Sánchez-Dehesa, J. (2013). Three-dimensional axisymmetric cloak based on the cancellation of acoustic scattering from a sphere, *Phys. Rev. Lett.*, **110**, 124301.
68. Scherrer, G., Hofman, M., Śmigaj, W., Kadic, M., Chang, T.-M., Mélique, X., Lippens, D., Vanbésien, O., Cluzel, B., de Fornel, F., Guenneau, S., and Gralak, B. (2013). Photonic crystal carpet: manipulating wave fronts in the near field at 1.55 μm, *Phys. Rev. B*, **88**, 115110.
69. Schittny, R., Bückmann, T., Kadic, M., and Wegener, M. (2013). Elastic measurements on macroscopic three-dimensional pentamode metamaterials, *Appl. Phys. Lett.*, **103**, 231905.

70. Schittny, R., Kadic, M., Bückmann, T., and Wegener, M. (2014). Invisibility cloaking in a diffusive light scattering medium, *Science*, **345**, 427–429.

71. Schittny, R., Kadic, M., Guenneau, S., and Wegener, M. (2013). Experiments on transformation thermodynamics: molding the flow of heat, *Phys. Rev. Lett.*, **110**, 195901.

72. Schittny, R., and Wegener, M. (2014). Metamaterialien: Physikalische tarnkappen, *Spektrum der Wissenschaft*, **1**, 80.

73. Schmied, R., Halimeh, J.-C., and Wegener, M. (2010). Conformal carpet and grating cloaks, *Opt. Express*, **18**, 24361–24367.

74. Schurig, D., Mock, J., Justice, B., Cummer, S., Pendry, J., Starr, A., and Smith, D. (2006). Metamaterial electromagnetic cloak at microwave frequencies, *Science*, **314**, 977–980.

75. Sigmund, O. (1995). Tailoring materials with prescribed elastic properties, *Mechanics of Materials*, **20**, 351–368.

76. Stenger, N., Wilhelm, M., and Wegener, M. (2012). Experiments on elastic cloaking in thin plates, *Phys. Rev. Lett.*, **108**, 014301.

77. Urzhumov, Y., Landy, N., Driscoll, T., Basov, D., and Smith, D. R. (2013). Thin low-loss dielectric coatings for free-space cloaking, *Opt. Lett.*, **38**, 10, 1606–1608.

78. Valentine, J., Li, J., Zentgraf, T., Bartal, G., and Zhang, X. (2009). An optical cloak made of dielectrics, *Nat. Mater.*, **8**, 569–571.

79. Vasquez, F., Milton, G., Onofrei, D., and Seppecher, P. (2011). Transformation elastodynamics and active exterior acoustic cloaking, URL http://arxiv.org/abs/1105.1221.

80. von Freymann, G., Ledermann, A., Thiel, M., Staude, I., Essig, S., Busch, K. and Wegener, M. (2010). Three-dimensional nanostructures for photonics, *Adv. Funct. Mater.*, **20**, 1038–1052.

81. Xu, H., Shi, X., Gao, F., Sun, H., and Zhang, B. (2014). Ultrathin three-dimensional thermal cloak, *Phys. Rev. Lett.*, **112**, 054301.

82. Yang, F., Mei, Z., Jin, T., and Cui, T. (2012). DC electric invisibility cloak, *Phys. Rev. Lett.*, **109**, 053902.

83. Yang, F., Mei, Z., Yang, X., Jin, T., and Cui, T. (2013). A negative conductivity material makes a dc invisibility cloak hide an object at a distance, *Adv. Funct. Mater.*, **23**, 4306–4310.

84. Zhang, B., Luo, Y., Liu, X., and Barbastathis, G. (2011). Macroscopic invisibility cloak for visible light, *Phys. Rev. Lett.*, **106**, 033901.

85. Zhang, J., Liu, L., Luo, Y., Zhang, S., and Mortensen, N. A. (2011). Homogeneous optical cloak constructed with uniform layered structures, *Opt. Express*, **19**, 8625–8631.

86. Zhang, S., Xia, C., and Fang, N. (2011). Broadband acoustic cloak for ultrasound waves, *Phys. Rev. Lett.*, **106**, 024301.
87. Zhou, X., Hu, G., and Lu, T. (2008). Elastic wave transparency of a solid sphere coated with metamaterials, *Phys. Rev. B*, **77**, 024101.
88. Zhu, W., Ding, C., and X., Z. (2010). A numerical method for designing acoustic cloak with homogeneous metamaterials, *Appl. Phys. Lett.*, **97**, 131902.

Chapter 11

Transformation Multiphysics

Massimo Moccia,[a] Giuseppe Castaldi,[a] Salvatore Savo,[b] Yuki Sato,[b] and Vincenzo Galdi[a]

[a]*Department of Engineering, Waves Group, University of Sannio, I-82100 Benevento, Italy*

[b]*Rowland Institute at Harvard, Harvard University, Cambridge, MA 02142, USA*

sato@rowland.harvard.edu, vgaldi@unisannio.it

In artificially engineered materials (also known as metamaterials), fascinating phenomena such as a negative index of refraction and a negative modulus can emerge and defy our common intuition. Such responses of metamaterials are strongly dictated by their engineered internal structures. As conventional media owe their properties to the average response from an ensemble of atoms and molecules, in metamaterials, each structural unit plays the role of an atom. This emergent nature provides researchers with great flexibility in tailoring the material's response to the external fields and waves of interest. Many significant advances in this rapidly moving field have been made utilizing transformation optics (TO) theory. In a dauntingly large phase space of material parameters and their arrangement, the TO formalism provides a clear-cut recipe

Transformation Wave Physics: Electromagnetics, Elastodynamics, and Thermodynamics
Edited by Mohamed Farhat, Pai-Yen Chen, Sebastien Guenneau, and Stefan Enoch

ISBN 978-981-4669-95-5 (Hardcover), 978-981-4669-96-2 (eBook)
www.panstanford.com

that is based on the idea of representing the spatial manipulation of a phenomenon of interest with a coordinate transformation. While tying such a geometrical interpretation to required material parameters through linear algebraic operations, this concept provides a systematic approach for designing a single functionality in a given physical domain. In this chapter, we describe recent developments in our attempt to extend the TO formalism beyond a single functionality to independently manipulate multiple physical phenomena simultaneously. Through the example of designing a multifunctional shell that behaves as an electrical invisibility cloak and a thermal concentrator, we describe a step towards a framework that can be called transformation multiphysics [1].

11.1 Introduction and Background

11.1.1 *Coordinate-Transformation-Based Metamaterials*

Transformation optics (TO) [2, 3] is a method for the conceptual design of complex media for controlling various physical phenomena. Originally developed in the context of manipulating electromagnetic waves, one of the defining characteristics of the TO formalism is to view the bending of the trajectories of a particular physical phenomenon as warping of space. In this approach, nontrivial physical manipulations can often be written as mere coordinate transformations, through which material parameters can be computed to induce a particular functionality. As long as the governing equations are form-invariant, such an operation can be utilized, and in principle the resultant medium is immune to scattering and undesired interactions with the background media. The highly complex nature of the material parameters computed with this methodology almost invariably requires truncation, discretization, and approximation in general, which inevitably introduces those aforementioned losses back to the system. However, the technique has been instrumental in much of the recent development in the rapidly growing field of metamaterials. Starting with electromagnetic waves [4–6], the TO formalism has seen applications in the areas of other wave-based phenomena such as acoustics [7–12], liquid surface waves [13], and elastodynamics [14–19], as well as for electrostatic [20–23] and magnetostatic fields [24–29] and most recently for the manipulation of diffusive mass concentration transfer [30, 31] and heat flow [32–38]. We refer readers to all other chapters within this book and [39–41] for some of the recent reviews on the application of the TO formalism in various scientific fields.

11.1.2 *Beyond Single Functionalities*

As TO theory makes catalytic influence in a wide variety of fields, the terms such as transformation acoustics [8] and transformation thermodynamics [33] are coined to describe its reincarnation in a specific domain. Naturally one has to wonder whether or not the approach can be utilized to design a *single* composite material that exhibits *distinct* functionalities in those varying physical domains at the same time. Such an independent and yet simultaneous manipulation of multiple physical phenomena is a task that remains elusive with naturally occurring media. If artificially constructed materials can provide this flexibility, a completely new avenue could be paved for implementing materials and devices capable of carrying out complex operations.

Taking advantage of the versatility of the TO formalism, one may envision, for instance, an architected material that manipulates the incoming electromagnetic wave and the heat dissipated within the medium from irradiation at the same time or an acoustic material that manipulates the sound propagating through it while providing a particular electrical/magnetic capability such as concentration and rotation. The material properties that were treated separately in the past could be folded into the design process altogether to construct a multifunctional device. We devote this chapter to discuss our recent attempt in extending the TO paradigm to induce multifunctionality with metastructures [1].

We present and highlight the transformation multiphysics platform while working through the example of an artificial shell that can simultaneously behave as an electrical invisibility cloak and a thermal concentrator.

11.2 Models and Methods

11.2.1 *Transformation Media in Thermal and Electrical Domains*

In what follows, we focus on thermal and electrical bifunctionalities. Nonetheless, we stress that our approach is not restricted to this scenario, and may be applied to different physical domains.

The problem of interest is schematically illustrated in Fig. 11.1a. We consider an auxiliary space $\mathbf{r}' = (x', y', z')$ filled with an isotropic medium of thermal and electrical conductivities κ' and σ', and the corresponding stationary, sourceless heat and electrical conduction equations

$$\nabla \cdot \left(\kappa' \nabla T'\right) = 0, \quad \nabla \cdot \left(\sigma' \nabla V'\right) = 0, \tag{11.1}$$

with T' and V' denoting the temperature and electrical potential, respectively. Assuming, for simplicity, a homogeneous medium (i.e., κ' and σ' constant), temperature and potential differences at the two boundaries would give rise to heat flux and electrical current density directed along straight, parallel paths, as typical for natural materials (see Fig. 11.1a).

We now introduce two intertwined coordinate transformations to a new *curved-coordinate* space $\mathbf{r}$,

$$\mathbf{r}' = \begin{cases} \mathbf{F}_{\mathrm{t}}(\mathbf{r}), \\ \mathbf{F}_{\mathrm{e}}(\mathbf{r}), \end{cases} \tag{11.2}$$

inducing different local metric distortions in the thermal ($\mathbf{F}_{\mathrm{t}}$) and electrical ($\mathbf{F}_{\mathrm{e}}$) domains. Figure 11.1a shows the specific example of a thermal concentrator and an electrical invisibility cloak, which will be pursued. Thanks to the form invariance properties of Eq. 11.1, it is possible to relate the temperature and potential distributions in the transformed domains to the original quantities as follows [39]:

$$T(\mathbf{r}) = T'[\mathbf{F}_{\mathrm{t}}(\mathbf{r})], \quad V(\mathbf{r}) = V'[\mathbf{F}_{\mathrm{e}}(\mathbf{r})]. \tag{11.3}$$

More importantly, it is possible to attain the same distortion effects induced by the coordinate transformations also in a *flat*, Cartesian space $\mathbf{r} = (x, y, z)$ filled with an inhomogeneous, anisotropic transformation medium (see Fig. 11.1b) with constitutive tensors

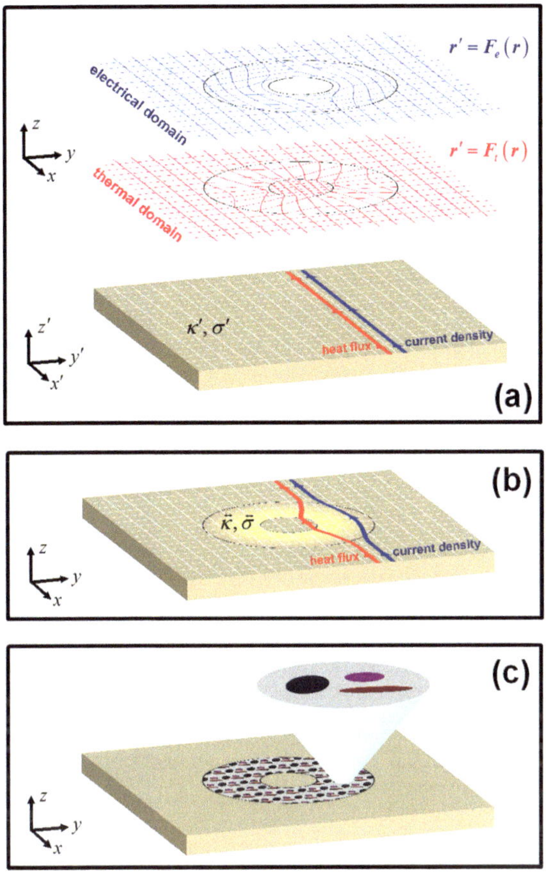

Figure 11.1 (a) Auxiliary space filled with an isotropic, homogeneous medium. Two coordinate transformations are applied, which induce different behaviors in the thermal (e.g., concentrator) and electrical (e.g., cloak) domains. (b) Equivalent interpretation in a flat-metric space filled with a transformation medium (see Eq. 11.4). The (otherwise straight) heat flux and current density paths are distorted in different fashions. (c) Metamaterial-based approximate implementation of the required nominal constitutive parameters via a mixture of inclusions of different shapes and materials (as qualitatively depicted in the magnified details) embedded in a host medium. Adapted from Ref. [1].

[39]

$$\overleftrightarrow{\kappa} = \kappa' \det\left(\overleftrightarrow{\Lambda}_{\rm t}\right) \overleftrightarrow{\Lambda}_{\rm t}^{-1} \cdot \overleftrightarrow{\Lambda}_{\rm t}^{-T}, \quad \overleftrightarrow{\sigma} = \sigma' \det\left(\overleftrightarrow{\Lambda}_{\rm e}\right) \overleftrightarrow{\Lambda}_{\rm e}^{-1} \cdot \overleftrightarrow{\Lambda}_{\rm e}^{-T}, \quad (11.4)$$

where the double arrow denotes tensor quantities,

$$\overleftrightarrow{\Lambda}_{\rm t} = \frac{\partial \mathbf{F}_{\rm t}}{\partial \mathbf{r}}, \quad \overleftrightarrow{\Lambda}_{\rm e} = \frac{\partial \mathbf{F}_{\rm e}}{\partial \mathbf{r}} \quad (11.5)$$

denote the Jacobian matrices associated with the two coordinate transformations, and the superscripts -1 and $-T$ indicate the inverse and inverse-transpose, respectively. Such a medium would exhibit *markedly different* thermal and electrical responses. This is schematically illustrated in Fig. 11.1b for the concentrator/cloak example chosen, with the heat flux concentrating in the inner region, and the current density circumventing that region.

11.2.2 *Joint Synthesis of Effective Parameters*

Unfortunately, it is generally impossible to find in nature the desired constitutive relationships in Eq. 11.4. Nevertheless, metamaterial engineering provides a powerful framework to approximate them to a certain extent. To the best of our knowledge, the only example of bifunctional device in the technical literature implements the same (cloak) functionality in both thermal and electrical domains [42], whereas most other studies [22, 36, 37] deal with the design of a single functionality (e.g., cloak or concentrator) in a single domain (e.g., thermal or electrical). Here, we are instead interested in prescribing *different* functionalities in *multiple* domains.

As typical in metamaterial engineering, our synthesis approach relies on the mixture of N different types of material inclusions embedded in a host medium (Fig. 11.1c). We characterize the host and inclusions by their thermal and electrical conductivities κ_n and σ_n, respectively, and filling fractions f_n, with $n = 0, 1, \ldots, N$, and the subscript 0 denoting the host medium. Inclusions are also characterized by depolarization tensors $\overleftrightarrow{\Gamma}_n$, which depend on their shape and orientation [43]. Our synthesis problem can be generally formulated as determining the material and structural compound parameters $\boldsymbol{\kappa} = \{\kappa_0, \kappa_1, \ldots, \kappa_N\}$, $\boldsymbol{\sigma} = \{\sigma_0, \sigma_1, \ldots, \sigma_N\}$, $\mathbf{f} =$

$\{f_0, f_1, \ldots, f_N\}, \overleftrightarrow{\mathbf{\Gamma}} = \left\{\overleftrightarrow{\Gamma}_1, \ldots, \overleftrightarrow{\Gamma}_N\right\}$ so that

$$\begin{cases} \overleftrightarrow{\kappa}_{\text{eff}}\left(\boldsymbol{\kappa}, \mathbf{f}, \overleftrightarrow{\mathbf{\Gamma}}\right) = \overleftrightarrow{\kappa}_{\text{nom}}, \\ \overleftrightarrow{\sigma}_{\text{eff}}\left(\boldsymbol{\sigma}, \mathbf{f}, \overleftrightarrow{\mathbf{\Gamma}}\right) = \overleftrightarrow{\sigma}_{\text{nom}}, \end{cases} \tag{11.6}$$

with $\overleftrightarrow{\kappa}_{\text{nom}}$ and $\overleftrightarrow{\sigma}_{\text{nom}}$ being the desired nominal constitutive tensors (arising from Eq. 11.4), and $\overleftrightarrow{\kappa}_{\text{eff}}$ and $\overleftrightarrow{\sigma}_{\text{eff}}$ the effective constitutive tensors which characterize the mixture. These effective parameters can be related to the host and inclusion parameters via approximate mixing formulae [43] (see also Appendix A for details). It is important to emphasize the *nonlinear* character of Eq. 11.6 (stemming from the mixing formulae), as well as the inherent *coupling* between the thermal and electrical domains via the structural compound parameters $\mathbf{f}$ and $\overleftrightarrow{\mathbf{\Gamma}}$. Moreover, in order to obtain physically feasible solutions, the search space should be constrained by the passivity requirements $\kappa_n \geq 0$ and $\sigma_n \geq 0$, as well as by the self-consistency conditions $\sum_{n=0}^{N} f_n = 1$, and unit-trace conditions $\text{tr}\left(\overleftrightarrow{\Gamma}_n\right) = 1$.

While solving Eq. 11.6 generally represents a formidable task, significant simplifications may arise in particular cases. For instance, if the *same* functionality is required in both domains, that is, $\mathbf{F}_{\text{t}} = \mathbf{F}_{\text{e}}$ (as in Ref. [42]), we obtain (from Eq. 11.4) $\overleftrightarrow{\kappa}_{\text{nom}}/\kappa' = \overleftrightarrow{\sigma}_{\text{nom}}/\sigma'$, which implies that the problems in Eq. 11.6 are *decoupled*, with only one synthesis needed. However, our scenario of interest is considerably more complex. In order to deal with two *distinct* functionalities, we exploit the concept of neutral inclusions from the theory of composites [44, 45]. More specifically, we consider inclusions that are matched with the host medium in one physical domain, so that they effectively operate only in the other domain. This assumption still allows to decouple the thermal and electrical syntheses in Eq. 11.6, but without constraining the two functionalities to be identical. Although these neutrality conditions may only be fulfilled approximately with natural material constituents, they may also be more precisely synthesized via multilayered inclusions [46, 47].

From now on, we focus on a two-dimensional scenario in the associated (ρ, ϕ, z) cylindrical coordinate system, which is better matched with our scenario of interest (see Fig. 11.1). Letting $\kappa_{\rho,\text{nom}}$,

$\kappa_{\phi,\mathrm{nom}}$, $\sigma_{\rho,\mathrm{nom}}$, $\sigma_{\phi,\mathrm{nom}}$ the nominal values of relevant constitutive-tensor components to be synthesized, and $\kappa_{\rho,\mathrm{eff}}$, $\kappa_{\phi,\mathrm{eff}}$, $\sigma_{\rho,\mathrm{eff}}$, $\sigma_{\phi,\mathrm{eff}}$ the corresponding effective parameters, we consider a three-phase mixture featuring two types of elliptic cylindrical inclusions with axes locally oriented along the cylindrical coordinates ρ and ϕ. Accordingly, our search parameter space comprises the constitutive parameters $\boldsymbol{\kappa} = \{\kappa_0, \kappa_1, \kappa_2\}$ and $\boldsymbol{\sigma} = \{\sigma_0, \sigma_1, \sigma_2\}$, filling fractions $\mathbf{f} = \{f_0, f_1, f_2\}$ (with $f_0 + f_1 + f_2 = 1$), and relevant depolarization tensor components $\Gamma_{1\rho} = 1 - \Gamma_{1\phi}$ and $\Gamma_{2\rho} = 1 - \Gamma_{2\phi}$. In particular, for an elliptical inclusion with axes A_ρ (along the ρ direction) and A_ϕ (along the ϕ direction), these latter components can be written as [43]

$$\Gamma_\rho = 1 - \Gamma_\phi = \frac{A_\phi}{A_\rho + A_\phi}. \tag{11.7}$$

Assuming the type 1 and type 2 inclusions as *thermally neutral* ($\kappa_1 = \kappa_0$) and *electrically neutral* ($\sigma_2 = \sigma_0$), respectively, and considering standard Maxwell–Garnett mixing formulae [43], we obtain for the effective parameters (see Appendix A for details):

$$\frac{\kappa_{\rho,\mathrm{eff}}}{\kappa_0} = \frac{\kappa_0 + (\kappa_2 - \kappa_0)\left[\Gamma_{2\rho}(1 - f_2) + f_2\right]}{\kappa_0 + \Gamma_{2\rho}(1 - f_2)(\kappa_2 - \kappa_0)}, \tag{11.8}$$

$$\frac{\kappa_{\phi,\mathrm{eff}}}{\kappa_0} = \frac{\kappa_0 + (\kappa_2 - \kappa_0)\left[\Gamma_{2\phi}(1 - f_2) + f_2\right]}{\kappa_0 + \Gamma_{2\phi}(1 - f_2)(\kappa_2 - \kappa_0)}, \tag{11.9}$$

$$\frac{\sigma_{\rho,\mathrm{eff}}}{\sigma_0} = \frac{\sigma_0 + (\sigma_1 - \sigma_0)\left[\Gamma_{1\rho}(1 - f_1) + f_1\right]}{\sigma_0 + \Gamma_{1\rho}(1 - f_1)(\sigma_1 - \sigma_0)}, \tag{11.10}$$

$$\frac{\sigma_{\phi,\mathrm{eff}}}{\sigma_0} = \frac{\sigma_0 + (\sigma_1 - \sigma_0)\left[\Gamma_{1\phi}(1 - f_1) + f_1\right]}{\sigma_0 + \Gamma_{1\phi}(1 - f_1)(\sigma_1 - \sigma_0)}. \tag{11.11}$$

Substitution of Eqs. 11.8–11.11 in Eq. 11.6 readily shows that the synthesis problem is now *effectively decoupled*, as the thermal parameters in Eqs. 11.8 and 11.9 only depend on the type 2 inclusions, while the electrical parameters in Eqs. 11.10 and 11.11 only depend on the type 1 inclusions. This allows for an analytical closed-form solution of the synthesis problem, whose general expression is given in Appendix A.

Here, we focus instead on the simplified expressions that are obtained in limit case $\kappa_2 \ll \kappa_0$ and $\sigma_1 \ll \sigma_0$,

$$\kappa_1 = \kappa_0 = \kappa_{\phi,\text{nom}} \frac{1 - \Gamma_{2\phi}(1 - f_2)}{(1 - \Gamma_{2\phi})(1 - f_2)},$$

$$\Gamma_{2\phi} = \frac{1 - \bar{\kappa}_{\text{nom}}(1 - 2f_2) - \sqrt{(1 - \bar{\kappa}_{\text{nom}})^2 + 4f_2^2\bar{\kappa}_{\text{nom}}}}{2(1 - f_2)(1 - \bar{\kappa}_{\text{nom}})}, \tag{11.12}$$

$$\sigma_2 = \sigma_0 = \sigma_{\phi,\text{nom}} \frac{1 - \Gamma_{1\phi}(1 - f_1)}{(1 - \Gamma_{1\phi})(1 - f_1)},$$

$$\Gamma_{1\phi} = \frac{1 - \bar{\sigma}_{\text{nom}}(1 - 2f_1) - \sqrt{(1 - \bar{\sigma}_{\text{nom}})^2 + 4f_1^2\bar{\sigma}_{\text{nom}}}}{2(1 - f_1)(1 - \bar{\sigma}_{\text{nom}})}, \tag{11.13}$$

with $\bar{\sigma}_{\text{nom}} = \sigma_{\phi,\text{nom}}/\sigma_{\rho,\text{nom}}$, $\bar{\kappa}_{\text{nom}} = \kappa_{\phi,\text{nom}}/\kappa_{\rho,\text{nom}}$, and the filling fractions f_1 and f_2 appearing as free parameters. Besides the inherent simplifications, the above choice is also consistent with realistic parameter designs (see Section 11.3.3 below). Nevertheless, other (asymptotic) simplifying assumptions are also possible (see Appendix A). Interestingly, the results in Eqs. 11.12 and 11.13 turn out to be *inherently feasible* as, for *arbitrary* values of the nominal anisotropy ratios $\bar{\kappa}_{\text{nom}}$ and $\bar{\sigma}_{\text{nom}}$, they always yield $\kappa_0 \geq 0$, $\sigma_0 \geq 0$, $0 < \Gamma_{1\phi} < 1$, and $0 < \Gamma_{2\phi} < 1$. However, as discussed in detail hereafter, the attainable anisotropy ratios are effectively restricted to *moderate* values by practical considerations and model consistency issues.

11.2.3 *Numerical Modeling*

For the numerical simulations of the thermal and electrical responses in our study, we rely on the finite-element-based commercial software package COMSOL Multiphysics 4.2 [48]. In particular, with reference to the square computational domain shown in Fig. 11.2, we use the "Heat Transfer" and "AC/DC" modules [48] for solving the stationary, sourceless heat and electrical conduction equations, $\nabla \cdot (\overleftrightarrow{\kappa} \cdot \nabla T) = 0$ and $\nabla \cdot (\overleftrightarrow{\sigma} \cdot \nabla V) = 0$, in the general presence of tensor, inhomogeneous thermal and electrical conductivities ($\overleftrightarrow{\kappa}$ and $\overleftrightarrow{\sigma}$, respectively), which reduce to

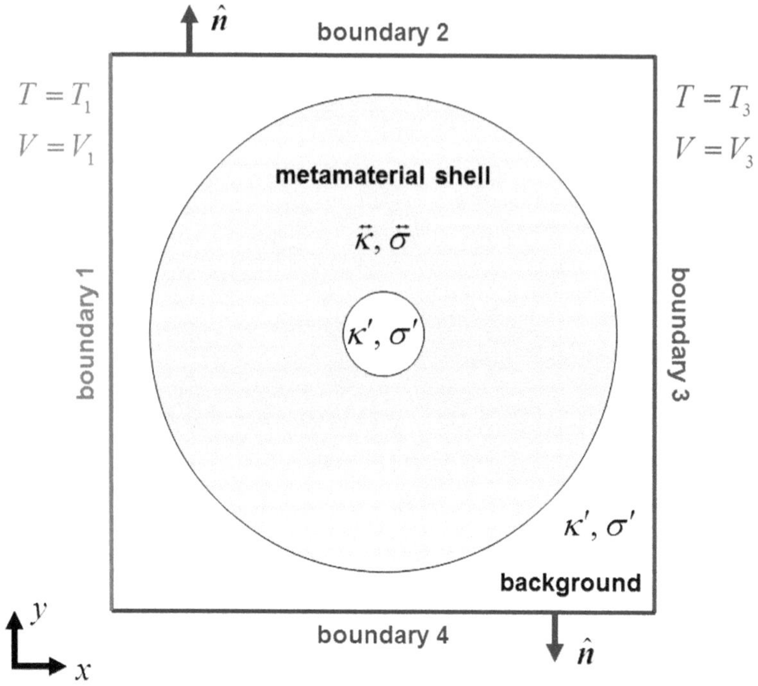

Figure 11.2 Schematic of the computational domain considered in the numerical simulations: a square of side length $L = 28$ cm filled with a background material with constitutive parameters κ' and σ', and a metamaterial annular shell of radii $R_1 = 2$ cm and $R_2 = 12$ cm. Also indicated are the boundary conditions enforced at the left and right boundaries (1 and 3, respectively), as well as the outward normal unit vectors involved in the thermal insulation and electrical insulation conditions enforced at boundaries 2 and 4. Adapted from Ref. [1].

scalar, piecewise-homogeneous quantities in the inclusion-based implementations. The computational domain is terminated by setting a temperature difference $\Delta T = T_3 - T_1$ and a potential difference $\Delta V = V_3 - V_1$ between the right and left boundaries (labeled as 3 and 1, respectively, in Fig. 11.2), and enforcing the thermal and electrical insulation conditions $\hat{\mathbf{n}} \cdot (\overleftrightarrow{\kappa} \cdot \nabla T) = 0$, $\hat{\mathbf{n}} \cdot (\overleftrightarrow{\sigma} \cdot \nabla V) = 0$, at the two remaining boundaries (labeled as 2 and 4, in Fig. 11.2), with $\hat{\mathbf{n}}$ denoting the outward normal unit vector.

We consider a discretization based on triangular meshing (using default criteria), and we rely on the SPOOLES direct solver, with default parameters [48]. For the more complex inclusion-based structures, resulting in about 15 million degrees of freedom, typical simulations (on a dedicated workstation with quad-core Intel Core i7 3.40 GHz processor, 16 GB RAM, running 64bit Windows8) may require up to 13 hours.

As meaningful observables, we show throughout the total heat flux $\overleftrightarrow{\kappa} \cdot \nabla T$ (W/m^2) and current density $\overleftrightarrow{\sigma} \cdot \nabla V$ (A/m^2), normalized by the enforced quantities $\kappa' \Delta T / L$ and $\sigma' \Delta V / L$, respectively (with L denoting the sidelength of the square computational domain; see Fig. 11.2). More specifically, we represent in false-color scale the magnitudes of these (vector) observables, and we show their local directions by superimposed streamlines.

11.3 Proof-of-Principle Example

11.3.1 *Thermal Concentrator and Electrical Cloak*

In order to apply the above synthesis procedure to the our scenario of interest (see Fig. 11.1), we introduce two (scalar) radial coordinate transformations

$$\rho' = \begin{cases} F_{\mathrm{t}}(\rho), \\ F_{\mathrm{e}}(\rho), \end{cases} \tag{11.14}$$

and we particularize Eq. 11.4 in terms of the relevant components [42]

$$\kappa_\rho = \frac{\kappa'^2}{\kappa_\phi} = \kappa' \frac{F_{\mathrm{t}}(\rho)}{\rho \dot{F}_{\mathrm{t}}(\rho)}, \quad \sigma_\rho = \frac{\sigma'^2}{\sigma_\phi} = \sigma' \frac{F_{\mathrm{e}}(\rho)}{\rho \dot{F}_{\mathrm{e}}(\rho)}, \tag{11.15}$$

where the overdot indicates differentiation with respect to the argument. Referring to Fig. 11.3a for illustration, an annular cylinder of radii $R_1 = 2$ cm and $R_2 = 12$ cm in the transformed space $\mathbf{r}$ is mapped in the auxiliary space $\mathbf{r}'$ onto an annular cylinder of radii $R_c > R_1$ and R_2 by the thermal domain transformation F_{t}, and onto a cylinder of radius R_2 by the electrical domain transformation F_{e}.

Accordingly, F_{t} yields a concentration effect (with $c = R_c/R_1 > 1$ denoting the concentration factor), whereas F_{e} yields an invisibility

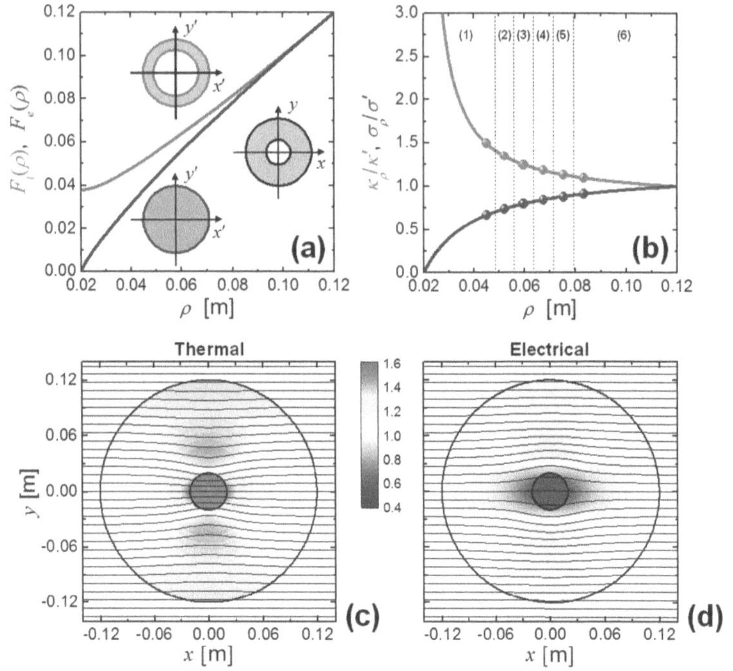

Figure 11.3 (a) Radial coordinate transformations implementing the ideal thermal concentrator (red) and electrical cloak (blue), and qualitative illustration of the mapping between auxiliary and transformed spaces. (b) Corresponding relevant constitutive tensor components $\kappa_\rho/\kappa' = \kappa'/\kappa_\phi$ (red) and $\sigma_\rho/\sigma' = \sigma'/\sigma_\phi$ (blue). Outside the annulus $R_1 < \rho < R_2$, the coordinate transformations reduce to the identity $F_{\mathrm{t,e}}(\rho) = \rho$, and the parameters coincide with those in the auxiliary space (κ' and σ'). The vertical dashed lines indicate the six-layer piecewise-constant radial discretization considered, with the markers representing the constant values assumed in each layer. (c, d) Numerically computed steady-state total heat flux and electrical current density magnitudes, respectively, with local directions indicated by the superimposed streamlines. Adapted from Ref. [1].

cloaking effect. We highlight that only the boundary values,

$$F_{\mathrm{e}}(R_1) = 0, \quad F_{\mathrm{t}}(R_1) = cR_1, \quad F_{\mathrm{t}}(R_2) = F_{\mathrm{e}}(R_2) = R_2, \quad (11.16)$$

are prescribed in order to achieve these effects, whereas the function behaviors in between are only partially constrained (see Appendix B for more details).

In our specific example, we found it expedient to exploit this degree of freedom by choosing the two mapping functions so that

$$\frac{\kappa_\rho}{\kappa_\phi} = \frac{\sigma_\phi}{\sigma_\rho}. \tag{11.17}$$

This choice, though not strictly necessary, arguably facilitates the implementation since it allows us to utilize two types of inclusions with identical shape (just rotated 90°) and filling fractions (see below and Appendix B for details).

The corresponding profiles of the constitutive parameters $\kappa_\rho = \kappa'^2/\kappa_\phi$ and $\sigma_\rho = \sigma'^2/\sigma_\phi$ are shown in Fig. 11.3b. It is worth pointing out that an *exact* implementation of the transformations would require *extreme* parameters (either zero or infinite) at the inner boundary $\rho = R_1$. Taking into account this and the aforementioned practical limitations, we approximate the continuous parameter distributions in terms of six-layer piecewise-constant profiles (indicated by dashed lines and markers in Fig. 11.3b), with parameters truncated so as to limit the anisotropy ratio $\kappa_\rho/\kappa_\phi = \sigma_\phi/\sigma_\rho$ to moderate values ≤ 2.5 (see also the discussion below). The corresponding thermal (concentrator) and electrical (cloak) responses are shown in Figs. 11.3c and 11.3d, respectively. We observe that the two responses differ substantially inside the transformation medium shell and in the inner region, whereas they essentially reduce to those observed in the unperturbed background medium (constant heat flux and current density, and straight, parallel streamlines) in the exterior region $\rho > R_2$. As expected, the thermal response (Fig. 11.3c) resembles that of a concentrator, with streamlines focusing toward the inner region. This results in an enhancement, by a factor of 1.53, of the enforced heat flux, in the inner region, by comparison with the case in the absence of the shell. On the other hand, the electrical response (Fig. 11.3d) resembles that of an (imperfect) invisibility cloak, with only little penetration of the streamlines in the inner region. This corresponds to a reduction, by a factor of 0.55, of the enforced current density in the inner region, by comparison with the case in the absence of the shell.

11.3.2 *Preliminary Ideal Parameter Design*

For each sampled value in the six-layer nominal profiles in Fig. 11.3b, we extract the scaled conductivities κ_ρ/κ', κ_ϕ/κ', σ_ρ/σ' and σ_ϕ/σ'

Table 11.1 Geometrical, structural, and constitutive parameters of the ideal parameter synthesis, from the piecewise-constant nominal profiles in Fig. 11.3b, assuming neutral inclusions ($\kappa_1 = \kappa_0$, $\sigma_2 = \sigma_0$) with $\kappa_2 = 0$ and $\sigma_1 = 0$

	Radii (cm)		Host	Inclusions Type 1 ($\kappa_1 = \kappa_0$, $\sigma_1 = 0$)		Inclusions Type 2 ($\kappa_2 = 0$, $\sigma_2 = \sigma_0$)		Fractions
Layer	R_{in}	R_{out}	$\kappa_0/\kappa' = \sigma_0/\sigma'$	$\Gamma_{1\phi}$	$A_{1\rho}/A_{1\phi}$	$\Gamma_{2\phi}$	$A_{2\rho}/A_{2\phi}$	$f_1 = f_2$
1	2	4.86	1.875	0.121	0.138	0.879	7.255	0.18
2	4.86	5.60	1.630	0.147	0.172	0.853	5.813	0.15
3	5.60	6.36	1.416	0.144	0.168	0.856	5.969	0.10
4	6.36	7.14	1.307	0.158	0.188	0.842	5.311	0.08
5	7.14	7.94	1.221	0.166	0.199	0.834	5.035	0.06
6	7.94	12	1.150	0.160	0.191	0.840	5.246	0.04

(with κ' and σ' denoting the background parameters), from which we can compute the nominal anisotropy ratios $\bar{\kappa}_{\text{nom}}$ and $\bar{\sigma}_{\text{nom}}$. We then choose the filling fractions $f_1 = f_2$ taking into account the aforementioned assumptions and limitations (see also Appendixes A and B). Since the transformation media to synthesize tend to become more isotropic towards the exterior layers (see Appendix B), we assume a gradually decreasing law with $f_1 = f_2 \leq 0.2$ (see also our discussion below). Assuming neutral inclusions ($\kappa_1 = \kappa_0$ and $\sigma_2 = \sigma_0$) with $\kappa_2 \ll \kappa_0$ and $\sigma_1 \ll \sigma_0$ (for simplicity, we set $\kappa_2 = 0$ and $\sigma_1 = 0$), Table 11.1 shows the parameters κ_0/κ', σ_0/σ', $\Gamma_{1\phi} = 1 - \Gamma_{1\rho}$, and $\Gamma_{2\phi} = 1 - \Gamma_{2\rho}$ computed from Eqs. 11.12 and 11.13, as well as the axis ratios calculated from the depolarization factors [43],

$$\frac{A_\rho}{A_\phi} = \frac{\Gamma_\phi}{1 - \Gamma_\phi} = \frac{\Gamma_\phi}{\Gamma_\rho}. \tag{11.18}$$

As anticipated (see also Appendix B), the particular choice in Eq. 11.17 yields $\Gamma_{2\phi} = 1 - \Gamma_{1\phi} = \Gamma_{2\rho}$, that is, two types of inclusions with identical shape (just rotated 90°), and, from Eqs. 11.9 and 11.10, $\kappa_0/\kappa' = \sigma_0/\sigma'$. Table 11.1 contains all the information needed for an inclusion-based implementation, in terms of geometrical, structural, and constitutive parameters for the host medium and the two types of inclusions in each layer.

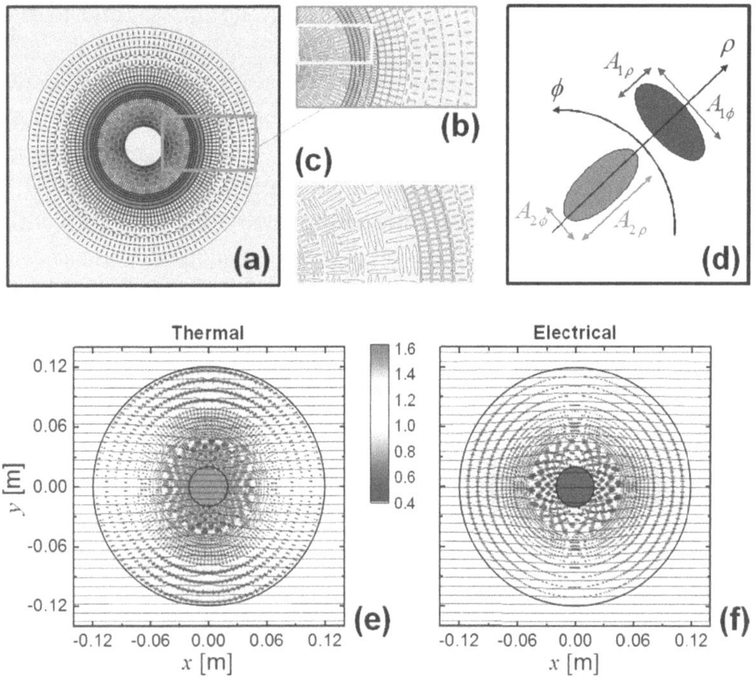

Figure 11.4 (a) Geometry of the ideal material inclusion-based implementation (details in Table 11.1) of the piecewise-constant constitutive parameter distributions in Fig. 11.3b. (b, c) Magnified details of the inclusions. (d) Schematic of the generic type 1 (blue) and type 2 (red) elliptical inclusions. (e, f) Corresponding thermal and electrical responses, respectively, as in Figs. 11.3c and 11.3d, assuming a background medium with $\kappa' = 1$ W/(mK) and $\sigma' = 1$ S/m. Adapted from Ref. [1].

Figure 11.4a shows one such possible implementation (with magnified details shown in Figs. 11.4b and 11.4c), generated via heuristic placement of the elliptical inclusions, oriented along the local ρ (and ϕ) directions (see Fig. 11.4d), in different host materials. Within this framework, working with the two types of inclusions having same shape (just rotated 90°) and filling fractions significantly facilitates their spatial arrangement. It should be emphasized that, although we took particular care in ensuring a uniform and spatially efficient packing and avoiding dense concentrations, the procedure is not fully optimized, and further improvement is possible.

The corresponding thermal and electrical responses are shown in Figs. 11.4e and 11.4f, respectively, assuming a background medium with $\kappa' = 1$ W/(mK) and $\sigma' = 1$ S/m. A generally good agreement with the nominal parameter predictions in Figs. 11.3c and 11.3d can be observed, both in the qualitative behaviors of the streamlines, and in the quantitative figures of merit.

11.3.3 *Realistic Parameter Design*

The reliance on host materials with strictly prescribed constitutive parameters (see Table 11.1) and neutral inclusions still renders the above design somehow idealized.

Via a heuristic matching with a list of realistic material parameters [49], we tried to approximate such design by resorting to only five realistic material constituents (detailed in Table 11.2), while maintaining the same geometrical and structural parameters as those in Table 11.1 and Fig. 11.4a–c. Also in this case, the procedure is not fully optimized, and there is room for further improvements. Nonetheless, the obtained configuration serves as a proof of principle demonstration of the practical feasibility of our proposed transformation multiphysics approach.

The corresponding thermal and electrical responses are shown in Figs. 11.5a and 11.5b, respectively, and, in spite of the imperfect fulfillment of the neutral inclusion conditions, they are in good agreement with both the nominal parameter predictions (Figs. 11.3c and 11.3d) and the previous ideal parameter synthesis (Figs. 11.4e

Table 11.2 Realistic materials (and corresponding thermal and electrical conductivities [49]) considered for approximating the ideal parameter synthesis in Table 11.1 and Fig. 11.4

	Material	κ (W/mK)	σ (S/m)
Layers 1 and 2	PG momentive	300	2×10^5
Layers 3 and 4	P-75S 2k	185	1.43×10^5
Layers 5 and 6, background	P-55S 2k	120	1.18×10^5
Type 1 inclusions	Aluminum nitride	190	10^{-11}
Type 2 inclusions	Silver conductive epoxy	1.75	1.4×10^5

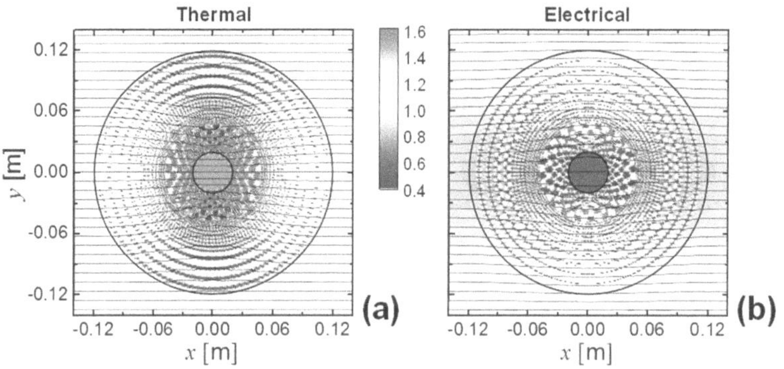

Figure 11.5 (a, b) As in Figs. 11.4e and 11.4f, respectively, but assuming the realistic material parameters in Table 11.2. Adapted from Ref. [1].

and 11.4f). In particular, in the inner region, the enforced heat flux is enhanced by a factor of 1.45 (concentrator), while the enforced current density is reduced by a factor of 0.52 (cloak).

11.4 Discussion

11.4.1 *Comparison with Conventional Material Shell*

The effects of our bifunctional metamaterial shell can be better understood by comparing its thermal and electrical responses with a reference configuration based on a conventional material. For example, we consider a shell of same size made of stainless steel ($\kappa = 16.3$ W/(mK), $\sigma = 1.42 \cdot 10^6$S/m) [50], and immersed in the same P-55S 2k ($\kappa = 120$ W/(mK), $\sigma = 1.18 \cdot 10^5$ S/m) background medium as in Fig. 11.5, with identical boundary conditions. The corresponding thermal and electrical responses are shown in Figs. 11.6a and 11.6b, respectively. By comparison with our bifunctional metamaterial shell, a similar reduction (by a factor of 0.58) of the current density in the inner region can be observed in the electrical response, which is however accompanied by a sensible reduction (by a factor of 0.3) of the heat flux. Moreover, significant perturbations in the area surrounding the shell are observed in both cases. Conversely, our metamaterial shell enhances the heat flux in

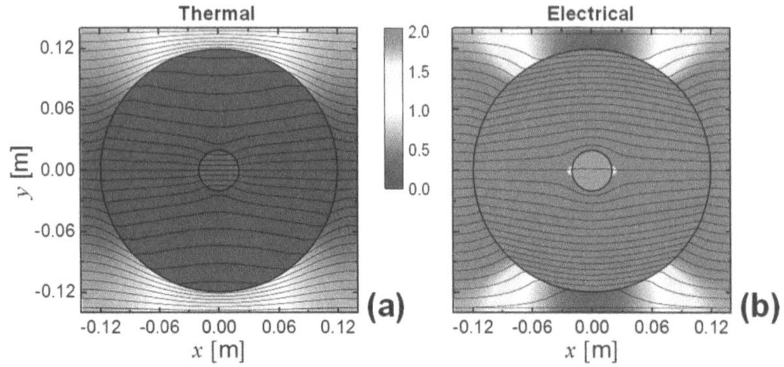

Figure 11.6 As in Fig. 11.5 but for a shell (of same size) made of stainless steel ($\kappa = 16.3$ W/(mK), $\sigma = 1.42 \cdot 10^6$S/m) [1, 50].

the inner region and reduces the current density, without sensible effects in the exterior background region.

11.4.2 *Realistic Bounds*

As previously mentioned, although arbitrary values of the nominal anisotropy ratios $\bar{\kappa}_{\text{nom}}$ and $\bar{\sigma}_{\text{nom}}$ would always yield physically feasible parameters in our synthesis procedure (see Eqs. 11.12 and 11.13), we need to account for certain practical limitations. In particular, extreme anisotropy ratios should be avoided, as they require extreme values of the depolarization factors (and hence *needle-shaped* inclusions, with $A_\rho \ll A_\phi$ or $A_\phi \ll A_\rho$, which may be difficult to arrange) and/or high values of the filling fractions (and hence significantly dense mixtures, for which the assumed Maxwell–Garnett mixing formulae may be inadequate [43]).

On these bases, it makes therefore sense to estimate some realistic bounds on the anisotropy ratios, and the associated constraints in the attainable coordinate transformations. In our design procedure, we found that inclusion arrangements that are both spatially efficient and adequately modeled by the Maxwell–Garnett mixing formulae (see Appendix A) are typically obtained with values of the depolarization factors of $0.1 \leq \Gamma_{1,2\phi} \leq 0.9$ and of the filling fractions $f_{1,2} \leq 0.2$. Figure 11.7a shows the anisotropy ratio $\kappa_{\rho,\text{eff}}/\kappa_{\phi,\text{eff}}$ or $\sigma_{\rho,\text{eff}}/\sigma_{\phi,\text{eff}}$ attainable from Eqs. 11.8–11.11 by

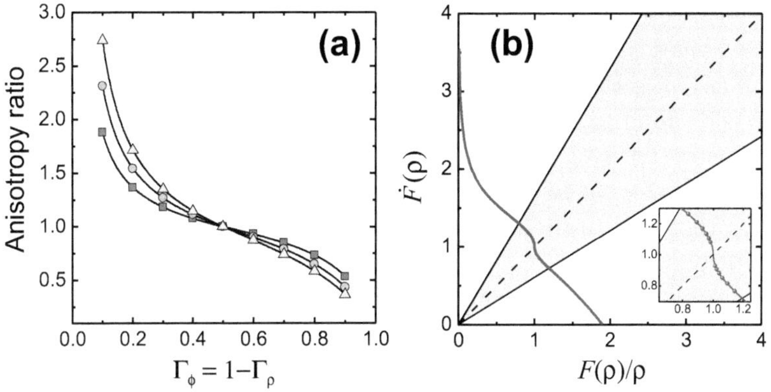

Figure 11.7 (a) Anisotropy ratio $\kappa_{\rho,\text{eff}}/\kappa_{\phi,\text{eff}}$ or $\sigma_{\rho,\text{eff}}/\sigma_{\phi,\text{eff}}$ attainable from Eqs. 11.8–11.11 (for the case of neutral inclusions $\kappa_1 = \kappa_0$ and $\sigma_2 = \sigma_0$, with $\kappa_2 \ll \kappa_0$ and $\sigma_1 \ll \sigma_0$) as a function of the depolarization factor Γ_ϕ, for three representative filling fraction values $f_1 = f_2 = 0.1,\ 0.15,\ 0.2$ (square, circle, and triangle markers, respectively). (b) Representation of the anisotropy ratio bounds in the space $\dot{F}(\rho)$ vs. $F(\rho)/\rho$, assuming $f_1 = f_2 = 0.2$. The cyan-shaded angular sector represents the allowed region, and the dashed bisector the identity transformation. Also shown are the curves pertaining to the concentrator (red) and cloak (blue) coordinate transformations in Fig. 11.3a. The inset shows a magnified view of the allowed portions of these curves, with the markers corresponding to the discretized samples in Fig. 11.3b. Adapted from Ref. [1].

varying the depolarization factors and filling fractions within these ranges, for the case of neutral inclusions ($\kappa_1 = \kappa_0$ and $\sigma_2 = \sigma_0$) with $\kappa_2 \ll \kappa_0$ and $\sigma_1 \ll \sigma_0$. It is worth highlighting that the syntheses in the thermal and electrical domains are effectively decoupled by the neutral inclusion assumption, and thus the results in Fig. 11.7a hold for either the thermal parameters ($\kappa_{\rho,\text{eff}}/\kappa_{\phi,\text{eff}}$, assuming $\Gamma_{2\phi}$ on the abscissa) or the electrical parameters ($\sigma_{\rho,\text{eff}}/\sigma_{\phi,\text{eff}}$, assuming $\Gamma_{1\phi}$ on the abscissa). These results confirm the anticipated trends (with the extreme values attained at the extrema of the allowed range for Γ_ϕ, and improving for higher filling fractions), and also quantify the attainable anisotropy ratios to moderate values ranging from $\sim$0.25 to $\sim$2.5.

These bounds can be translated to the space of implementable coordinate transformations $F(\rho)$ by noting from Eq. 11.15 that the anisotropy ratios directly affect the function $F(\rho)/[\rho\dot{F}(\rho)]$, so that it makes sense to consider the two-dimensional space $\dot{F}(\rho)$ vs. $F(\rho)/\rho$, as shown in Fig. 11.7b. In this space, fixed anisotropy ratios correspond to straight lines passing through the origin, with the slope decreasing for higher values of the anisotropy ratio $\kappa_{\rho,\mathrm{eff}}/\kappa_{\phi,\mathrm{eff}}$ (or $\sigma_{\rho,\mathrm{eff}}/\sigma_{\phi,\mathrm{eff}}$). In particular, the identity transformation $F(\rho) = \rho$ corresponds to the (dashed) bisector. Accordingly, for fixed filling fractions ($f_1 = f_2 = 0.2$), by extracting from Fig. 11.7a the maximum and minimum anisotropy, we can identify an angular sector (cyan shaded in Fig. 11.7b) containing all the possible $\dot{F}(\rho)$ vs. $F(\rho)/\rho$ combinations that are compatible with the assumed parameter constraints. For better understanding, Fig. 11.7b also shows the curves pertaining to the ideal concentrator (red) and cloak (blue) coordinate transformations in Fig. 11.3a, only a portion of which (magnified in the inset, with the markers corresponding to the discretized samples in Fig. 11.3b) falls within the allowed region.

From the above analysis, we can derive useful indications and guidelines for more general syntheses, not necessarily limited to the specific concentrator/cloak example.

11.5 Conclusions and Perspectives

The concept of manipulating electrical and thermal currents independently has relevance to many sectors. For example, as more components are packed into 3D integrated circuits to enhance performance in a smaller form factor, many sorts of physical phenomena at varying length and timescales become involved to significantly complicate the removal of heat. The multifunctional approach in one way or the other may indeed become a necessity for their thermal management. The involved scales could range from individual materials to a component to a whole device package. The field of thermoelectric materials may also be an interesting candidate suitable for the application of multifunctional concept presented here. Currently one of the fundamental bottlenecks for energy harnessing via thermoelectric conversions stems from the difficulty associated with controlling electrical and thermal transport independently in a given medium. This has forced many researchers to work with nanoscale materials, creating scalability issues [51–55]. Design protocols that link thermal and electrical responses could potentially allow one to manipulate the figure of merit and power conversion factors in ways unexplored in the past.

Furthermore, looking ahead to the design of advanced functional materials with some degree of self-regulation, the incorporation of ingredients such as vanadium dioxide that drastically alter their characteristics with the application of external stimuli will be deemed critical [56]. In such contexts, the transformation multiphysics platform may prove useful in, for instance, designing a metamaterial whose electrical anisotropy can be reconfigured with thermal stimulus for autonomous control of electrical current, while independently routing heat flux in an effective and predetermined fashion.

It is important to note that, as illuminating as the bifunctional electrical/thermal metamaterial case study may be, it is simply one of many possible scenarios that can be imagined. The concept of inducing multifunctionality with the coordinate transformation based formalism is general enough that phenomena in various electrical, magnetic, thermal, and acoustic domains may be effectively linked together. Furthermore it may be feasible to incorporate target functionalities that appear when the system is at static equilibrium in one domain while mixing them with transient functionalities in other domains. Possibilities extend far beyond the specific combinations of phenomena and their niche applications considered here.

Various different fields that incorporate the TO formalism in their respective artificial material engineering have all quickly started to mature over the past decade. A suitable time has certainly arrived for weaving those disciplines to construct multifunctional materials. With much possibility in the specific combinations of physical phenomena to manipulate and the types of functionalities to induce, one of the novelties of the transformation multiphysics formalism may be that it acts as an open source protocol that one can keep building upon to incorporate various different phenomena of interest.

Acknowledgments

S. Savo and Y. Sato acknowledge support from the Rowland Institute at Harvard University.

Appendix A

Details on Effective Medium Formulation

In our approach, we rely on simple Maxwell–Garnett mixing formulae [43] for modeling the effective thermal and electrical constitutive parameters of a multiphase mixture composed of N types of inclusions embedded in a host material. For the cylindrical scenario of interest, with inclusions aligned along the local ρ (and ϕ) direction (see Fig. 11.4d), the relevant constitutive tensor components can be written as [43]

$$\frac{\kappa_{\rho,\mathrm{eff}}}{\kappa_0} = 1 + \frac{\sum\limits_{n=1}^{N} \dfrac{f_n\left(\kappa_n - \kappa_0\right)}{\kappa_0 + \Gamma_{n\rho}\left(\kappa_n - \kappa_0\right)}}{1 - \sum\limits_{n=1}^{N} \dfrac{f_n \Gamma_{n\rho}\left(\kappa_n - \kappa_0\right)}{\kappa_0 + \Gamma_{n\rho}\left(\kappa_n - \kappa_0\right)}},$$

$$\frac{\kappa_{\phi,\mathrm{eff}}}{\kappa_0} = 1 + \frac{\sum\limits_{n=1}^{N} \dfrac{f_n\left(\kappa_n - \kappa_0\right)}{\kappa_0 + \Gamma_{n\phi}\left(\kappa_n - \kappa_0\right)}}{1 - \sum\limits_{n=1}^{N} \dfrac{f_n \Gamma_{n\phi}\left(\kappa_n - \kappa_0\right)}{\kappa_0 + \Gamma_{n\phi}\left(\kappa_n - \kappa_0\right)}}, \tag{A.1}$$

$$\frac{\sigma_{\rho,\mathrm{eff}}}{\sigma_0} = 1 + \frac{\sum\limits_{n=1}^{N} \dfrac{f_n\left(\sigma_n - \sigma_0\right)}{\sigma_0 + \Gamma_{n\rho}\left(\sigma_n - \sigma_0\right)}}{1 - \sum\limits_{n=1}^{N} \dfrac{f_n \Gamma_{n\rho}\left(\sigma_n - \sigma_0\right)}{\sigma_0 + \Gamma_{n\rho}\left(\sigma_n - \sigma_0\right)}},$$

$$\frac{\sigma_{\phi,\mathrm{eff}}}{\sigma_0} = 1 + \frac{\sum\limits_{n=1}^{N} \dfrac{f_n\left(\sigma_n - \sigma_0\right)}{\sigma_0 + \Gamma_{n\phi}\left(\sigma_n - \sigma_0\right)}}{1 - \sum\limits_{n=1}^{N} \dfrac{f_n \Gamma_{n\phi}\left(\sigma_n - \sigma_0\right)}{\sigma_0 + \Gamma_{n\phi}\left(\sigma_n - \sigma_0\right)}}, \tag{A.2}$$

with all parameters already defined in the main text. In particular, $\Gamma_{n\rho}$ and $\Gamma_{n\phi} = 1 - \Gamma_{n\rho}$ denote the components of the diagonal depolarization tensor $\overleftrightarrow{\boldsymbol{\Gamma}}_n$ pertaining to the generic nth inclusion. Equations A.1 and A.2 establish the nonlinear character of the general synthesis problem in Eq. 11.6, as well as the coupling between the thermal and electrical domains, as previously highlighted.

This model relies on the calculation of the static polarizabilities of the inclusions, assumed embedded in an *infinite* host medium, and is therefore adequate for *sparse* mixtures. In the case of *densely packed* inclusions, for which the medium effectively seen outside the generic inclusion is different from the host medium, more refined models can be applied [43] which typically require the numerical solution of implicit equations. Accordingly, to avoid the dense-packing conditions and guarantee the applicability of the Maxwell–Garnett mixing formulae in Eqs. A.1 and A.2, we restrict the structural parameters of the mixtures to *moderate* anisotropy levels.

We highlight that in our three-phase mixture example we consider the simplest reduction ($N = 2$) of Eqs. A.1 and A.2 that still allows the joint and independent synthesis of different functionalities in the two physical domains. For better understanding, here we also consider the simpler two-phase mixture, that is, with only one type of inclusions embedded in a host medium, for which Eq. A.1 can be particularized ($N = 1$) as follows

$$\frac{\kappa_{\rho,\text{eff}}}{\kappa_0} = 1 + \frac{\dfrac{f_1(\kappa_1 - \kappa_0)}{\kappa_0 + \Gamma_{1\rho}(\kappa_1 - \kappa_0)}}{1 - \dfrac{f_1\Gamma_{1\rho}(\kappa_1 - \kappa_0)}{\kappa_0 + \Gamma_{1\rho}(\kappa_1 - \kappa_0)}} = \frac{\bar{\kappa}_1 - f_0\Gamma_{1\phi}(\bar{\kappa}_1 - 1)}{1 + f_0\Gamma_{1\rho}(\bar{\kappa}_1 - 1)}, \quad \text{(A.3)}$$

$$\frac{\kappa_{\phi,\text{eff}}}{\kappa_0} = 1 + \frac{\dfrac{f_1(\kappa_1 - \kappa_0)}{\kappa_0 + \Gamma_{1\phi}(\kappa_1 - \kappa_0)}}{1 - \dfrac{f_1\Gamma_{1\phi}(\kappa_1 - \kappa_0)}{\kappa_0 + \Gamma_{1\phi}(\kappa_1 - \kappa_0)}} = \frac{\bar{\kappa}_1 - f_0\Gamma_{1\rho}(\bar{\kappa}_1 - 1)}{1 + f_0\Gamma_{1\phi}(\bar{\kappa}_1 - 1)}, \quad \text{(A.4)}$$

with $\bar{\kappa}_1 = \kappa_1/\kappa_0$, and the second equalities stemming from the consistency conditions $f_0 + f_1 = 1$ and $\Gamma_{1\rho} + \Gamma_{1\phi} = 1$. Similarly, Eq. A.2 yield

$$\frac{\sigma_{\rho,\text{eff}}}{\sigma_0} = 1 + \frac{\dfrac{f_1(\sigma_1 - \sigma_0)}{\sigma_0 + \Gamma_{1\rho}(\sigma_1 - \sigma_0)}}{1 - \dfrac{f_1\Gamma_{1\rho}(\sigma_1 - \sigma_0)}{\sigma_0 + \Gamma_{1\rho}(\sigma_1 - \sigma_0)}} = \frac{\bar{\sigma}_1 - f_0\Gamma_{1\phi}(\bar{\sigma}_1 - 1)}{1 + f_0\Gamma_{1\rho}(\bar{\sigma}_1 - 1)}, \quad \text{(A.5)}$$

$$\frac{\sigma_{\phi,\text{eff}}}{\sigma_0} = 1 + \frac{\dfrac{f_1(\sigma_1 - \sigma_0)}{\sigma_0 + \Gamma_{1\phi}(\sigma_1 - \sigma_0)}}{1 - \dfrac{f_1\Gamma_{1\phi}(\sigma_1 - \sigma_0)}{\sigma_0 + \Gamma_{1\phi}(\sigma_1 - \sigma_0)}} = \frac{\bar{\sigma}_1 - f_0\Gamma_{1\rho}(\bar{\sigma}_1 - 1)}{1 + f_0\Gamma_{1\phi}(\bar{\sigma}_1 - 1)}, \quad \text{(A.6)}$$

with $\bar{\sigma}_1 = \sigma_1/\sigma_0$. It can be observed that, for the extreme values $\Gamma_{1\phi} = 0$ and $\Gamma_{1\phi} = 1$, Eqs. A.3–A.6 reduce to the expressions pertaining to radial and angular multilayers [43], respectively. While these latter typically work well in the synthesis of single-functionality (e.g., cloak, concentrator) metamaterial structures in a given physical domain [20, 22–27, 29, 33, 36, 37], Eqs. A.3–A.6 do not generally provide sufficient degrees of freedom to design *different* anisotropic behaviors in the thermal and electrical domains. For example, if we choose the parameters $\bar{\kappa}_1$, f_0 and $\Gamma_{1\phi}$ in Eqs. A.3 and A.4 so that $\kappa_{\rho,\text{eff}}/\kappa_{\phi,\text{eff}} > 1$, it will be impossible to attain $\sigma_{\rho,\text{eff}}/\sigma_{\phi,\text{eff}} < 1$ from Eqs. A.5 and A.6, irrespective of the choice of the (passive) material parameter $\bar{\sigma}_1$. For general values of $\Gamma_{1\phi}$, this can be verified using symbolic algebra tools [57] and, among other things, it implies the impossibility to jointly synthesize a thermal concentrator (which requires $\kappa_\rho/\kappa_\phi > 1$) and an electrical cloak (which requires $\sigma_\rho/\sigma_\phi < 1$). These constraints may be relaxed in the presence of *negative-conductivity* material constituents, which could be in turn synthesized as metamaterials [23].

Maintaining the passivity requirements, the above limitations can be also overcome by considering a *three-phase* mixture, featuring two types of inclusions embedded in a host medium, for which ($N = 2$) we now obtain from Eqs. A.1 and A.2:

$$\frac{\kappa_{\rho,\text{eff}}}{\kappa_0} = 1 + \frac{\dfrac{f_1(\kappa_1 - \kappa_0)}{\kappa_0 + \Gamma_{1\rho}(\kappa_1 - \kappa_0)} + \dfrac{f_2(\kappa_2 - \kappa_0)}{\kappa_0 + \Gamma_{2\rho}(\kappa_2 - \kappa_0)}}{1 - \dfrac{f_1\Gamma_{1\rho}(\kappa_1 - \kappa_0)}{\kappa_0 + \Gamma_{1\rho}(\kappa_1 - \kappa_0)} - \dfrac{f_2\Gamma_{2\rho}(\kappa_2 - \kappa_0)}{\kappa_0 + \Gamma_{2\rho}(\kappa_2 - \kappa_0)}}, \quad \text{(A.7)}$$

$$\frac{\kappa_{\phi,\text{eff}}}{\kappa_0} = 1 + \frac{\dfrac{f_1(\kappa_1 - \kappa_0)}{\kappa_0 + \Gamma_{1\phi}(\kappa_1 - \kappa_0)} + \dfrac{f_2(\kappa_2 - \kappa_0)}{\kappa_0 + \Gamma_{2\phi}(\kappa_2 - \kappa_0)}}{1 - \dfrac{f_1\Gamma_{1\phi}(\kappa_1 - \kappa_0)}{\kappa_0 + \Gamma_{1\phi}(\kappa_1 - \kappa_0)} - \dfrac{f_2\Gamma_{2\phi}(\kappa_2 - \kappa_0)}{\kappa_0 + \Gamma_{2\phi}(\kappa_2 - \kappa_0)}}, \quad \text{(A.8)}$$

$$\frac{\sigma_{\rho,\text{eff}}}{\sigma_0} = 1 + \frac{\dfrac{f_1(\sigma_1 - \sigma_0)}{\sigma_0 + \Gamma_{1\rho}(\sigma_1 - \sigma_0)} + \dfrac{f_2(\sigma_2 - \sigma_0)}{\sigma_0 + \Gamma_{2\rho}(\sigma_2 - \sigma_0)}}{1 - \dfrac{f_1\Gamma_{1\rho}(\sigma_1 - \sigma_0)}{\sigma_0 + \Gamma_{1\rho}(\sigma_1 - \sigma_0)} - \dfrac{f_2\Gamma_{2\rho}(\sigma_2 - \sigma_0)}{\sigma_0 + \Gamma_{2\rho}(\sigma_2 - \sigma_0)}}, \quad \text{(A.9)}$$

$$\frac{\sigma_{\phi,\mathrm{eff}}}{\sigma_0}=1+\frac{\dfrac{f_1(\sigma_1-\sigma_0)}{\sigma_0+\Gamma_{1\phi}(\sigma_1-\sigma_0)}+\dfrac{f_2(\sigma_2-\sigma_0)}{\sigma_0+\Gamma_{2\phi}(\sigma_2-\sigma_0)}}{1-\dfrac{f_1\Gamma_{1\phi}(\sigma_1-\sigma_0)}{\sigma_0+\Gamma_{1\phi}(\sigma_1-\sigma_0)}-\dfrac{f_2\Gamma_{2\phi}(\sigma_2-\sigma_0)}{\sigma_0+\Gamma_{2\phi}(\sigma_2-\sigma_0)}}. \quad (A.10)$$

These formulae allow, in principle, to achieve different anisotropy ratios in the thermal and electrical domains, although the general synthesis in Eq. 11.6 remains very complicated. A significant simplification is obtained by assuming neutral inclusions [44, 45], matched with the background medium in one physical domain. For instance, assuming $\kappa_1=\kappa_0$ (i.e., thermally neutral type 1 inclusions) and $\sigma_2=\sigma_0$ (i.e., electrically neutral type 2 inclusions), Eqs. A.7–A.10 reduce to the forms in Eqs. 11.8–11.11, for which the synthesis problem is effectively decoupled. By substituting Eqs. 11.8–11.11 in Eq. 11.6, and letting $\kappa_{\rho,\mathrm{nom}}$, $\kappa_{\phi,\mathrm{nom}}$, $\sigma_{\rho,\mathrm{nom}}$, and $\sigma_{\phi,\mathrm{nom}}$ the relevant components of the nominal parameters to synthesize, we obtain analytical closed-form solutions

$$\Gamma_{2\phi}=\frac{(1-\bar{\kappa}_2)\left[1-\bar{\kappa}_{\mathrm{nom}}(1-2f_2)\right]\pm\sqrt{\begin{array}{l}(1+\bar{\kappa}_2)^2(1-\bar{\kappa}_{\mathrm{nom}})^2\\+4f_2^2\bar{\kappa}_{\mathrm{nom}}(1-\bar{\kappa}_2)^2\end{array}}}{2(1-\bar{\kappa}_2)(1-f_2)(1-\bar{\kappa}_{\mathrm{nom}})}, \quad (A.11)$$

$$\Gamma_{1\phi}=\frac{(1-\bar{\sigma}_1)\left[1-\bar{\sigma}_{\mathrm{nom}}(1-2f_1)\right]\pm\sqrt{\begin{array}{l}(1+\bar{\sigma}_1)^2(1-\bar{\sigma}_{\mathrm{nom}})^2\\+4f_1^2\bar{\sigma}_{\mathrm{nom}}(1-\bar{\sigma}_1)^2\end{array}}}{2(1-\bar{\sigma}_1)(1-f_1)(1-\bar{\sigma}_{\mathrm{nom}})}, \quad (A.12)$$

$$\begin{aligned}\kappa_0=\kappa_1&=\kappa_{\phi,\mathrm{nom}}\frac{1+\Gamma_{2\phi}(1-f_2)(\bar{\kappa}_2-1)}{1+\Gamma_{2\phi}(1-f_2)(\bar{\kappa}_2-1)+f_2(\bar{\kappa}_2-1)}\\&=\frac{2\kappa_{\rho,\mathrm{nom}}\left[1+\bar{\kappa}_2+(1-\bar{\kappa}_2)f_2\right]}{(1+\bar{\kappa}_2)(1+\bar{\kappa}_{\mathrm{nom}})\pm\sqrt{\begin{array}{l}(1+\bar{\kappa}_2)^2(1-\bar{\kappa}_{\mathrm{nom}})^2\\+4f_2^2\bar{\kappa}_{\mathrm{nom}}(1-\bar{\kappa}_2)^2\end{array}}},\end{aligned} \quad (A.13)$$

$$\begin{aligned}\sigma_0=\sigma_2&=\sigma_{\phi,\mathrm{nom}}\frac{1+\Gamma_{1\phi}(1-f_1)(\bar{\sigma}_1-1)}{1+\Gamma_{1\phi}(1-f_1)(\bar{\sigma}_1-1)+f_1(\bar{\sigma}_1-1)}\\&=\frac{2\sigma_{\rho,\mathrm{nom}}\left[1+\bar{\sigma}_1+(1-\bar{\sigma}_1)f_1\right]}{(1+\bar{\sigma}_1)(1+\bar{\sigma}_{\mathrm{nom}})\pm\sqrt{\begin{array}{l}(1+\bar{\sigma}_1)^2(1-\bar{\sigma}_{\mathrm{nom}})^2\\+4f_1^2\bar{\sigma}_{\mathrm{nom}}(1-\bar{\sigma}_1)^2\end{array}}}.\end{aligned} \quad (A.14)$$

where $\bar{\kappa}_2 = \kappa_2/\kappa_0$, $\bar{\sigma}_1 = \sigma_1/\sigma_0$, $\bar{\kappa}_{\text{nom}} = \kappa_{\rho,\text{nom}}/\kappa_{\phi,\text{nom}}$, $\bar{\sigma}_{\text{nom}} = \sigma_{\rho,\text{nom}}/\sigma_{\phi,\text{nom}}$, and the $\pm$ sign is consistently chosen so as to ensure the passivity ($\kappa_0 \geq 0, \sigma_0 \geq 0$) and model consistency ($0 < \Gamma_{1\phi} < 1$, $0 < \Gamma_{2\phi} < 1$) conditions. The above solutions contain as free parameters the normalized conductivities $\bar{\kappa}_2$ and $\bar{\sigma}_2$ (subject to the passivity conditions $\bar{\kappa}_2 \geq 0, \bar{\sigma}_2 \geq 0$) as well as the fractions f_1 and f_2 (subject to $0 < f_1 < 1$, $0 < f_2 < 1$, and $f_0 + f_1 + f_2 = 1$). By setting $\bar{\kappa}_2 = \bar{\sigma}_1 = 0$ in Eqs. A.11–A.14 we readily obtain the simplified expressions in Eqs. 11.12 and 11.13. Alternatively, in the opposite asymptotic limit $\bar{\kappa}_2 \gg 1$ and $\bar{\sigma}_2 \gg 1$, we obtain

$$\kappa_1 = \kappa_0 = \kappa_{\phi,\text{nom}} \frac{\Gamma_{2\phi}\left(1 - f_2\right)}{\Gamma_{2\phi}\left(1 - f_2\right) + f_2}, \tag{A.15}$$

$$\sigma_2 = \sigma_0 = \sigma_{\phi,\text{nom}} \frac{\Gamma_{1\phi}\left(1 - f_1\right)}{\Gamma_{1\phi}\left(1 - f_1\right) + f_1}, \tag{A.16}$$

with $\Gamma_{1\phi}$ and $\Gamma_{2\phi}$ still given by Eqs. 11.12 and 11.13. It is also possible to derive approximations valid in mixed limits, such as $\bar{\kappa}_2 \ll 1$ and $\bar{\sigma}_2 \gg 1$, or $\bar{\kappa}_2 \gg 1$ and $\bar{\sigma}_2 \ll 1$.

Appendix B

Details on Coordinate Transformations

As previously mentioned, the boundary values of the mapping functions in Eq. 11.16 are critical to establish the thermal concentration and electrical cloak effects, whereas the function behaviors in between are only subjected to the continuity requirement (so as to avoid additional boundary conditions) as well as to the passivity related constraints $F_\text{t}/\dot{F}_\text{t} \geq 0$, $F_\text{e}/\dot{F}_\text{e} \geq 0$ (see Eq. 11.15). In our specific example, we choose an algebraic cloak transformation (blue curve in Fig. 11.3a)

$$F_\text{e}\left(\rho\right) = R_2 \left(\frac{\rho - R_1}{R_2 - R_1}\right)^\gamma, \quad \gamma > 0, \tag{B.1}$$

which satisfies the required boundary conditions (16), and maps a cylinder of radius R_2 in the auxiliary space onto an annular cylinder of radii R_1 and R_2 in the transformed space; this creates a hole of radius R_1, with no image in the auxiliary space. For $\gamma = 1$, we recover the standard (linear) cloak transformation [3]. Here, we consider instead a different value

$$\gamma = 1 - \frac{R_1}{R_2}, \tag{B.2}$$

which yields

$$\dot{F}_e(R_2) = \frac{F_e(R_2)}{R_2} = 1. \tag{B.3}$$

From Eq. 11.15, this implies

$$\sigma_\rho(R_2) = \sigma_\phi(R_2) = \sigma', \tag{B.4}$$

i.e., the gradual tendency (for $\rho \rightarrow R_2$) of the transformation medium to an isotropic material matched with the background medium. As anticipated, while not strictly necessary, this assumption turns out to be expedient in the inclusion-based implementation.

We then exploit the degrees of freedom in the choice of the concentrator-type mapping function, by enforcing the condition in Eq. 11.17. From Eqs. 11.12 and 11.13, this assumption (i.e., $\bar{\kappa}_{\text{nom}} = 1/\bar{\sigma}_{\text{nom}}$) together with $f_1 = f_2$, yields $\Gamma_{2\phi} = 1 - \Gamma_{1\phi} = \Gamma_{1\rho}$, that is, two types of inclusions with identical shape (just rotated 90°). Once again, albeit not strictly necessary, the above assumption allows more efficient packing strategies in the inclusion-based implementation. From Eq. 11.6, the condition in Eq. 11.17 yields the differential equation

$$\frac{\rho^2 \dot{F}_e(\rho)}{F_e(\rho)} \dot{F}_t(\rho) - F_t(\rho) = 0, \tag{B.5}$$

with the boundary condition $F_t(R_2) = r_2$, which for the cloak transformation in Eq. B.1 can be solved analytically in closed form, yielding the mapping function

$$F_t(\rho) = R_2 \left(\frac{\rho}{R_2}\right)^{\frac{1}{\gamma}} \exp\left[\frac{R_1(R_2 - \rho)}{\gamma R_2 \rho}\right] \tag{B.6}$$

considered in our study (red curve in Fig. 11.3a). From Eq. B.6, the concentration factor is readily obtained

$$c = \frac{F_t(R_1)}{R_1} = e\left(\frac{R_1}{R_2}\right)^{\frac{R_1}{R_2 - R_1}} > 1, \tag{B.7}$$

and the concentrator functionality is verified.

References

1. Moccia, M., Castaldi, G., Savo, S., Sato, Y., and Galdi, V. (2014). Independent manipulation of heat and electrical current via bifunctional metamaterials, *Phys. Rev. X*, **4**, 021025.
2. Leonhardt, U. (2006). Optical conformal mapping, *Science,* **312**, 1777.
3. Pendry, J.B., Schurig, D., and Smith, D.R. (2006). Controlling electromagnetic fields, *Science*, **312**, 1780.
4. Chen, H., Chan, C.T., and Sheng, P. (2010). Transformation optics and metamaterials, *Nat. Mater.*, **9**, 387.
5. Schurig, D., Mock, J.J., Justice, B.J., Cummer, S.A., Pendry, J.B., Starr, A.F., and Smith, D.R. (2006). Metamaterial electromagnetic cloak at microwave frequencies, *Science*, **314**, 977.
6. Werner, D.H., and Kwon, D.H. (eds.) (2013). *Transformation Electromagnetics and Metamaterials* (Springer, Berlin-Heildelberg).
7. Chen, H., and Chan, C.T. (2007). Acoustic cloaking in three dimensions using acoustic metamaterials, *Appl. Phys. Lett.*, **91,** 183518.
8. Chen, H., and Chan, C.T. (2010). Acoustic cloaking and transformation acoustics, *J. Phys. D: Appl. Phys.*, **43**, 113001.
9. Craster, R.V., and Guenneau, S. (eds.) (2013). *Acoustic Metamaterials* (Springer Netherlands, Dordrecht).
10. Cummer, S.A., and Schurig, D. (2007). One path to acoustic cloaking, *New J. Phys.*, **9**, 45.
11. Norris, A.N. (2008). Acoustic cloaking theory, *Proc. R. Soc. A*, **464**, 2411.
12. Zhang, S., Xia, C., and Fang, N. (2011). Broadband acoustic cloak for ultrasound waves, *Phys. Rev. Lett.*, **106**, 024301.
13. Farhat, M., Enoch, S., Guenneau, S., and Movchan, A.B. (2008). Broadband cylindrical acoustic cloak for linear surface waves in a fluid, *Phys. Rev. Lett.*, **101**, 134501.
14. Brun, M., Guenneau, S., and Movchan, A.B. (2009). Achieving control of in-plane elastic waves, *Appl. Phys. Lett.*, **94**, 061903.
15. Farhat, M., Guenneau, S., and Enoch, S. (2009). Ultrabroadband elastic cloaking in thin plates, *Phys. Rev. Lett.*, **103**, 024301.
16. Milton, G.W., Briane, M., and Willis, J.R. (2006). On cloaking for elasticity and physical equations with a transformation invariant form, *New J. Phys.*, **8**, 248.
17. Norris, A.N., and Shuvalov, A.L. (2011). Elastic cloaking theory, *Wave Motion*, **48**, 525.

18. Norris, A.N., and Parnell, W.J. (2012). Hyperelastic cloaking theory: transformation elasticity with pre-stressed solids, *Proc. R. Soc. A*, **468**, 2881.
19. Stenger, N., Wilhelm, W., and Wegener, M. (2012). Experiments on elastic cloaking in thin plates, *Phys. Rev. Lett.*, **108**, 014301.
20. Greenleaf, A., Lassas, M., and Uhlmann, G. (2003). Anisotropic conductivities that cannot be detected by EIT, *Physiol. Meas.*, **24**, 413.
21. Greenleaf, A., Lassas, M., and Uhlmann, G. (2003). On uniqueness for Calderon's inverse problem, *Math. Res. Meas.*, **10**, 685.
22. Yang, F., Mei, Z.L., Jin, T.Y., and Cui, T.J. (2012). DC electric invisibility cloak, *Phys. Rev. Lett.*, **109**, 053902.
23. Yang, F. Mei, Z.L., Yang, X.Y., Jin, T.Y., and Cui, T.J. (2013). A negative conductivity material makes a DC invisibility cloak hide an object at a distance, *Adv. Funct. Mater.*, **23**, 4306.
24. Gömöry, F., Solovyov, M., Šouc, J., Navau, C., Prat-Camps, J., and Sanchez, A. (2012). Experimental realization of a magnetic cloak, *Science*, **335**, 1466.
25. Magnus, F., Wood, B., Moore, J., Morrison, K., Perkins, G., Fyson, J., Wiltshire, M.C.K., Caplin, D., Cohen, L.F., and Pendry, J.B. (2008). A d.c. magnetic metamaterial, *Nat. Mater.*, **7**, 295.
26. Narayana, S., and Sato, Y. (2012). DC magnetic cloak, *Adv. Mater.*, **24**, 71.
27. Sanchez, A., Navau, C., Prat-Camps, J., and Chen, D.X. (2011). Antimagnets: controlling magnetic fields with superconductor–metamaterial hybrids, *New J. Phys.*, **13**, 093034.
28. Sun, F., and He, S. (2013). Static magnetic field concentration and enhancement using magnetic materials with positive permeability, *Prog. Electromagn. Res.*, **142**, 579.
29. Wood, B., and Pendry, J.B. (2007). Metamaterials at zero frequency, *J. Phys.: Condens. Matter*, **19**, 076208.
30. Guenneau, S., and Puvirajesinghe, T.M. (2013). Fick's second law transformed: one path to cloaking in mass diffusion, *J. R. Soc. Interface*, **10**, 20130106.
31. Zeng, L., and Song, R. (2013). Controlling chloride ions diffusion in concrete, *Sci. Rep.*, **3**, 3359.
32. Dede, E.M., Nomura, T., Schmalenberg, P., and Lee, J.S. (2013). Heat flux cloaking, focusing, and reversal in ultra-thin composite considering conduction-conduction effects, *Appl. Phys. Lett.*, **103**, 063501.

33. Guenneau, S., Amra, C., and Veynante, D. (2012). Transformation thermodynamics: cloaking and concentrating heat flux, *Opt. Express*, **20**, 8207.
34. Han, T., Bai, X., Gao, D., Thong, J.T.L., Li, B., and Qiu, C.W. (2014). Experimental demonstration of a bilayer thermal cloak, *Phys. Rev. Lett.*, 054302.
35. Ma, Y., Lan, L., Jiang, W., Sun, F., and He, S. (2013). A transient thermal cloak experimentally realized through a rescaled diffusion equation with anisotropic thermal diffusivity, *NPG Asia Mater.*, **5**, e73.
36. Narayana, S., and Sato, Y. (2012). Heat flux manipulation with engineered thermal materials, *Phys. Rev. Lett.*, **108**, 214303.
37. Schittny, R., Kadic, M., Guenneau, S., and Wegener, M. (2013). Experiments on transformation thermodynamics: molding the flow of heat, *Phys. Rev. Lett.*, **110**, 195901.
38. Xu, H., Shi, X., Gao, F., Sun, H., and Zhang, B. (2014). Ultrathin three-dimensional thermal cloak, *Phys. Rev. Lett.*, **112**, 054301.
39. Kadic, M., Bückmann, T., Schittny, R., and Wegener, M. (2013). Metamaterials beyond electromagnetism, *Rep. Progr. Phys.*, **76**, 126501.
40. Maldovan, M. (2013). Sound and heat revolutions in phononics, *Nature*, **503**, 209.
41. Wegener, M. (2013). Metamaterials beyond optics, *Science*, **342**, 939.
42. Li, J.Y., Gao, Y., and Huang, J.P. (2010). A bifunctional cloak using transformation media, *J. Appl. Phys.*, **108**, 074504.
43. Sihvola, A.H. (1999). *Electromagnetic Mixing Formulas and Applications* (IET Press, London).
44. Mansfield, E.H. (1953). Neutral holes in plane sheet: reinforced holes which are elastically equivalent to the uncut sheet, *Q. J. Mech. Appl. Math.*, **6**, 370.
45. Milton, G.W. (2002). *The Theory of Composites* (Cambridge University Press, Cambridge, UK).
46. He, X., and Wu, L. (2013). Thermal transparency with the concept of neutral inclusion, *Phys. Rev. E*, **88**, 033201.
47. Jarczyk, P., and Mityushev, V. (2012). Neutral coated inclusions of finite conductivity, *Proc. R. Soc. A*, **468**, 954.
48. COMSOL Group, *COMSOL Multiphysics: Version 4.2* (COMSOL, Stockholm, 2011).
49. http://indico.cern.ch/event/99464/material/slides/2?contribId=22
50. http://www.goodfellow.com/E/Stainless-Steel-AISI-302.html

51. Dresselhaus, M.S., Chen, G., Tang, M.Y., Yang, R.G., Lee, H., Wang, D.Z., Ren, Z.F., Fleurial, J.P., and Gogna, P. (2007). New directions for low-dimensional thermoelectric materials, *Adv. Mat.*, **19**, 1043.
52. Li, J.F., Liu, W.S., Zhao, L.D., and Zhou, M. (2010). High- performance nanostructured thermoelectric materials, *NPG Asia Mat.*, **2**, 152.
53. Maldovan, M. (2013). Narrow low-frequency spectrum and heat management by thermocrystals, *Phys. Rev. Lett.*, **110**, 025902.
54. Minnich, A.J., Dresselhaus, M.S., Ren, Z.F., and Chen, G. (2009). Bulk nanostructured thermoelectric materials: current research and future prospects, *Energy Env. Sci.*, **2**, 466.
55. Yu, J.K., Mitrovic, S., Tham, D., Varghese, J., and Heath, J.R. (2010). Reduction of thermal conductivity in phononic nanomesh structures, *Nat. Nanotechnol.*, **5**, 718.
56. Savo, S., Zhou, Y., Castaldi, G., Moccia, M., Galdi, V., Ramanathan, S., and Sato, Y. Reconfigurable anisotropy and functional transformations with VO2-based metamaterial electric circuits, *arXiv*:1405.7743.
57. Wolfram Research. (2010). *Mathematica, Version 8.0* (Wolfram, Champaign, IL).

Chapter 12

Time Reversal of Linear and Nonlinear Water Waves

A. Chabchoub,[a,b] A. Maurel,[c] V. Pagneux,[d] P. Petitjeans,[e] A. Przadka,[e] and M. Fink[c]

[a]*Department of Ocean Technology Policy and Environment, Graduate School of Frontier Sciences, University of Tokyo, Kashiwa, Chiba 277-8563, Japan*
[b]*Department of Mechanical Engineering, School of Engineering, Aalto University, FI-02150 Espoo, Finland*
[c]*Institut Langevin; ESPCI Paris, PSL Research University; CNRS; UPMC-Paris 6, Sorbone University; Univ. UPD-Paris 7; 1 rue Jussieu, 75005 Paris, France*
[d]*Laboratoire d'Acoustique de l'Université du Maine, UMR CNRS 6613, Avenue Olivier Messiaen, 72085 Le Mans, France*
[e]*Physique et Mécanique des Milieux Hétérogènes; ESPCI Paris, PSL Research University; CNRS; UPMC-Paris 6, Sorbone University; Univ. UPD-Paris 7; 10 rue Vauquelin, 75005 Paris, France*

12.1 Introduction

Time reversal (TR) of acoustic, elastic, and electromagnetic waves has been extensively studied in recent years [1, 2]. In a standard TR experiment, waves generated by a source are first measured by an array of antennas positioned around the source and then time reversed and simultaneously rebroadcasted by the same

Transformation Wave Physics: Electromagnetics, Elastodynamics, and Thermodynamics
Edited by Mohamed Farhat, Pai-Yen Chen, Sebastien Guenneau, and Stefan Enoch

ISBN 978-981-4669-95-5 (Hardcover), 978-981-4669-96-2 (eBook)
www.panstanford.com

antenna array. Due to the time invariance of the wave process, the reemitted energy will focus back on the original source, whatever the complexity of the propagation medium [3, 4]. The present paper concentrates on the application of TR to the focusing and manipulation of water waves both in linear and nonlinear regimes. The problem is not as trivial as that with acoustic waves. Let us cite Richard Feynman [5], "Water waves that are easily seen by everyone and which are usually used as an example of waves in elementary courses are the worst possible example, because they are in no respects like sound and light; they have all the complications that waves can have." Water waves are scalar waves; that refers to the evolution of small perturbation of the height of fluid under the action of gravity and surface tension. They are dispersive by nature, nonlinear when generated with standard wave makers, and they experience strong damping at the scale of laboratory experiments. The effect of dispersion on the TR process has already been studied in TR experiments for guided elastic waves [6, 7]; these waves are dispersionless in free space, and the dispersion is due only to the reflection on the boundaries of the waveguide. In the case of water waves, the dispersion is intrinsic but preserves the TR invariance (obviously, not taking the damping into account). The effect of the nonlinearities has been experimentally studied in Ref. [8] for acoustic waves where it has been shown that the TR invariance is preserved as long as nonlinearities do not create dissipation, that is, as long as the propagation distance is smaller than the shock distance. In the case of water waves, the effect of nonlinearities has to be treated. The evolution dynamics in time and space of nonlinear wave trains in deep water can be modeled using the focusing nonlinear Schrödinger equation (NLSE). We will show the implication of the TR invariance on the NLSE and we will demonstrate a way to experimentally focus, both in time and space, rogue waves using the principles of TR mirrors.

12.2 Surface Gravity Water Waves

In this chapter the mathematical and physical preliminaries on water waves are introduced. It relies on the many excellent monographs that exist on the subject, see for instance [9–14].

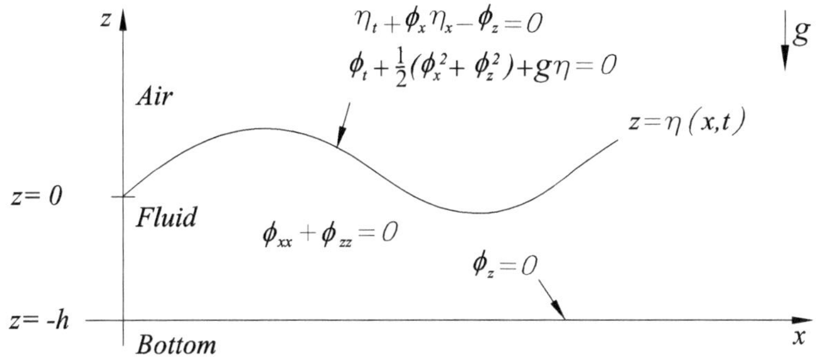

Figure 12.1 Water waves with elevation perturbation η around $z = 0$ and bottom at $z = -h(x, y)$.

We consider the propagation of water waves with local depth $h(x, y)$ for a fluid at rest (Fig. 12.1). We start with the incompressible flow assumption. There are two criteria to be satisfied to assume incompressibility: low Mach number and a timescale associated to the flow much smaller than the sound timescale. The first is obviously highly justified in the context of water waves, since the velocity of the fluid is much smaller than the speed of sound in water $c_{\text{sound}} \simeq 1500$ m/s. The timescale associated to water wave propagation scales as the inverse of the wave speed c_w, so that the second condition reads $c_w/c_{\text{sound}} \ll 1$; as we will see later $c_w \leq \sqrt{gh}$ with g the gravity constant, which tell us that it is sufficient to have $h < 200$ km to assume incompressibility, which is evidently always verified [9].

For incompressible flows, Navier–Stokes equations take the form

$$\begin{cases} \dfrac{\partial \mathbf{u}}{\partial t} + \mathbf{u}\nabla\mathbf{u} = -\dfrac{1}{\rho}\nabla P + \mathbf{g} + \nu\Delta\mathbf{u}, \\ \nabla\mathbf{u} = 0, \end{cases} \tag{12.1}$$

where $\nu = \mu/\rho$ is the kinematic viscosity. Here, we also assume a small effect of the viscosity. This means that we neglect the condition of no slip at the bottom $z = -h$ and that we neglect the shear stress at the free surface $z = \eta$. These boundary layer effects would be a correction responsible of the wave damping [9, 10].

Next, introducing the vorticity $\boldsymbol{\omega} = \nabla \times \mathbf{u}$, we get the Euler equation

$$\begin{cases} \dfrac{\partial \mathbf{u}}{\partial t} + \nabla \left(\dfrac{\mathbf{u}^2}{2} \right) - \mathbf{u} \times \boldsymbol{\omega} = -\dfrac{1}{\rho} \nabla P + \mathbf{g} + \nu \Delta \mathbf{u}, \\ \nabla \mathbf{u} = 0, \end{cases} \tag{12.2}$$

If we assume that the flow is initially irrotational $\omega(t = 0) = 0$, Kelvin's theorem ensures that the flow remains irrotational for subsequent time. This allows to define a velocity potential $\phi(\mathbf{x}, t)$ such that

$$\mathbf{u} = \nabla \phi. \tag{12.3}$$

Euler equations simplify into

$$\begin{cases} \nabla \left(\dfrac{\partial \phi}{\partial t} + \dfrac{\mathbf{u}^2}{2} + \dfrac{P}{\rho} + gz \right) = 0, \\ \nabla \mathbf{u} = \Delta \phi = 0. \end{cases} \tag{12.4}$$

The first equation in Eq. 12.4 leads to the Bernoulli equation

$$\frac{\partial \phi}{\partial t} + \frac{\mathbf{u}^2}{2} + \frac{P}{\rho} + gz = F(t) \tag{12.5}$$

Next, because only the gradient of ϕ has a physical meaning, it can be shifted by any function of time. Thus, applying Eq. 12.5 at the free surface and the shift $\phi \to \phi + P_0 \ t/\rho - \int dt \ F(t)$, with P_0 the ambient atmospheric pressure, we get

$$\frac{\partial \phi}{\partial t} + \frac{\mathbf{u}^2}{2} + g\eta = 0, \quad \text{at } z = \eta. \tag{12.6}$$

Two other boundary conditions have to be accounted for. The first is at the free surface $z = \eta$ and it links the motion of the free surface to the velocity of the fluid. To do this, we use the principle that "what happens at the free surface stays at the free surface." More precisely, the free surface $F(\mathbf{x}, t) = z - \eta(x, y, t) = 0$ is assumed to be a material surface, yielding $(\partial_t + (\mathbf{u}\nabla)) F = 0$. Then, using $(\partial_t + (\mathbf{u}\nabla)) z = v_z = \partial_z \phi$, we obtain

$$\partial_z \phi = \partial_t \eta + \partial_x \phi \ \partial_x \eta + \partial_y \phi \ \partial_y \eta, \quad \text{at } z = \eta. \tag{12.7}$$

The boundary condition at the bottom is the most straightforward to obtain. Since we neglect the viscous boundary layer, it simply states that there is no normal velocity (nonpenetrable condition)

$$\mathbf{n} \times \nabla\phi = 0, \quad \text{at } z = -h, \tag{12.8}$$

with $\mathbf{n}$ the vector normal to the bottom $z = -h$.

Eventually, the problem to be solved is reduced to

$$\begin{cases} \Delta\phi = 0, & \text{in the bulk,} \\ \partial_t\phi + \dfrac{\mathbf{u}^2}{2} + g\eta = 0, & \text{at } z = \eta, \\ \partial_z\phi = \partial_t\eta + \partial_x\phi\ \partial_x\eta + \partial_y\phi\ \partial_y\eta, & \text{at } z = \eta, \\ \mathbf{n} \times \nabla\phi = 0, & \text{at } z = -h(x, y). \end{cases} \tag{12.9}$$

12.2.1 *Linear Approximation*

Here, we denote $k = 2\pi/\lambda$ with λ the typical wavelength of the water waves, a the typical wave amplitude and H the typical depth scale. Linear approximation is submitted to the condition of small generalized Ursell number [11]

$$U_r = \frac{ka}{\tanh(kh)^3} \ll 1. \tag{12.10}$$

In the shallow water approximation, $kh \ll 1$, it becomes the Ursell number, and in the deep water approximation, $kh \gg 1$, it becomes simply the sea slope or steepness ka. Below, we present the water wave equations for transient behaviors (time domain) and for forced behaviors (harmonic regime in the frequency domain).

12.2.1.1 Equations in the time domain

For transient behaviors we keep the time differentiation. Due to linear approximation, the system (12.9) is simplified into

$$\begin{cases} \Delta\phi = 0, & \text{in the bulk,} \\ \partial_t\phi + g\eta = 0, & \text{at } z = 0, \\ \partial_z\phi = \partial_t\eta, & \text{at } z = 0, \\ \mathbf{n} \times \nabla\phi = 0, & \text{at } z = -h(x, y), \end{cases} \tag{12.11}$$

where the explicit nonlinear terms have been suppressed and where the boundary condition at $z = \eta$ has been linearized to $z = 0$. This system can be arranged in terms of ϕ only

$$\begin{cases} \Delta\phi = 0, & \text{in the bulk,} \\ \partial_{tt}\phi + g\partial_z\phi = 0, & \text{at } z = 0, \\ \mathbf{n} \times \nabla\phi = 0, & \text{at } z = -h(x, y). \end{cases} \tag{12.12}$$

12.2.1.2 Harmonic regime and flat bottom

In the harmonic regime, the time dependence is written $\exp(-\mathrm{i}\,\omega t)$ and it will be omitted in the following equation. For a flat bottom, the depth is a constant $h(x, y) = h$, and the boundary condition at the bottom become $\partial_z\phi = 0$, so that the system (12.13) becomes

$$\begin{cases} \Delta\phi = 0, & \text{in the bulk,} \\ \partial_z\phi = \dfrac{\omega^2}{g}\phi, & \text{at } z = 0, \\ \partial_z\phi = 0, & \text{at } z = -h. \end{cases} \tag{12.13}$$

In this simple case, we can look for a modal solution in the form $\phi(x, y, z) = \exp(\mathrm{i}\,\mathbf{kx})\, f(z)$, where $\mathbf{x}$ denotes now the horizontal position vector $\mathbf{x} = (x, y)$. In what follows, we consider a 1D propagation, and the function f is found to satisfy

$$f'' - k^2 f = 0, \quad \text{with } f'(-h) = 0, \text{ and } f'(0) = \frac{\omega^2}{g} f(0). \tag{12.14}$$

The dispersion relation than comes from this eigenvalue problem is

$$\omega^2 = gk \tanh kh. \tag{12.15}$$

The associated eigenfunctions are of the form $f(z) = \cosh k(z + h)$. For a given frequency, the dispersion relation has one real solution associated to propagating surface waves and an infinity of imaginary solutions associated to evanescent waves.

Deep-water waves are characterized by having small wavelengths compared to the water depth, that is, $kh \gg 1$. Since for this case $\tanh kh \approx 1$, the linear dispersion relation for deep-water waves is

$$\omega = \sqrt{gk}. \tag{12.16}$$

Hence, the phase velocity of the waves is

$$c_{\mathrm{pd}} = \frac{\omega}{k} = \sqrt{\frac{g}{k}}, \tag{12.17}$$

while the regular surface elevation of amplitude a, a wave frequency of ω, and a wave number of k is determined by

$$\eta\,(x,\,t) = a\cos\,(kx - \omega t)\,. \tag{12.18}$$

Moreover, slowly modulated deep-water periodic-wave train envelopes propagate with the group velocity

$$c_{\mathrm{g}} = \frac{\mathrm{d}\,\omega}{\mathrm{d}\,k} = \frac{1}{2}\sqrt{\frac{g}{k}} = \frac{\omega}{2k} = \frac{1}{2}c_{\mathrm{pd}}. \tag{12.19}$$

For the deep-water case, it is possible to approximate the evolution of linear wave packets within the framework of envelope evolution equation. The latter can be derived heuristically by expanding the dispersion relation (Eq. 12.16) about $k = k_0$ [15, 16]

$$\omega\,(k) = \omega(k_0) + \left(\frac{\partial\omega}{\partial k}\right)(k - k_0) + \frac{1}{2}\,(k - k_0)^2\left(\frac{\partial^2\omega}{\partial k^2}\right)\Big|_{k=k_0}. \tag{12.20}$$

Now, we define the slowly varying wave number and wave frequency to be $K := k - k_0$ and $\Omega = \omega - \omega_0$, respectively. Thus, Eq. 12.20 becomes

$$\Omega - \left(\frac{\partial\omega}{\partial k}\right)K - \frac{1}{2}K^2\left(\frac{\partial^2\omega}{\partial k^2}\right)\Big|_{k=k_0} = 0. \tag{12.21}$$

Applying the Fourier transform gives

$$E(K,\,\Omega) = \mathcal{F}\left[E\,(x,\,t)\right] = \int_{-\infty}^{+\infty} E\,[x,\,t]\exp\left[\mathrm{i}\,(\Omega t - Kx)\right]\mathrm{d}\,x\,\mathrm{d}\,t, \tag{12.22}$$

while the inverse Fourier transform provides

$$\begin{aligned} E(x,\,t) &= \mathcal{F}^{-1}\left[E\,(K,\,\Omega)\right] \\ &= \frac{1}{(2\pi)^2}\int_{-\infty}^{+\infty} E\,[K,\,\Omega]\exp\left[-\,\mathrm{i}\,(\Omega t - Kx)\right]\mathrm{d}\,K\,\mathrm{d}\,\Omega, \end{aligned} \tag{12.23}$$

It is straightforward to verify from Eqs. 12.22 and 12.23 that

$$E_x = \mathrm{i}\,K\mathcal{F}^{-1}\left[E\,(K,\,\Omega)\right];\; E_t = -\,\mathrm{i}\,\Omega\mathcal{F}^{-1}\left[E\,(K,\,\Omega)\right]; \tag{12.24}$$

Applying the operator (Eq. 12.21) on the envelope function E and using the derivative expression (Eq. 12.24), we obtain its linear evolution dynamics in space and time

$$-\mathrm{i}\left(E_t+\frac{\omega}{2k}E_x\right)+\frac{\omega}{8k^2}E_{xx}=0. \tag{12.25}$$

Equation 12.25 is also referred to as the linear Schödinger equation. For a given linear solution of Eq. 12.25, the corresponding surface elevation to first order of approximation is determined by

$$\eta(x,t)=\mathrm{Re}\left(E\left(x,t\right)\cdot\exp\left[\mathrm{i}\left(kx-\omega t\right)\right]\right). \tag{12.26}$$

For shallow-water waves, the wavelength is large compared with the water depth. Therefore, $kh \ll 1$. Hence, $\tanh kh \approx kh$, which means that the dispersion relation for shallow-water waves is

$$\omega=\sqrt{gk^2h}. \tag{12.27}$$

Then, the phase velocity is

$$c_{\text{ps}}=\frac{\omega}{k}=\sqrt{gh}. \tag{12.28}$$

Since shallow-water waves are nondispersive, the group velocity is equal to the phase velocity.

Whatever the depth (deep water, shallow water, or intermediate depth) the phase velocity is

$$c_{\text{ps}}=\frac{\omega}{k}=\sqrt{gh}\sqrt{\frac{\tanh kh}{kh}}. \tag{12.29}$$

Then, since $\tanh a \leq a$, it shows that the shallow-water expression of the phase velocity is a maximum, in the sense that we always have

$$c_{\text{ps}}\leq\sqrt{gh}. \tag{12.30}$$

12.2.1.3 2D equation in the harmonic regime for a flat bottom

The vertical mode that we have defined previously is useful to obtain a 2D equation for water waves, analogous to the Helmholtz 2D equation for acoustic waves. We know from (12.13) that the equation for the potential in the bulk is

$$\frac{\partial^2\phi}{\partial x^2}+\frac{\partial^2\phi}{\partial y^2}+\frac{\partial^2\phi}{\partial z^2}=0. \tag{12.31}$$

Thus, for a geometry that is separable in (x, y) and z, it is possible to decompose the potential ϕ with the vertical mode $f(z)$ by writing

$$\phi(x, y, z) = f(z)\Psi(x, y).$$

Then, from the previous equation, we obtain

$$\frac{\partial^2 \Psi}{\partial x^2} + \frac{\partial^2 \Psi}{\partial y^2} + k^2 \Psi = 0. \tag{12.32}$$

This is the 2D Helmholtz equation and it is exact—in the linear regime—for separable geometries, that is, with a flat bottom and lateral vertical walls. The boundary conditions at these vertical walls correspond to vanishing normal velocity and thus to

$$\partial_n \Psi = 0. \tag{12.33}$$

Besides, the relation between the potential at the surface and the elevation perturbation, $\partial_z \phi(x, y, 0) = -\mathrm{i}\,\omega\eta(x, y)$, allows us to express the 2D problem for η:

$$\frac{\partial^2 \eta}{\partial x^2} + \frac{\partial^2 \eta}{\partial y^2} + k^2 \eta = 0 \tag{12.34}$$

with Neumann boundary conditions on the vertical walls

$$\frac{\partial \eta}{\partial n} = 0. \tag{12.35}$$

Eventually, it appears that, in the harmonic regime for separable geometries, there is a perfect analogy between water waves and acoustic waves (as well as with transverse electromagnetic waves).

12.2.1.4 Time reversal invariance in the linear regime

From the previous considerations that have shown the analogy with 2D acoustics we can conclude immediately that water waves possess the TR invariance when the geometry is separable.

However, there is simpler and more general argument. We remind the linear equations in the time domain

$$\begin{cases} \Delta\phi = 0, & \text{in the bulk,} \\ \partial_{tt}\phi + g\partial_z\phi = 0, & \text{at } z = 0, \\ \mathbf{n}\cdot\nabla\phi = 0, & \text{at } z = -h(x, y). \end{cases} \tag{12.36}$$

It is valid in any geometry, with any shape of the bottom, and it is obviously invariant with respect to the symmetry $t \to -t$. It means that linear water waves are TR invariant in any geometry (as long as the damping is neglected).

12.2.2 *Nonlinear Regime*

The propagation of nonlinear waves in one spatial dimension and in deep-water condition can be described within the framework of weakly nonlinear Stokes waves [17]. These waves have the property to have flatter troughs and sharper peaks. Another fundamental property is their instability to side-band perturbation [18]. The dynamics of these waves in both, stationary and unstable regime can be accurately described by nonlinear Schrödinger-type evolution equations [19, 20]. Being an integrable evolution equation, exact envelope model solutions of the NLSE can be used to model exact localized structures on the water surface [21].

12.2.2.1 Stokes waves and modulation instability

Using the perturbation Ansatz in the small steepness parameter $\varepsilon := ka$, Stokes found weakly nonlinear periodic solutions, which satisfy the nonlinear governing equations of an ideal fluid [17]. The weakly nonlinear surface elevation to second order in steepness is determined by

$$\eta(x, t) = a\cos(kx - \omega t) + \frac{1}{2}ka^2\cos\left[2(kx - \omega t)\right] + \ldots \quad (12.37)$$

The dispersion relation is then corrected to

$$\omega = \sqrt{gk}\left(1 + \frac{a^2k^2}{2}\right). \quad (12.38)$$

Considering the corrected dispersion relation (Eq. 12.38), we can notice the dependency of the wave velocity with respect to the amplitude. This amplitude dependency may engender a nonlinear focusing of Stokes waves under certain conditions. This instability of periodic deep-water wave trains is also referred to as the Benjamin–Feir instability [18], the side-band instability or the Bespalov–Talanov instability [22]. This instability has been discovered at the same time by various researcher [12]. A geometric condition for the instability was already provided by Lighthill [9]. Benjamin and Feir [18] investigated the stability of Stokes waves theoretically and experimentally, while Zakharov [19] derived the same result using a Hamiltonian approach. In the same work he derived an equation, which describes the evolution in time and space of slowly modulated weakly nonlinear wave trains in deep water, the NLSE, which will be discussed in the next section. Benjamin and Feir examined the second-order Stokes solution to side-band disturbances [18, 23, 28], where the disturbed surface elevation has the form

$$\tilde{\eta}(x,t) = \eta(x,t) + \epsilon(x,t), \tag{12.39}$$

while the side-band disturbance is given by

$$\begin{aligned}\epsilon(x,t) = {} & \epsilon_{+} \exp(\gamma t) \cos\left[k(1+\kappa)x - \omega(1+\delta)t\right] + \\ & \epsilon_{-} \exp(\gamma t) \cos\left[k(1-\kappa)x - \omega(1-\delta)t\right].\end{aligned} \tag{12.40}$$

Here, $\epsilon_{\pm}$ denote small amplitudes of the perturbation, whereas $\delta = \Omega/\omega$ and $\kappa = K/k$ are small perturbations fractions of the wave frequency and wave number, respectively. After substitution of the perturbed wave equation $\tilde{\eta}(x,t)$ in the governing equations (11) and performing a linear stability analysis of the obtained dynamical system provides to the instability growth rate

$$\gamma = \frac{1}{2}\delta\sqrt{2k^2a^2 - \delta^2}\omega \tag{12.41}$$

Therefore, the disturbances grow exponentially with time for real values of γ. That is, whenever γ is real, the instability exists in a limited range of modulation frequencies:

$$0 < \delta < \sqrt{2}ka. \tag{12.42}$$

Therefore, for a given wave steepness ε of a regular deep-water wave train, there exists an unstable frequency range, centered around the main frequency ω, for which the growth rate γ is real and positive. In this case, disturbances grow exponentially with time. The side-band instability was also experimentally observed in a large wave facility shortly after its theoretical discovery [24, 25].

12.2.2.2 Nonlinear Schrödinger equation and doubly localized breather-type solutions

An alternative to describe the exact dynamics of Stokes waves is provided by the theory of nonlinear wave packet evolution equations. The simplest equation for nonlinear wave packets of this type is known as the NLSE and can be derived heuristically by extending the results Taylor expansion, presented for the linear case, by including the dependency of the dispersion relation with respect to the amplitude, as described in Eq. 12.38. That is,

$$\omega(k) = \omega(k_0) + \left(\frac{\partial \omega}{\partial k}\right)(k-k_0) + \frac{1}{2}(k-k_0)^2\left(\frac{\partial^2 \omega}{\partial k^2}\right)\Big|_{k=k_0} + \left(\frac{\partial \omega}{\partial |a|^2}\right)\Big|_{|a|^2=0} |a|^2 . \tag{12.43}$$

Therefore, Eq. 12.25 is corrected to

$$-\mathrm{i}\left(E_t + \frac{\omega}{2k}E_x\right) + \frac{\omega}{8k^2}E_{xx} + \frac{\omega k^2}{2}|E|^2 E = 0, \tag{12.44}$$

which the NLSE. Now, weakly nonlinear surface elevation to first order of approximation is determined by

$$\eta(x,t) = \mathrm{Re}\left(E(x,t)\cdot \exp\left[\mathrm{i}(kx-\omega t)\right]\right). \tag{12.45}$$

The latter equation can be derived more rigorously from the governing equations, using the multiple scales technique, see [10, 26]. The interesting fact about the NLSE, is that its stationary solution $E_S(x,t) = a\exp\left(-\mathrm{i}\frac{a^2k^2}{2}\omega t\right)$ describes a second-order Stokes waves and same instability condition (Eq. 12.42) can be derived, by perturbing the stationary Stokes envelope $E_S(x,t)$ by a modulation frequency Ω and a modulation wave number K. Therefore, the considered perturbed envelope function becomes

$$\tilde{E_S}(x,t) = E_S(x,t)\left[1 + \varepsilon_- \exp\left(-\mathrm{i}K\left(x - \frac{\omega}{2k}t\right) - \mathrm{i}\Omega t\right) + \varepsilon_+ \exp\left(\mathrm{i}K\left(x-\frac{\omega}{2k}t\right) + \mathrm{i}\Omega t\right)\right]. \tag{12.46}$$

The criterion (Eq. 12.42) is then obtained after inserting the perturbed Stokes envelope in the NLSE, linearizing and following the modulation dispersion relation

$$\Omega^2 = \frac{\omega^2}{8k^2}\left(\frac{K}{8k^2} - k^2a^2\right)K^2, \tag{12.47}$$

see [10, 11, 27] for details. An exponential growth takes place, when the modulation frequency is imaginary. This is the case, when

$$0 < \frac{K}{k} < 2\sqrt{2}ka. \tag{12.48}$$

Since for deep water $c_g = \frac{\Omega}{K} = \frac{\omega}{2k}$, thus, $\delta = \frac{\Omega}{\omega} = \frac{1}{2}\frac{K}{k} = \frac{1}{2}\kappa$, Eq. 12.48 is clearly equivalent to Eq. 12.42. The modulation instability can be discussed within the framework of exact solutions of the NLSE. Besides stationary wave envelope soliton solutions, the NLSE admits a family of breather solution on a constant Stokes background, known to model strong localizations of Stokes waves [29–33]. For simplicity, we scale the NLSE to a dimensional form [12]

$$\mathrm{i}\,\psi_T + \psi_{XX} + 2\,|\psi|^2\,\psi = 0. \tag{12.49}$$

The solution, which describes the modulation instability dynamics in space and time is the family of space-periodic Akhmediev solutions [29], expressed as

$$\psi_A\,(X,\,T) = \frac{\cosh\,(\Omega T\,-\,2\,\mathrm{i}\,\varphi) - \cos\,(\varphi)\cos\,(pX)}{\cosh\,(\Omega T) - \cos(\varphi)\cos\,(pX)}\exp\,(2\,\mathrm{i}\,T), \tag{12.50}$$

where $\Omega = 2\sin(2\varphi)$, $p = 2\sin(\varphi)$ and $\varphi \in \mathbb{R}$. In the limiting case of infinite modulation period, which occurs when the breather parameter $\varphi \to 0$, is known as the Peregrine solution [33]

$$\psi_1\,(X,\,T) = \left(-1 + \frac{4 + 16\,\mathrm{i}\,T}{1 + 4X^2 + 16T^2}\right)\exp\,(2\,\mathrm{i}\,T), \tag{12.51}$$

Here, the modulation wavelength $K \approx 0$ and the growth rate becomes algebraic and is not of exponential nature anymore. This latter solution is very particular, since it is localized in both space and time and it amplifies the amplitude of the carrier theoretically by an exact factor of three. A higher-order solution of this kind is referred to as the Akhmediev–Peregrine solution [29, 34, 35]. This doubly localized solution is characterized by a maximal amplitude amplification of five (see Fig. 12.2 for an illustration) and is defined by

$$\psi_2\,(X,\,T) = \left(1 + \frac{G + \mathrm{i}\,H}{D}\right)\exp\,(2\,\mathrm{i}\,T), \tag{12.52}$$

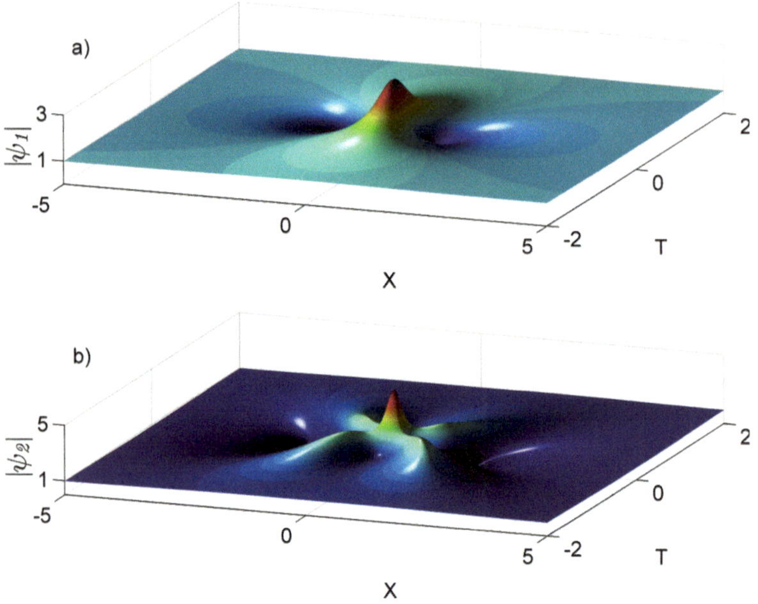

Figure 12.2 (a) First-order doubly localized rational solution (Peregrine breather), which at $X = T = 0$ amplifies the amplitude of the carrier by a factor of 3. (b) Second-order doubly localized rational solution (Akhmediev–Peregrine breather), which at $X = T = 0$ amplifies the amplitude of the carrier by a factor of 5.

while

$$G = -\left(X^2 + 4T^2 + \frac{3}{4}\right)\left(X^2 + 20T^2 + \frac{3}{4}\right) + \frac{3}{4} \tag{12.53}$$

$$H = 2T\left(3X^2 - 4T^2 - 2\left(X^2 + 4T^2\right)^2 + \frac{15}{8}\right) \tag{12.54}$$

$$D = \frac{1}{3}\left(X^2 + 4T^2\right)^3 + \frac{1}{4}\left(X^2 - 12T^2\right)^2 + \frac{3}{64}\left(12X^2 + 176T^2 + 1\right). \tag{12.55}$$

All these basic breather solutions, which describe the strong focusing of a regular background have been only recently observed in Kerr media, water waves and plasma [36–42]. Breathers attracted the significant scientific interest, since they describe the backbone dynamics of the modulation instability and therefore, of rogue wave as well.

12.2.2.3 Time reversal invariance in the nonlinear regime

The TR invariance of Stokes waves can be discussed within the framework of the NLSE. In fact, considering the weakly nonlinear evolution equation (Eq. 12.44), we can notice that the time-dependent part of the NLSE equation, contains a term in $i\frac{\partial E}{\partial t}$. Therefore, it is straightforward to see that if $E\,(x, t)$ is a solution, then its complex conjugate $E^*\,(x, -t)$ is also a solution. Consequently, both surface elevations $\eta\,(x, t)$ and $\eta\,(x, -t)$ describe two possible evolutions for the weakly nonlinear water wave problem. Consequently, a TR mirror can be used to create the time-reversed wave field $\eta\,(x, -t)$ in the whole propagating medium. For this 1D problem it is sufficient to measure the wave field $\eta\,(x, t)$ at one unique point x_M and to rebroadcast the time-reversed signal $\eta\,(x_M, -t)$ from this mirror point in order to observe the solution $\eta\,(x, -t)$ in the whole medium. If this approach is valid for breather dynamics, it would be a confirmation of the TR invariance of strongly localized waves and therefore for nonlinear waves as well.

12.3 Experiments of Time Reversal

In this section we describe recent experiments, confirming the TR invariance of linear [43] and nonlinear surface gravity waves [44].

12.3.1 *Time Reversal of Linear Water Waves*

In this section we will start reporting the results of the TR of water waves in the linear regime. The experiment is conducted in a water tank cavity to take advantage of multiple reflections on the boundaries. The influence of the number of channels in the TR mirror is studied and it allows us to show that a small number of channels is sufficient to obtain the TR refocusing owing to the reverberating effect of the cavity.

The reverberating tank is filled with water with depth at rest $H = 10$ cm. The dimension of the rectangular tank is 53×38 cm^2 with obstacles placed in order to break the spatial symmetries (see Fig. 12.3). The waves are generated by using a vertical conical vibrator and recorded by using an optical method (note that

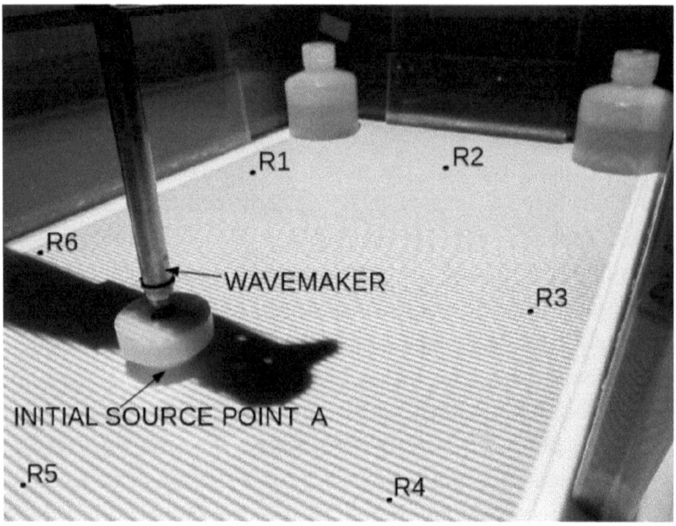

Figure 12.3 Water tank used in the experiment. Point A is at the initial source position. Note the fringes used for the Fourier transform profilometry technique. Points R_1 to R_6 correspond to the positions of the different channels of the TR mirror. At each of these channel positions the wave is measured and can be reemitted after the TR operation. Reprinted (figure) with permission from Ref. [43]. Copyright (2012) by the American Physical Society.

this differs from TR experiments in acoustics that use the same transducer to record and to regenerate the wave). We perform a typical one channel TR experiments in two steps. In the first step of emission, a wave packet is generated from a conical vibrator that can be considered as a point source. In a second step, the signal recorded at a receiver point is time-reversed and reemitted.

The time-reversed wave is expected to refocus spatially at the source point and refocus temporally at the recompression time. The key point to ensure TR refocusing is the number of cavity modes that the wave packet has been able to excite in the first step of the TR.

After a few experimental trials, the authors found the central frequency $f_0 = 5$ Hz the best to obtain good refocusing in time and space. This is a compromise between the bulk dissipation that grows with a frequency increase and the bottom friction that becomes significant while decreasing frequency (note: no significant

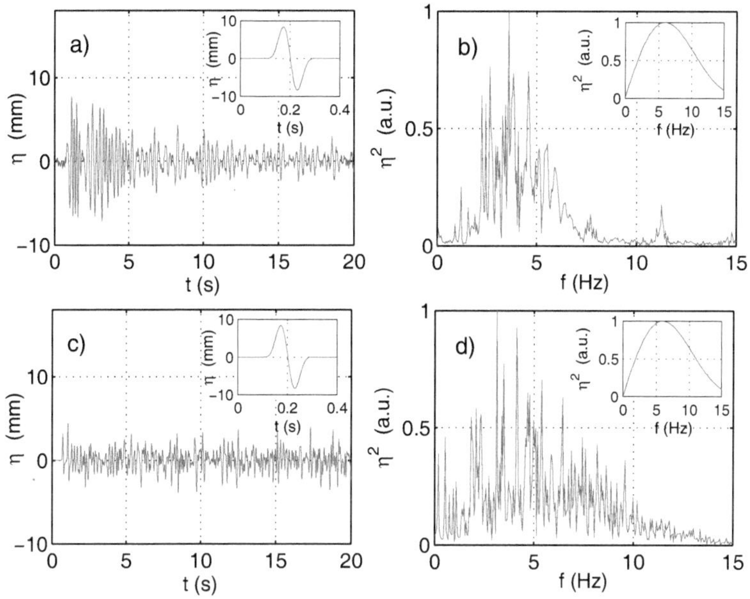

Figure 12.4 (a) Experimental measurement of the temporal evolution of the surface elevation, $\eta(\mathbf{r_1}, t)$ during the forward propagation after emission from point A. The inset shows the signal emitted from point A. (b) Corresponding spectrum. The inset shows the spectrum of the signal emitted from point A. (c, d) The same representation from numerical simulations of the wave equation. Reprinted (figure) with permission from Ref. [43]. Copyright (2012) by the American Physical Society.

peaks in the low-frequency region in Fig. 12.4b). Figure 12.4a shows the signal recorded at one point (point R_1 in Fig. 12.3) when a one-period sinusoidal pulse centered at f_0 Hz is generated at the initial source position. The duration of the signal is typically 20 s, corresponding both to reverberating effects and linear dispersion effects. This latter is given by the linear dispersion relation for water wave propagation (taking into account the effects of finite depth H and surface tension γ):

$$\omega^2 = \left(gk + \frac{\gamma}{\rho}k^3\right)\tanh kH, \qquad (12.56)$$

where k denotes the complex wave number, g the gravity acceleration and ρ the water density. The wavelength at the central frequency f_0 is $\lambda = 6$ cm. The magnitude of nonlinearity of the waves based on the maximum measured gradient of surface elevation was found to be $\epsilon = 0.13$. The attenuation is such that the wave can propagate over roughly 100 wavelengths, that is, about 10 to 20 times the length scale of the cavity. This is consistent with the 20 s duration of the time signal recorded at one point in the cavity (Fig. 12.4a) since the phase velocity at the central frequency is 0.33 m/s corresponding to 12–17 reflections from the boundaries. The spectrum of the signal recorded during the direct propagation is shown in Fig. 12.4b. It presents several peaks (one can count roughly 20 peaks) corresponding to the eigenmodes of the cavity that have been excited. Although the initial pulse covers a broadband frequency range [0 15] Hz, the signal recorded is limited to frequencies smaller than about 10 Hz. We have checked that this is an effect of the attenuation: direct numerical simulations of the 2D wave equation in the same geometry but omitting the attenuation give a spectrum with around 100 cavity modes excited in the whole range 0 15 Hz (Fig 12.4c,d).

We now investigate the refocusing. The perturbation of the surface elevation $\eta(\mathbf{r}, t)$ is measured in time and in space during the wave propagation using an optical method (FTP for Fourier transform profilometry) that has been adapted recently to water wave measurements [45–48]. FTP is used to measure the whole pattern of surface elevation $\eta(\mathbf{r}, t)$ at *each time* of the reverse propagation. This has been done in one-channel experiments ($N = 1$). Although the spatial refocusing and temporal recompression are visible (Fig. 12.5), it is not possible to distinguish the converging wave fronts before the recompression and the diverging wave fronts after recompression in this one channel experiments (for a movie, see the supplementary material).

To improve the refocusing, it is possible to increase the number of channels. In a TR experiments with multiple channels, the signal emitted from the source point is recorded at N receiver points. The TR signal are then reemitted simultaneously from the N receiver points. If the N receiver points are uncorrelated, it is meant to improve the refocusing since the wave experiences many different trajectories in the cavity. In our experiment, rather than using N wave generators to reemit the signal, the N channel TR have been done with just one wave generator. This is possible by exploiting the

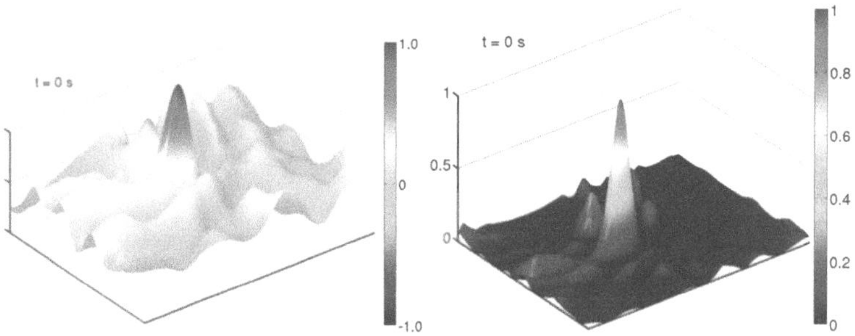

Figure 12.5 Measured amplitude, $\eta(\mathbf{r}, t = 0)$, and intensity, $\eta(\mathbf{r}, t = 0)^2$ of the time-reversed wave around the initial source position (point A) where the wave is refocused at the recompression time ($t = 0$) in a one-channel TR experiment ($N = 1$). The spatial range around the refocusing point A is 34×26 cm^2. Reprinted (figure) with permission from Ref. [43]. Copyright (2012) by the American Physical Society.

linearity of the problem which implies that the wave field excited by N generators is equal to the sum of the N wave fields excited by each generator alone. We have checked this linearity by comparing the time signal recorded at the refocalization point when using 2 channels and when summing the two signals obtained in two single channel experiments (see supplementary material). The temporal signals in both cases coincide, with less than 10% discrepancy in a $10/f_0$ time window centered at the recompression time. Six different positions of the receiver point have been used (points R_1 to R_6 in Fig. 12.3). Figure 12.6 shows a time sequence of the reverse propagation for the 6 channel TR. As expected, the peak at the recompression time is much higher than in the one channel TR, confirming that the channels are uncorrelated. Besides, the converging and diverging wave fronts, before and after the recompression time are visible.

To gain insights into the study of the quality of the refocusing, we want to inspect both the effect of the number of channels and the effect of the damping. To characterize the quality of the spatial refocusing, we define a peak-to-noise ratio (PNR) at the recompression time as the ratio of the maximum intensity at the focal spot to the mean side-lobe intensity. Experimental results are shown in Fig. 12.7 (black points). With a single channel, the PNR is typically 20, a value that is related to the number of cavity modes that can be resolved from the spectrum at the receiver

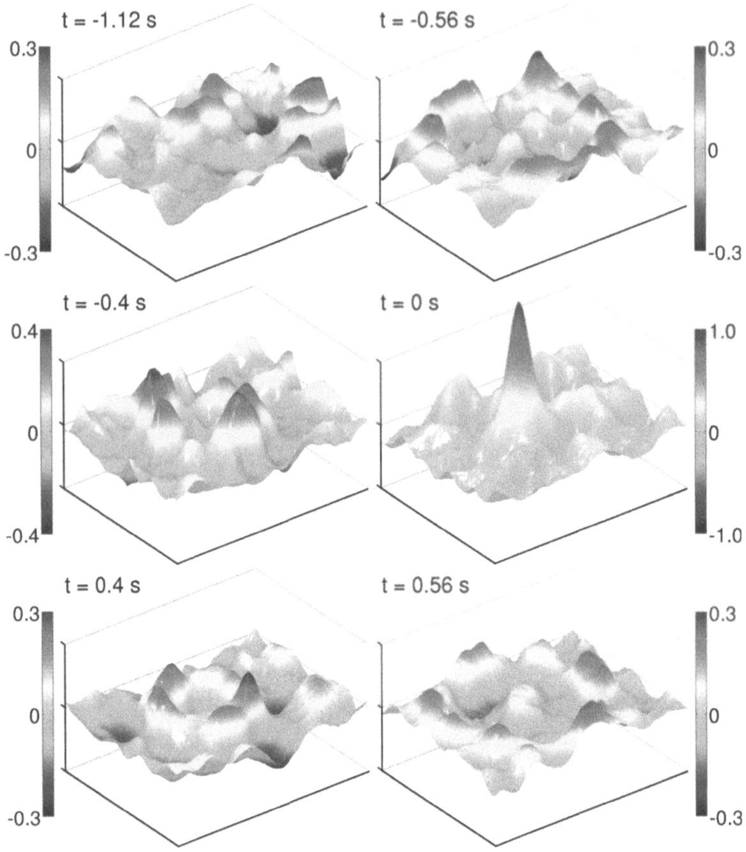

Figure 12.6 Space–time-resolved experimental measurements of the surface elevation $\eta(\mathbf{r}, t)$ during the refocusing of the time-reversed wave. In this case, $N = 6$ channels (points R_1 to R_6) reemit the time-reversed signals. The recompression time is at $t = 0$ s. Converging and diverging wave fronts appear respectively for negative and positive times. The spatial range around the refocusing point A is 34×26 cm^2. Reprinted (figure) with permission from Ref. [43]. Copyright (2012) by the American Physical Society.

point in Fig. 12.4. With several channels, the PNR increases linearly with the number N of channels [49]. Although this behavior is expected without damping, it was not obvious to be verified with the typical range of damping of our experiment. The insets of Fig. 12.7

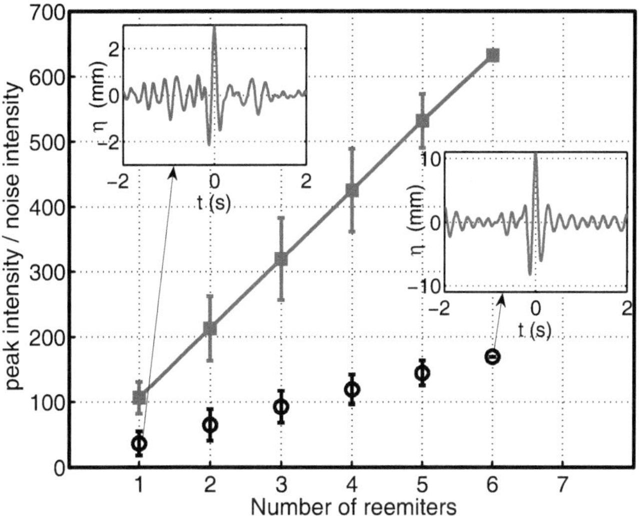

Figure 12.7 Experimentally measured peak-to-noise ratio as a function of the number N of channels (black points). For comparison, the blue squares show numerical results with negligible damping. Note the standard deviation accounting for the sensitivity of the refocusing to the position of the reemission point. The insets present the experimental temporal refocalization while using one and six channels. Reprinted (figure) with permission from Ref. [43]. Copyright (2012) by the American Physical Society.

show the temporal recompression for $N = 1$ and $N = 6$ at the refocusing point A. The refocusing is clearly visible in the one-channel TR experiment but with higher temporal side lobes than with 6 channels. Note that these temporal signals allow also to define a PNR and we observed that PNR either defined in space or in time have roughly the same values. Varying the damping is more difficult. To perform experiments where the damping effect is negligible would necessitate much larger size of the cavity (e.g., the size of a swimming pool) because the attenuation per wavelength decreases with the frequency [9]. Therefore, numerical simulations have been used to model the case with negligible damping. The results are shown in Fig. 12.7 (blue square) where computations have been done by taking the same protocol as in the experiment. The same trends as in the experiment are observed: i) for $N = 1$, the PNR is equal to 100 and is given by the number of excited cavity modes that can be resolved in the spectrum in Fig. 12.4, ii) the PNR increases linearly with N. With about 20 excited modes in our laboratory experiments, versus the 100 modes obtained in the numerics, it appears that the refocusing can be significantly reduced because of the attenuation occurring at that laboratory scale.

Our experiments illustrate the feasibility of a few channel TR refocusing for gravity capillary waves. This has been performed in a well-controlled laboratory context that allows quantitative measurements simultaneously in time and space. At this laboratory scale, with centimetric wavelengths, the quality of the refocusing is limited by the damping due to viscous effects but it is not suppressed. At larger scales, viscous damping highly decreases and numerical simulations show that the refocusing is greatly improved. Thus, this paves the way to applications in the context of water waves in the sea, with very small damping, where very high quality of refocusing is expected.

12.3.2 *Time Reversal of Nonlinear Water Waves*

Next, an experimental demonstration of TR of nonlinear waves is presented. Due to the strong nonlinear focusing of NLSE breathers, the doubly localized Peregrine and the higher-order Akhmediev–Peregrine breathers solutions will be used for the demonstration in one spatial direction of wave propagation. Following the 1D NLSE (Eq. 12.44), the experiments, performed in a unidirectional wave basin, are first started by generating the maximal amplitude of hydrodynamic and doubly localized breather by the wave maker, which is considered to be the source of the propagation. Its position is labeled by x_S. The attenuated breather-type wave field is then measured at the mirror position, labeled by x_M, after a specific propagation distance. The collected signal is then reversed in time, providing new initial conditions to a wave generator. As for the linear experiment, if the TR symmetry is valid, one should expect the refocusing and the perfect reconstruction of the initial maximal breather compression, after reemitting the time-reversed signal. Due to the experimental limitations with respect to the generation of the time-reversed signal at the mirror position, we will use the spatial reciprocity of the NLSE in order to reemit the time-reversed surface elevation signal of the attenuated breather from the source and observe the refocusing at the mirror position x_M, instead of rebroadcasting the reversed wave field from x_M and expecting refocusing at the source position x_S.

The experiments are performed in a unidirectional and 15 m long wave flume facility with a constant depth of $h = 1$ m. A computer-controlled and hydraulic single flap generates surface gravity Stokes waves at one end of the flume at a position we refer to be the source of pulse generation x_S. A capacitance wave gauge measures the surface elevation along the flume at a position of interest with a sampling frequency of 500 Hz. The wave gauge is placed and fixed 9 m from the wave generator, keeping enough distance from the absorbing beach, which is installed at the other end of the facility in order to avoid wave reflections, which would obviously distort the dynamics of the generated waves. The position of the wave gauge is considered to be the mirror position x_M. The experimental setup is depicted in Fig. 12.8.

In order to provide ideal experimental conditions, the walls of the flume were therefore properly cleaned and the water was filtered accordingly before performing the experiments in order to minimize the dissipation effects, which have a significant influence on NLSE localized structures, propagating in a water wave flume.

First, the dimensional doubly localized wave profiles of the breathers, satisfying Eq. 12.44, at their maximal amplitude amplification are generated. These initial wave profile amplitudes are amplified by a factor of three for the Peregrine and of five for the Akhmediev–Peregrine solution. The parameters of the corresponding carrier wave have to be carefully chosen in order to

Figure 12.8 Schematic upper view of the unidirectional wave basin. The single flap, driven by a hydraulic cylinder, is installed at the right end of the wave flume at x_S. The wave gauge is located at x_M, 9 m from the wave maker. The absorbing beach is displayed at the left end of the wave facility.

avoid initial wave breaking of obviously strongly nonlinear focused waves at x_S and in order to reach satisfactory attenuation after 9 m of propagation. Consequently, the carrier wave parameters, determined by the steepness ε and the amplitude a, have to be chosen accordingly. Both parameters are set to be $\varepsilon = 0.09$ and $a = 3$ mm for the Peregrine as well as $\varepsilon = 0.03$ and $a = 1$ mm for the Akhmediev–Peregrine solution. These chosen steepness values are far from the experimentally determined wave breaking thresholds [50]. The steepness ε and the amplitude a of the Stokes background are sufficient to determine the carrier parameters. Trivially, the wave number is $k = \frac{\varepsilon}{a}$. The linear dispersion relation of deep water determine the wave frequency to be $\omega = \sqrt{gk}$, while denotes the gravitational acceleration and is $g = 9.81\,\mathrm{m}\cdot\mathrm{s}^{-2}$. The initial conditions are then trivially determined by

$$\eta(x^*, t) = \mathrm{Re}\left(E\left(x^*, t\right)\cdot\exp\left[\mathrm{i}\left(kx^* - \omega t\right)\right]\right), \qquad (12.57)$$

evaluated at $x^* = 0$, in order to satisfy the maximal breather compression. Figure 12.9 shows the initial conditions applied to the flap at the Position x_S.

After generating the Peregrine and the Akhmediev–Peregrine breather at their maximal wave amplitude of 0.9 cm and 0.5 cm, respectively, we collect the wave profiles after having being declined in amplitude, as expected and predicted by theory, by use of the wave gauge at x_M, that is, 9 m from the flap Position x_S. Here, we notice already a deviation from theoretical weakly nonlinear wave profiles, due to the strongly nonlinear nature of the initially generated wave forms. These discrepancies can be explained by higher-order NLSE, known as the modified NLSE [20]. The corresponding data are shown in Fig. 12.10a,b.

In the third step these recorded breather signals are revised in time. These time-reversed signals provide now new initial conditions to the flap in order to initiate the fourth and penultimate stage of the TR experiment. The latter are depicted in Fig. 12.10c,d. If the considered breather dynamics is indeed TR invariant, it is expected to observe the refocusing of the maximal wave compressions after reemitting the latter time-reversed and slightly modulated signals to the wave maker. Here, we take advantage of the spatial reciprocity of the NLSE equation, thus, we can reemit the

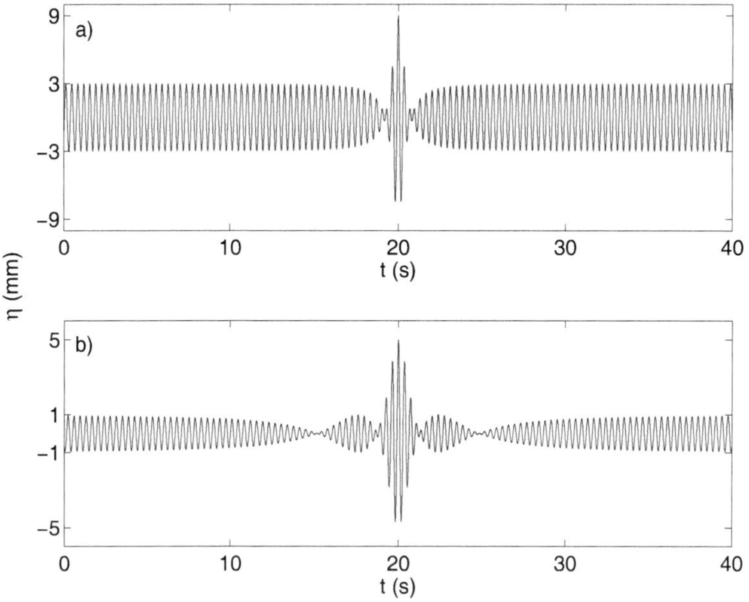

Figure 12.9 Initial conditions provided by theory (Eq. 12.45) at the position of maximal amplification, that is, at $X = 0$, applied to the wave flap at Position x_S. (a) The carrier parameters for the Peregrine breather-type water wave profile are $a_0 = 0.3$ cm and $\varepsilon = 0.09$. (b) The carrier parameters for the Akhmediev–Peregrine breather-type water wave profile are $a_0 = 0.1$ cm and $\varepsilon = 0.03$.

time-reversed signal from the point x_S and observe the refocusing at point x_M, rather than rebroadcasting the reversed wave field from x_M and expecting refocusing at the position x_S. At the fifth and last step of the experiments, we measure the surface elevations related to the time-reversed initial conditions, again 9 m from the flap at position x_M, that is, keeping the wave gauge at the same position. These refocused measurements are compared to the theoretical surface elevations. The corresponding wave profiles are shown in Fig. 12.11.

These results are a clear demonstration of the TR invariance of the hydrodynamic breathers. Clearly, Fig. 12.11 provides an accurate refocusing and reconstruction of the breather surface elevation of the corresponding NLSE solution at its maximal compression, as already presented in Fig. 12.9. The results are in a very good

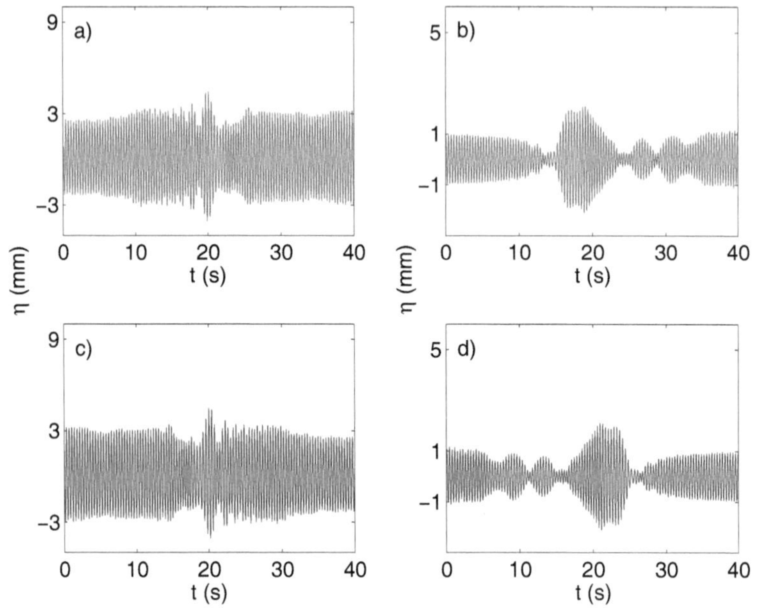

Figure 12.10 (a) Surface elevation of water wave profile of the attenuated Peregrine breather, recorded at x_M, located 9 m form the flap. (b) Surface elevation of water wave profile of the attenuated Akhmediev–Peregrine breather, recorded at x_M, located 9 m form the flap. (c) Time-reversed signal of the measurements shown in (a), providing new initial conditions to the flap and reemitted at x_S. (d) Time-reversed signal of the measurements shown in (b), providing new initial conditions to the flap and reemitted at x_S.

agreement with the theoretical predictions, expected at this position within the framework of NLSE. In fact, the maximal water surface amplitude is of 9 mm and of 5 mm for the Peregrine breather and the Akhmediev–Peregrine breather, respectively, which correspond to theoretical values of amplitude amplifications, related to these NLSE solutions at the maximal stage of breather compression and as generated in the first step of the experiment. These observations prove the TR invariance of strongly nonlinear water waves. In addition, these experimental results confirm the accuracy of the NLSE in describing the unidirectional and complex evolution dynamics of rogue waves, taking into account the complex phase-shift dynamics,

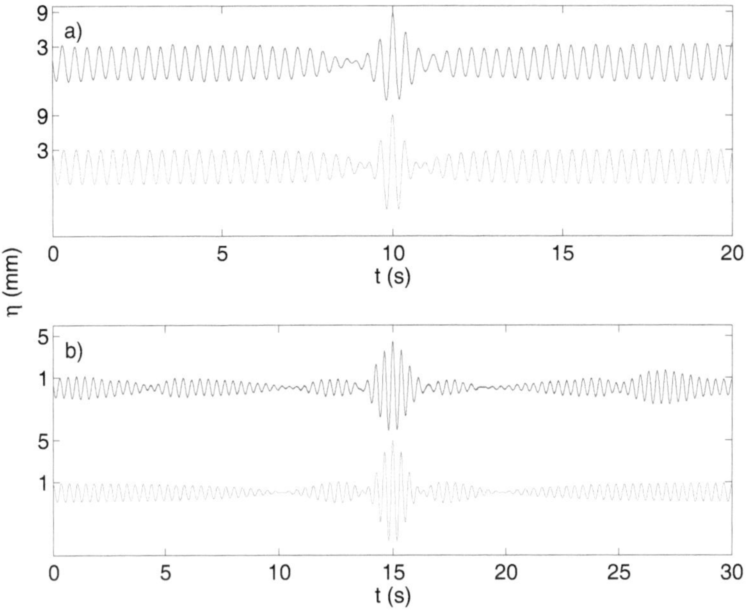

Figure 12.11 (a) Comparison of the Peregrine surface profile measured 9 m from the paddle at x_M, while starting its propagation from time-reversed initial conditions (blue upper line) with the expected theoretical NLSE prediction at the same position (red bottom line). (b) Comparison of the Akhmediev–Peregrine surface profile measured 9 m from the paddle at Position x_M, while starting its propagation from time-reversed initial conditions (blue upper line) with the expected theoretical NLSE prediction at the same position (red bottom line).

related to the modulation instability process. Nevertheless, some discrepancies between theory and experiment can be also noted in terms of asymmetric wave profile shape of the experimental observations, as can be noticed in Fig. 12.11. The latter are due to higher-order nonlinearities (Stokes drift) and to higher-order dispersion effects, not taken into account in the NLSE approach, as well as to occurring experimental imperfections, including dissipation and wave reflection and most importantly noise, naturally and always existing while performing experiments in wave basins. Nevertheless, the experimental observations confirm the possibility to reconstruct strongly localized, thus, strongly nonlinear waves through TR. Therefore, we emphasize that this technique may be used in order to construct new TR invariant localized structures, described by nonlinear evolution equations, which considerably amplify the amplitude of a wave field, thus, again also in the case of strong nonlinearity.

12.4 Discussion and Outlook

Experiments on water wave pulse field reconstruction in the linear and nonlinear regime using TR, provide not only a confirmation of the TR invariance of the hydrodynamic wave motion, but also open new field of possible applications in several dispersive media. This includes, for example, superfluidity [51, 52] and optical fibers [15, 53]. The experiments, involving the doubly localized breather solutions of the NLSE show also limitations, which should be addressed. First, discrepancies with respect to the theoretical NLSE predictions can be noticed. These can be easily explained in view of the natural limitation of the NLSE. A successive analysis could be performed in order to verify the TR invariance, also for higher-order NLSE-type evolution equation [20, 56] as well as within the framework of fully nonlinear equations [55]. This is to accurately characterize possible applications and limitations of the TR method for nonlinear, very steep, and nonbreaking water waves. Another interesting point, which should be addressed from a chaotic dynamics point view is the influence of random noise on the TR reconstruction of NLSE breathers. Next, we inspect the numerical instability of the TR scheme, by perturbing the initial conditions of the same nonlinear experiments within the framework of the NLSE. We used the split-step Fourier method [57, 58] to illustrate the influence of the random noise on the decaying and TR reconstruction dynamics of both, Peregrine and Akhmediev–Peregrine dynamics for the same wave parameters as used in the laboratory experiments. The initial conditions have been multiplied by $[1 + N\Gamma(t)]$ [59]. Here, Γ is a normally distributed random function, whereas N denotes the noise level, which is chosen to be 10^{-1} and 10^{-2}, respectively. The numerical results involving the Peregrine breather for the same laboratory parameters over a

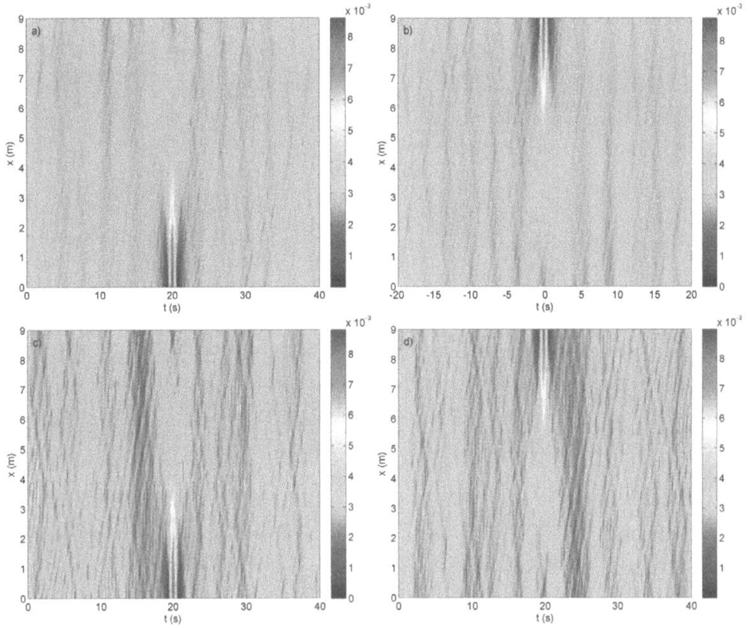

Figure 12.12 Numerical TR experiment of the Peregrine breather dynamics for $\varepsilon = 0.09$ and $a = 3$ mm for two different noise levels using the split-step Fourier method. (a) Decay of the Peregrine pulse from its maximal compression during its evolution over 9 m for a noise level of 10%. (b) Growth of the Peregrine pulse, starting from the time-reversed signal of the experiment (a) at 9 m. (c) Decay of the Peregrine pulse from its maximal compression during its evolution over 9 m for a noise level of 20%. (d) Growth of the Peregrine pulse, starting from the time-reversed signal of the experiment (c) at 9 m.

distance of 9 m are shown in Fig. 12.12a,b for a noise level of $N = 10\%$ and Fig. 12.12c,d for a considerable noise level of $N = 20\%$.

It is noticeable that TR techniques work very well in reconstructing both doubly localized Peregrine-type extreme waves, since the maximal compression of the envelope could be recovered, even in the presence of significant noise. Remarkably, this is also the case for the Akhmediev–Peregrine solution that has a much larger amplitude amplification value and therefore, a more significant nonlinear Stokes wave nature, as shown in Fig. 12.13 (same noise level N as in the previous figure).

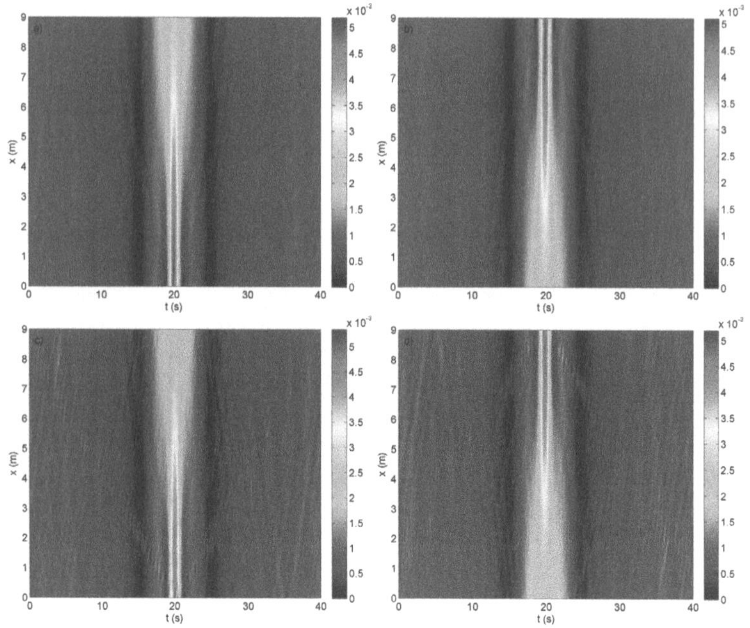

Figure 12.13 Numerical TR experiment of the Akhmediev–Peregrine breather dynamics for $\varepsilon = 0.03$ and $a = 1$ mm for two different noise levels using the split-step Fourier method. (a) Decay of the Akhmediev–Peregrine pulse from its maximal compression during its evolution over 9 m for a noise level of 10%. (b) Growth of the Peregrine pulse, starting from the time-reversed signal of the experiment (a) at 9 m. (c) Decay of the Akhmediev–Peregrine pulse from its maximal compression during its evolution over 9 m for a noise level of 20%. (d) Growth of the Akhmediev–Peregrine pulse, starting from the time-reversed signal of the experiment (c) at 9 m.

Another possible application is the validation of TR reconstruction of nonlinear waves, propagating in two spatial directions [60, 61]. Deep-water evolutions equations, such as the Davey–Stewartson equation [62] or the shallow water KP-I [63] could be used for this investigation. Obviously, experiments for such type of equations are much more difficult to perform. Furthermore, new possible solutions may be derived numerically in the limit of TR convergence.

12.5 Conclusion

To summarize, we have discussed the TR invariance of surface gravity water waves in the linear and nonlinear regime. After introducing the governing equations of an ideal fluid and emphasizing the conditions for the TR invariance, we have presented recent experiments, validating the TR approach in reconstructing linear and 2D water pulses, whereas two examples of nonlinear breather-type pulses have been refocused in a 1D water wave flume. The experimental results provide a clear confirmation of the TR invariance of water waves and emphasize novel applications in other dispersive media, such as fiber optics, plasma and Bose–Einstein condensates as well as in remote sensing to name only few. The effect of dynamical noise on the nonlinear breather wave propagation has been discussed numerically as well. The reported simulations, based on the split-step Fourier method are promising and motivate further analytical, numerical and experimental work in analyzing the hydrodynamics and complex propagation properties of linear and nonlinear waves, using the TR technique.

References

1. Fink, M. (1997). *Phys. Today*, **3**, 34.
2. Lerosey, G., de Rosny, J., Tourin, A., Derode, A., Montaldo, G., and Fink, M. (2004). *Phys. Rev. Lett.*, **92**, 193904.
3. Fink, M., Cassereau, D., Derode, A., Prada, C., Roux, P., Tanter, M., Thomas, J.-L., and Wu, F. (2000). *Rep. Prog. Phys.*, **63**, 1933.
4. Draeger, C., and Fink, M. (1997). *Phys. Rev. Lett.*, **79**, 407.
5. Feynman, R.P., Leighton, R.B., and Sands, M. (1963). *The Feynman Lectures on Physics*, Vol. I, Chaps. 51–54 (Addison-Wesley, Reading, MA).
6. Roux, P. and Fink, M. (2000). *J. Acoust. Soc. Am.*, **107**, 2418–2429.
7. Ing, R.K. and Fink, M. (1998). *IEEE Trans. Ultrason. Ferroelectr. Freq. Control*, **45**, 1032.
8. Tanter, M., Thomas, J.-L., Coulouvrat, F., and Fink, M. (2001). *Phys. Rev. E*, **64**, 016602.
9. Lighthill, J. (2001). *Waves in Fluids* (Cambridge University Press).

10. Mei, C.C., Stiassnie, M.M., and Yue, D.K.P. (2005). *Theory and Applications of Ocean Surface Waves*, Advanced Series on Ocean Engineering, Vol. 23.
11. Osborne, A. (2010). *Nonlinear Ocean Waves & the Inverse Scattering Transform*, Vol. 97 (Academic Press).
12. Kharif, C., Pelinovsky, E., and Slunyaev, A. (2009). *Rogue Waves in the Ocean* (Springer).
13. Dalrymple, R.A., and Dean, R.G. (1991). *Water Wave Mechanics for Engineers and Scientists* (Prentice Hall).
14. Linton, C.M., and McIver, P. (2001). *Handbook of Mathematical Techniques for Wave/Structure Interactions* (CRC Press).
15. Remoissenet, M. (1999). *Waves Called Solitons* (Springer).
16. Debnath, L. (1994). *Nonlinear Water Waves* (Academic Press).
17. Stokes, G.G. (1847). On the theory of oscillatory waves, *Trans. Camb. Philos. Soc.*, **8**, 441–455.
18. Benjamin, T.B. and Feir, J.E. (1967)The disintegration of wave trains on deep water. Part 1. Theory, *J. Fluid Mech.*, **27**, 417–430.
19. Zakharov, V.E. (1968). Stability of periodic waves of finite amplitude on a surface of deep fluid, *J. Appl. Mech. Tech. Phys.*, **9**, 190–194.
20. Dysthe, K.B. (1979). Note on the modification of the nonlinear Schrödinger equation for application to deep water waves, *Proc. R. Soc. A*, **369**, 105–114.
21. Zakharov, V.E. and Shabat, A.B. (1972). Exact theory of two-dimensional self-focusing and one-dimensional self-modulation of waves in nonlinear media, *J. Exp. Theor. Phys.*, **34**, 62–69.
22. Bespalov, V.I., and Talanov, V.I. (1966). Filamentary structure of light beams in nonlinear liquids, *JETP Lett.*, **3**, 307–310.
23. Lake, B.M., and Yuen, H.C. (1977). Note on some nonlinear water-wave experiments and the comparison of data with theory, *J. Fluid Mech.*, **83**, 75–81.
24. Benjamin, T.B. (1967). Instability of periodic wavetrains in nonlinear dispersive systems, *Proc. R. Soc. A.*, **299**, 59–75.
25. Zakharov, V.E. and Ostrovsky, L.A. (2009). Modulation instability: the beginning, *Physica D*, **238**, 540–548.
26. Hasimoto, H., and Ono, H. (1972). Nonlinear modulation of gravity waves, *J. Phys. Soc. Jpn.*, **33**, 805–811.
27. Yuen, H.C., and Lake, B.M. (1982). Nonlinear dynamics of deep-water gravity waves, *Adv. Appl. Mech.*, **22**, 67–229.

28. Tulin, M.P., and Waseda, T. (1999). Laboratory observations of wave group evolution, including breaking effects, *J. Fluid Mech.*, **378**, 197–232.
29. Akhmediev, N., Eleonskii, V.M., and Kulagin, N. (1985). Generation of periodic sequence of picosecond pulses in an optical fibre: exact solutions, *J. Exp. Theor. Phys.*, **61**, 894–899.
30. Akhmediev, N., and Korneev, V.I. (1986). Modulation instability and periodic solutions of the nonlinear Schrödinger equation, *Theor. Math. Phys.*, **69**, 1089–1093.
31. Kuznetsov, E.A. (1977). Solitons in a parametrically unstable plasma, *Sov. Phys. Dokl.*, **22**, 507–508.
32. Ma, Y.C. (1979). The perturbed plane wave solutions of the cubic nonlinear Schrödinger equation, *Stud. Appl. Math.*, **60**, 43–58.
33. Peregrine, D.H. (1983). Water waves, nonlinear Schrödinger equations and their solutions, *J. Aust. Math. Soc. Ser. B*, **25**, 16–43.
34. Akhmediev, N., Ankiewicz, A., and Soto-Crespo, J.M. (2009). Rogue waves and rational solutions of the nonlinear Schrödinger equation, *Phys. Rev. E*, **80**, 026601.
35. Akhmediev, N., Ankiewicz, A., and Taki, M. (2009). Waves that appear from nowhere and disappear without a trace, *Phys. Lett. A*, **373**, 675–678.
36. Dudley, J.M., Genty, G., Dias, F., Kibler, B., and Akhmediev, N. (2009). Modulation instability, Akhmediev breathers and continuous wave supercontinuum generation, *Opt. Express*, **17**, 21497.
37. Kibler, B., Fatome, J., Finot, C., Millot, G., Dias, F., Genty, G., Akhmediev, N., and Dudley, J.M. (2010). The Peregrine soliton in nonlinear fibre optics, *Nat. Phys.*, **6**, 790–795.
38. Chabchoub, A., Hoffmann, N.P., and Akhmediev, N. (2011). Rogue wave observation in a water wave tank, *Phys. Rev. Lett.*, **106**, 204502.
39. Chabchoub, A., Hoffmann, N., Onorato, M., and Akhmediev, N. (2012). Super rogue waves: observation of a higher-order breather in water waves, *Phys. Rev. X*, **2**, 011015.
40. Bailung, H., Sharma, S.K., and Nakamura, Y. (2011). Observation of Peregrine solitons in a multicomponent plasma with negative ions, *Phys. Rev. Lett.*, **107**, 255005.
41. Onorato, M., Proment, D., Clauss, G., and Klein, M. (2013). Rogue waves: from nonlinear Schrödinger breather solutions to sea-keeping test, *PLOS ONE*, **8**, e54629.

42. Chabchoub, A., Kibler, B., Dudley, J.M., and Akhmediev, N. (2014). Hydrodynamics of periodic breathers, *Phil. Trans. R. Soc. A*, **372**, 20140005.
43. Przadka, A., Feat, S., Petitjeans, P., Pagneux, V., Maurel, A., and Fink, M. (2012). TR of water waves, *Phys. Rev. Lett.*, **109**, 064501.
44. Chabchoub, A., and Fink, M. (2014). Time-reversal generation of rogue waves, *Phys. Rev. Lett.*, **112**, 124101.
45. Maurel, A., Cobelli, P., Pagneux, V., and Petitjeans, P. (2009). *J. Appl. Opt.*, **48**, 380; Cobelli, P., Maurel, A., Pagneux, V., and Petitjeans, P. (2009). *Exp. Fluids*, **46**, 1037.
46. Recent improvements of the technique for signal processing and the choice of painting particles can be found, respectively, in G. Lagubeau et al., in preparation; Przadka, A., Cabane, B., Pagneux, V., Maurel, A., and Petitjeans, P. (2012). *Exp. Fluids*, **52**, 519.
47. Cobelli, P., Petitjeans, P., Maurel, A., Pagneux, V., and Mordant, N. (2009). *Phys. Rev. Lett.*, **103**, 204301.
48. Cobelli, P., Pagneux, V., Maurel, A., and Petitjeans, P. (2009). *Europhys. Lett.*, **88**, 20006; (2011). *J. Fluid Mech.*, **666**, 445–476.
49. Draeger, C., and Fink, M. (1999). *J. Acoust. Soc. Am.*, **105**, 611.
50. Chabchoub, A., Hoffmann, N., Onorato, M., Slunyaev, A., Sergeeva, A., Pelinovsky, E., and Akhmediev, N. (2012). Observation of a hierarchy of up to fifth-order rogue waves in a water tank, *Phys. Rev. E*, **86**, 056601.
51. Gross, E.P. (1961). Structure of a quantized vortex in boson systems, *Il Nuovo Cimento*, **20**, 454–457.
52. Pitaevskii, L.P. (1961). Vortex lines in an imperfect Bose gas, *Sov. Phys. JETP*, **13**, 451–454.
53. Dudley, J.M., Dias, F., Erkintalo, M., and Genty, G. (2014). Instabilities, breathers and rogue waves in optics, *Nat. Photon.*, **8**, 755–764.
54. Craig, W., Guyenne, P., and Sulem, C. (2010). Hamiltonian approach to nonlinear modulation of surface water waves, *Wave Motion*, **47**, 552–563.
55. (2013). On the highest non-breaking wave in a group: fully nonlinear water wave breathers versus weakly nonlinear theory, *J. Fluid Mech.*, **735**, 203–248.
56. Craig, W., Guyenne, P., and Sulem, C. (2010). Hamiltonian approach to nonlinear modulation of surface water waves, *Wave Motion*, **47**, 552–563.
57. Hardin, R.H., and Tappert, F.D. (1973). *SIAM Rev. Chronicle*, **15**, 423.

58. Fisher, R.A., and Bischel, W.K. (1975). *Appl. Phys. Lett.*, **23**, 661 (1973); *J. Appl. Phys.*, **46**, 4921.
59. Chen, S., Soto-Crespo, J.M., and Grelu, P. (2014). Dark three-sister rogue waves in normally dispersive optical fibers with random birefringence, *Opt. Express*, **22**, 27632–27642.
60. Waseda, T., Kinoshita, T., and Tamura, H. (2009). Evolution of a random directional wave and freak wave occurrence, *J. Phys. Oceanogr.*, **38**, 621–639.
61. Toffoli, A., Gramstad, O., Trulsen, K., Monbaliu, J., Bitner-Gregersen, E., and Onorato, M. (2010). Evolution of weakly nonlinear random directional waves: laboratory experiments and numerical simulations, *J. Fluid. Mech.*, 313–336.
62. Davey, A., and Stewartson, K. (1974). On three dimensional packets of surface waves, *Proc. R. Soc. A*, **338**, 101–110.
63. Kadomtsev, B.B., and Petviashvili, V.I. (1970). On the stability of solitary waves in weakly dispersive media, *Sov. Phys. Dokl.*, **15**, 539–541.

Index